A Min Tjoa (ed.)

Information and Communication Technologies in Tourism 1997

Proceedings of the International Conference in Edinburgh, Scotland, 1997

SpringerWienNewYork

Univ.-Prof. Dipl.-Ing. Dr. A Min Tjoa
Institut für Softwaretechnik
Technische Universität Wien, Vienna, Austria

Printed in Austria

Printed by Novographic, Ing. Wolfgang Schmid, A-1230 Wien
Printed on acid-free and chlorine-free bleached paper
Graphic design: Ecke Bonk

With 86 Figures

ISBN 3-211-82963-6 Springer-Verlag Wien New York

Preface

For the fourth time now, experts in tourism from various countries come to attend the ENTER conference in order to inform themselves and others about the current developments in the usage of information and communication technologies.

The ENTER conference is thought as a platform for the exchange of ideas, experiences, opinions, and visions among scientists and practitioners. The visions presented at the last three ENTER conferences have triggered many requirements of important on-going and planned projects in the application of communication and information technologies in tourism.

The scope of the papers of this proceedings covers the most recent and relevant topics in our area of interest. The sessions are primarily devoted to intelligent agents and systems, the future role of global (reservation) systems, the new chances of database applications due to the most recent technological developments, and above all the role of the Internet (and Intranet).

I would like to express my cordial thanks to all institutions actively supporting this event, namely:

- The Edinburgh & Lothian Tourist Board
- The International Federation of Information Technology in Tourism (IFITT)

Several people spent numerous hours organizing the scientific programme of ENTER. The names of most of them will appear in the following pages.

I have to express my thanks to my friend Hannes Werthner, the president of IFITT, for his continuous interest and help. My special thanks are dedicated to the members of the programme committee who have given valuable inputs to the shape of ENTER '97 and above all for their careful reviewing of the submissions. I have to thank Dr. Roger Carter and his team (Celia Christie, Abigail Stevens, Mifamwy Thomas) for their patience and support; they were always present with their help whenever it was needed.

Above all I have to cordially thank my secretary Ms. Rosanna Löw who supported me during all the time with extra-ordinary patience and engagement during the whole life-cycle of the genesis of this proceedings, beginning from the management of the paper submissions to the last steps of editing this book. It was her help which made the edition of this book possible.

A Min Tjoa (Scientific Chairperson)
Institute of Software Technology
University of Technology Vienna

Contents

4. Strategic Concepts, Realization Tools II

5. Implementation Issues

6. The Role of Internet and World Wide Web I

7. The Role of Internet and World Wide Web II

8. The Role of Advanced Information Technology in Tourism

9. Advanced Information Systems Technology

10. Strategy Workshop

Index of Authors

Programme Committee

Fourth International Conference on Information and Communication Technologies in Tourism, Edinburgh, Scotland, January 22-24, 1997

AANONSEN Kari	Norwegian Computing Center, Norway
AIVALIS Constantin	Technological Education Foundation, Greece
BAUKNECHT Kurt	University of Zürich, Switzerland
BENTLEY Robert	Sustainable Tourism Development Associates, Spain
BRANDES Wolfram	Arthur D. Little, Germany
BUHALIS Dimitros	Westminster University, Great Britain
DJUNAEDI Achmad	Gadjah Mada University, Indonesia
EBNER Arno	TIS GmbH, Austria
FLEISCH Elgar	University of St. Gallen, Switzerland
FRANK Klaus	University of Worms, Germany
GO Frank	Technical University of Hong Kong
HITZ Martin	University of Vienna, Austria
JAFAR Jafari	University of Wisconsin, USA
KASPAR Claude	University of St. Gallen, Switzerland
KLEIN Stefan	University of St. Gallen, Switzerland
KUBICEK Herbert	University of Bremen, Germany
LANFRANCO Sam	York University, Canada
MATA-MONTERO Erick	Instituto Tecnologico de Costa Rica
MAZANEC Josef A.	University of Economics Vienna, Austria
MEIJS Chris	Wageningen Agricultural University, The Netherlands
MORENO-DIAZ Roberto	Instituto Tecnologico de Canarias, Spain
PAOLINI Paolo	Politecnico di Milano, Italy
PERONI Giovanni	CST Perugia, Italy
POLLOCK Anna	The Strategy Group, Great Britain
PRESS Larry	University of California, USA
RAFFERTY John	Board Failte, Ireland
REVELL Norman	University of Middlesex, Great Britain
RIBBERS Pieter	Tilburg University, The Netherlands
ROITHMAYR Friedrich	University of Innsbruck, Austria
SCHERTLER Walter	University of Trier, Germany
SHELDON Pauline	University of Hawaii, USA
SMERAL Egon	Austrian Institute for Economic Research, Austria

STOCK Oliveiro	Istituto per la Ricerca Scientifica Tecnologica, Italy
VAN DER PIJL John	Tilburg University, The Netherlands
VAN HOOF Hubert	Northern Arizona University, USA
WAGNER Roland	University of Linz, Austria
WERTHNER Hannes	University of Vienna, Austria
WIETRZYK Vlad	Macquarie University, Australia

A Meta-Data Driven Tourism Information System Prototype

Josef Küng, Richard Sikora, Roland R. Wagner, Wolfram Wöß

Research Institute for Applied Knowledge Processing (FAW),
Johannes Kepler University Linz
Altenbergerstraße 69, 4040 Linz, Austria, e-mail: kueng@ifs.uni-linz.ac.at

Abstract

In this contribution a concept and a prototype for a tourism information system based on meta-data will be introduced. With this approach it is possible to handle new kind of information, new data structures or new layouts better than in other information systems in which such changes have an impact on the software. With our system the adaptation can be done by the tourism organisation itself. Besides, the suitability of Microsoft Access for such meta-data based information systems was evaluated.

1 Introduction

The analysis of life cycles of data object types in tourism information systems showed, that these life cycles are often very short. Every season the market requires new offers. Therefore also the information systems have to be adapted accordingly. Often a simple data update is not enough. Also new object types have to be added or the structures of existing object types have to be altered. But changes in data structures have an impact on the software especially on the user interface. This adaptation of the database application can be done only by software engineers. Besides it is time and cost intensive.

At our institute, research work is done which tried to solve this problem by means of meta-data. Whenever new kinds of objects have to be inserted into the information system, only the meta-data has to be altered but not the software itself.

2 The Usage of Meta-Data

In the field of database design meta-data means the scheme which describes the conceptual and logical structure of all objects and entities of a database and the relationships between them. Data dictionaries, e.g., are a well known mechanism for storing meta-data.

In [3] two areas are identified in which meta-data are very useful: The system layer (storage and migration of data) and the application layer (i.e., searching). At the system layer meta-data is used to describe the object itself: structure, data types, semantic of attributes, etc. At the application layer the programs derive their behaviour from meta-data which are especially designed for the specific application areas, for instance for tourism information systems.

More than the existence of a data description for an information system an efficient user interface is important. Therefore it is necessary to consider both aspects - the data description and the user interface. More information about meta-data especially in geosciences can be found in [4]. Furthermore, efforts are taken in standardising meta-data [5].

In our prototype, both the structure of touristic object types and the user interface are supported with meta-data. A general entity-relationship diagram can be built as shown in Figure 1.

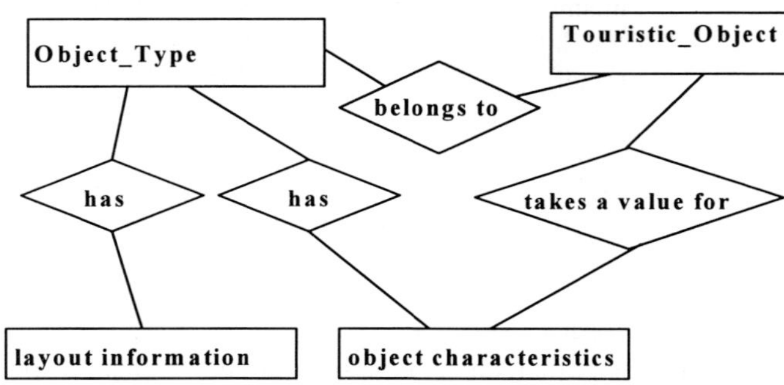

Fig. 1. A general ER-diagram for a tourism information system with meta-data

Meta-data for the user interface is depicted by „layout information“, meta-data for the structure of object types is indicated by „object characteristics“. By using a relational database system the relationship „takes a value for“ results in a very long table storing a separate record for each attribute value. To handle such a table is very complicated. Much effort has to be done in programming the queries, standard tools cannot be used and last but not least performance is much lower compared to the approach of storing a whole object in a single record.

Therefore we chose a mixed approach. That means, data itself is stored in usual tables of a relational database system. As a consequence, all tuning features of the database system are applicable. Besides, the meta-data describes the layout of the user interface and the mapping from object types to the tables in the database. With this concept almost the same performance as with a conventional database application could be reached.

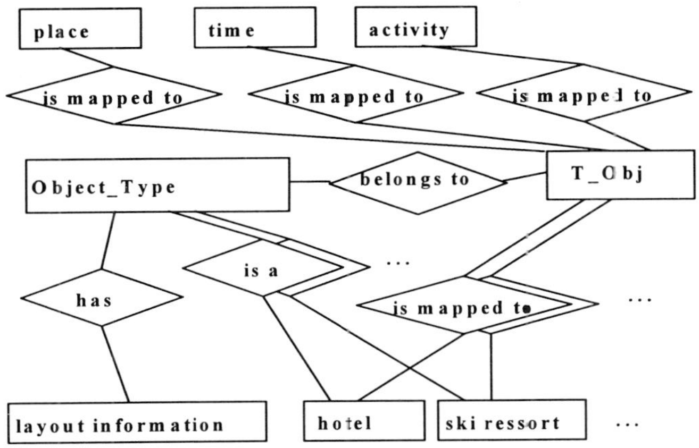

Fig. 2. ER-diagram of the „mixed approach"

Also the meta-data mechanisms of the database itself - i.e., the data dictionary - can be used for both describing the data structure and as an input for the searching facilities. Therefore the „object characteristics" are not visible in Figure 2.

The mapping of a touristic object to the places in which it is situated, to the times in which it is available and to the corresponding activities is modelled explicitly because each object can be described by these attributes and the behaviour of these objects is the same.

With this approach and based on the work of Werthner and Tjoa [1], [2] we realised the following prototype of a tourism information system for maintaining information about arbitrary touristic objects.

2 The Tourism Information System Prototype

Two questions gave the motivation for the development of this prototype:

a) Is the concept above suitable for information systems in general and for tourism information systems in particular.
b) To what extent does Microsoft Access support such kinds of information systems?

The functionality of the realised information system can be divided into three parts:

1) Administrator module - for defining object types (structure and layout)
2) Data acquisition module - for maintaining the data itself
3) Query module - for searching and finding

All these modules are already implemented completely so that the prototype is very close to production level.

2.1 Administrator Module

Since one of the main goals of our concept is that new object types can be defined by the tourism organisation itself, tools for this task are necessary - in our case the administrator module. It is realised like a wizard of Microsoft Access.

In a first step, object types, attributes and corresponding data types have to be entered. Possible data types are text, integer, real, currency, boolean, date/time, OLE-fields and memofields. With OLE-fields the whole OLE-facility of Microsoft Windows is available. Pictures, sounds, etc. are supported too. The links to the places, times and activities are installed automatically.

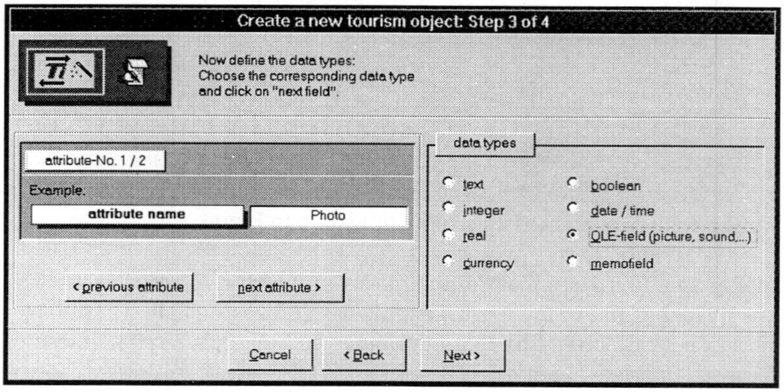

Fig. 3. Defining attributes and data types

The administrator module now generates the object type itself (resulting in an additional record within the table „Object_Type“ as illustrated in figure 2), a table with the defined attributes in which the objects of this type will be stored and the mapping between the object type and the table.

In a second step the layout of the object type has to be designed. The administrator module proposes one of two optional standard layouts (a single column, multiple columns). If not all attributes fit into a single window, the system automatically generates the necessary number of windows. Afterwards, the user can revise the layout as shown in Figure 4. For this task, the features of Microsoft Access are used directly.

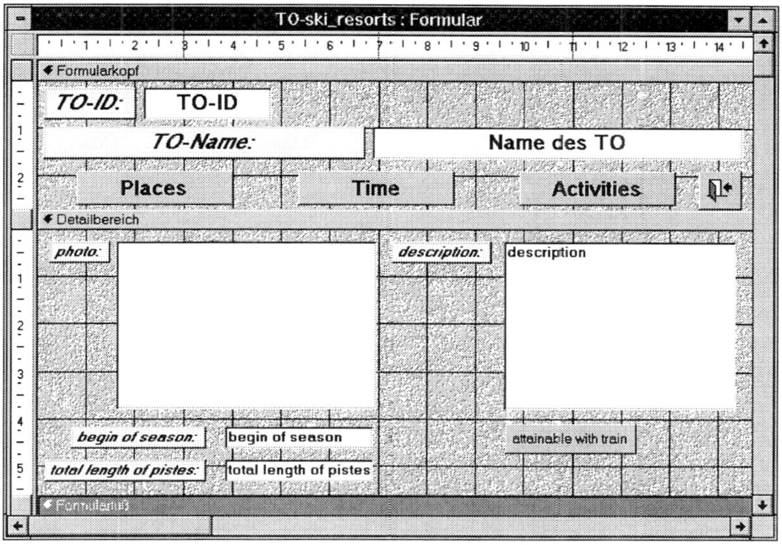

Fig. 4. Designing the layout of the user interface

2.2 Data Acquisition Module

After defining the object types, the system is ready to be filled with data - real touristic objects such as the Hilton Hotel in London or the ski resort St. Anton am Arlberg - an object of the type defined in the figures above:

Fig. 5. Entering data by using the dynamic interface

2.3 Query Module

The third and most component of our prototype is the query module. According to the output of the queries, four different types of queries can be distinguished:

Output	possible search criteria	Example
Objects	• Attributes regarding to the chosen object type • Places • Times • Activities	All objects of the type ski resort in Tyrol where the total length of pistes is more than 100 km.
Places	• Attributes regarding to the chosen object type (more than one object type is allowed) • Times • Activities	All places where ski resorts with more than 100 km total length of pistes and hotels with indoor swimming pool are available.
Times	• Attributes regarding to the chosen object type (more than one object type is allowed) • Places • Activities	All times in which swimming in an indoor pool with a length of 50 m is possible in the district of Innsbruck.
Activities	• Attributes regarding to the chosen object type (more than one object type is allowed) • Places • Times	Which activities are offered on January 10th in the district of Innsbruck.

In spite of the complexity of these query possibilities the user interface is easy to handle. When using places as a search criteria, a map or lists of possible values are available.

A special component of this module is the so called „query wizard". It is a general usable facility which interprets the meta-data, generates the user interface, composes a SQL-statement out of the user input, sends it to the database and displays the result of the query.

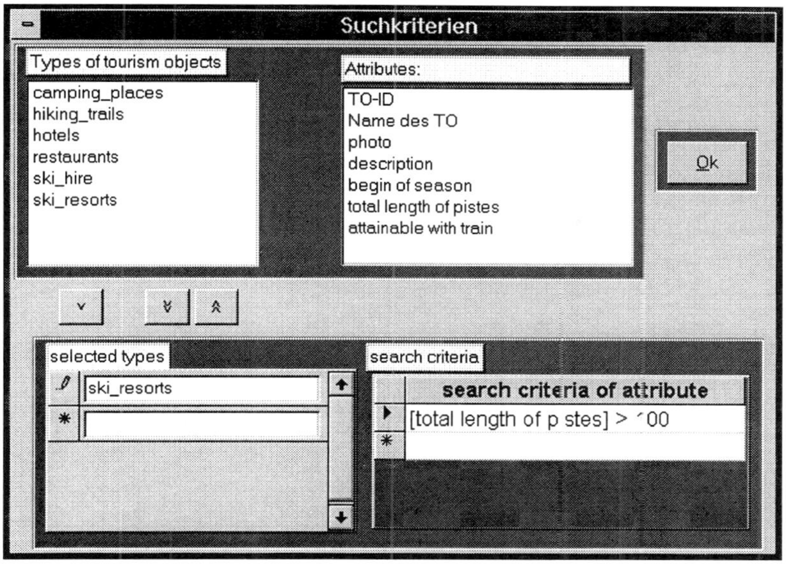

Fig. 6. One part of the „query wizard“

3 Results

The experiences we have gained from implementing our prototype are summarised in the following by means of two questions:

a) Is the concept above suitable for information systems in general and for touristic information systems in particular?

- The system allows the definition of object types - structure and layout - at runtime and without intervention of the system developer. The tourism organisation itself is able to do the maintenance of the object types. New releases of the software for each season are not necessary.

- The user interface is quite comfortable. Often such meta-data driven systems have considerable drawbacks in this field. Often only a few number of screen elements are possible or there exist restrictions in colour, styles, background and so on. In our prototype all features supported by Microsoft Access are available.

- Since our „mixed approach“ stores data in tables the performance of our information system is almost equal to those systems which do not make use of meta-data. Especially query execution time do not differ.

- In another system our institute implemented the idea of Figure 1 [6] in a restricted area. The comparison of the SQL-statements showed that less effort has to be taken in programming the access to the database. Also in this field the mixed approach has proven to be advantageous.

Consequently, the answer of the question above is that the concept is suitable for information systems in general, with no or only a few dynamic aspects. Each object type has the same behaviour. It allows storing objects as well as searching and displaying objects on the basis of meta-data.

b) To what extent does Microsoft Access support such kinds of information systems?

- The prototype shows that many features of Microsoft Access could be used directly, for instance the layout editor, window elements, etc.

- The system is implemented as an additional Microsoft Access wizard with the same look and feel.

- Meta-data of Microsoft Access could be used for the structure of object types and for the layout information.

In our opinion information systems can be realised very well with the tools of Microsoft Access, provided that the „mixed approach“ is chosen. Otherwise many of the features cannot be used and programming would take much more effort.

4 Conclusion and Further work

In this contribution we introduced a concept for meta-data based information systems based on relational databases. The concept was evaluated by means of a prototype.

Even if the prototype works well and has a very user friendly interface two things have to be mentioned. First, at this time data is also stored in a Microsoft Access database but we are working on transferring it to database systems with higher suitable for larger and possibly decentralised applications. Second, in the context of a distributed system each site is required to install the runtime module of Microsoft Access and the ODBC-connection to the database server.

References

[1] Werthner H. Design Principles of Tourist Information Systems. In: Klein, S. Schmid, B., Tjoa, AM., Werthner, H. (eds); *Information and Communication Technologies in Tourism*. Proceedings of the Internaltional Conference in Innsbruck, Austria, 1996. Springer, Wien, New York, pp. 70-77

[2] Tjoa AM., Werthner H. Interfacing WWW with Distributed Database Applications in the Field of Tourism. In: Klein, S. Schmid, B., Tjoa, AM., Werthner, H. (eds); *Information and Communication Technologies in Tourism*. Proceedings of the International Conference in Innsbruck, Austria, 1996. Springer, Wien, New York, pp. 78-85

[3] Sumpter, Raybyne M. Draft Whitepaper on Data Management. Developed from the first two IEEE Metadata Workshops, University of Texas, Austin, Lawrence Livermore National Laboratory, 1994 http://www.llnl.gov/liv_comp/metadata/papers/whitepaper-draft.html

[4]. Medyckyj-Scott, D.J., Newman, I. Ruggles, Walker D. (eds); *Metadata in the Geosciences* C. (1991) Group D Publications. Loughborough UK. ISBN 1-874152-00-4.

[5] Rubenstrunk K. Standardizing Metadata, In *DBMS Database & Client/Server Solutions*, Vol. 9, No. 2, Februar 1996, 44-48

[6] Steinegger G., Hohl R., Wagner R.R., Küng J., ProdIS: A Database Application for the Support of Computer Integrated Manufacturing. In: Tjoa. AM., Ramos, I. (eds); *Database and Expert Systems Applications*. Proceedings of the International Conference in Valencia, Spain, 1992. Springer, Wien, New York, pp. 379-383

Geographical Information System Based Decision Support System for Tourism Planning and Development

Vahap Tecim

Department of Management Science, The Management School, Lancaster University, Lancaster LA1 4YX, England

Abstract. During the last ten years, advances in information technology have had a dramatic influence on the tourism industry. This paper intends to present a geographical information system designed as a decision support system (DSS) for tourism planning and development. A user friendly computer system, called Tourism DSS, has been developed for the use of public and private organisations. The county of Cumbria has been chosen as a case study. The main functions of this DSS are (i) to display different kinds of geographic, socio-economic, environmental and tourism data of Cumbria that map onto a number of problems confronting the planner, in particular; (ii) to find potential sites for specific tourism investment under the diverse criteria relevant to the users; and (iii) to demonstrate the possible impacts of tourism investments.

1 Introduction

It has been suggested that many of the modeling techniques can be integrated under the Decision Support System (DSS) framework which has the potential to assist planners and decision makers in the selection of options in a changing environment [4]. This paper explores the way in which a geographical information system (GIS) based DSS can be used for tourism planning and development. The county of Cumbria, north west England, has been selected as a case study area for the following reasons in particular (i) issues surrounding tourism development in Cumbria, (ii) a wide range of organisations involved in tourism in Cumbria, and (iii) some key agencies involved in tourism in Cumbria have already been experiencing GIS.

The limitations of tourism models and use of GIS in tourism are first discussed. Then, the requirement for integrating GIS with DSS are investigated. Finally, a customised Tourism DSS computer system is demonstrated.

2 Limitation of Existing Tourism Models

Tourism is an inter-disciplinary and complex subject. The tourism industry is not well defined in the same way as other industries, for example the textile industry. Therefore it is not possible to transfer known theories and methodologies to the tourism industry [8]. Failure occurs in linking models to real tourism planning problems. As other literature [1, 2] demonstrates, there is little, if any, convincing evidence of models being used in tourism planning. The reasons for this failure include data availability and its accuracy, uncertainty, and the need to satisfy multiple objectives. Information technology has already been used in several ways to minimise these failures in the tourism industry [6]. One of the ways information technology is used is through a computer reservation system.

The positive and the negative impacts of tourism on a specific destination areas can be debated [7, 11] but how can these impacts be analysed and demonstrated? One aspect of the problem is that it is very difficult to measure all these impacts of tourism, because with tourism everything is integrated. A second more useful aspect is that in the last few years a huge volume of tourism statistics (data) has been published both as survey results or in official reports.

It has also become very difficult to deal with all those tourism related data which are exclusively related to location. Tourism is always associated with a geographical location, therefore typical tourism problems have an essentially geographic component. The difficulty of dealing with tourism data is that there are no specific models which can deal with the complexity of the problem. It is still not very clear how to break down the tourism problem in order to interpret tourism data in the context, and how the results of the analysis can be made available to the planners and researchers. For example, data collected on expenditure in restaurants includes not only tourist spending but also includes spending by local residents.

3 Capabilities and Limitations of GIS in Tourism Planning

Before the appearance of GIS, planners and researchers still had to examine large amounts of spatial data to help in taking rational decisions. This is typically done by using various spreadsheets or database software. The introduction of the digital tabular database greatly improved such decisions by allowing quick and efficient processing of large amounts of data. Planners also needed to consider the spatial dimension along with these tabular databases. With the advent of GIS, the attribute data in a tabular database could be linked to actual geographical locations and entities. GIS therefore met some of the needs which tourism researchers and planners were looking for, such as the ability to handle large volumes of data and the ability to represent multiple objectives geographically.

GIS has several capabilities in different subjects, but it has been rarely used to support tourism planning. However, tourism marketing [10] and landscape changes [5] are the two areas in which GIS is involved.

GIS is capable of integrating geographical data from various sources to provide information for effective decision making in tourism planning and development. Map overlay is the most useful tool for planning and decision making, because several maps can be laid over each other and intersect area arises in which different map features which can be seen very easily. This process can help users to identify areas which have the predetermined map features. Decision makers can also extract data from the database of GIS and then export the data to other modeling and analysis programs together with data from other database or specially conducted surveys. It has been used in information retrieval, development control, mapping, site selection, land use planning, land suitability analysis, and programming and monitoring. Recent advances in GIS aim to provide planners and decision makers with efficient tools with which to organise and structure the spatial decision making process.

GIS has been used usually only to a limited degree, because of the absence of specially trained GIS experts can deal with all the functions of GIS. For example, the Lake District National Park Authority has a GIS system and it is used only by GIS officer because it is too technical for other staff to use. The National Park Authority also uses GIS with limited capabilities mainly for monitoring landscape changes. There is an urgent need to design GIS for the non-expert user, even without making use of its all functions.

4 Requirement of GIS as a DSS in Tourism Planning

DSS is a method or an approach to solve some problems (semi-structured or unstructured) and GIS is a way of presenting data (using purpose designed software). Sprague and Carlson [9] define DSS as a computer-based system which can help decision makers confront ill-structured problems through direct interaction with data and analysis models. The choice of data and the format and design characteristics are chosen with a particular problem in mind. Therefore GIS can be designed as a DSS for specific problems. In this research GIS is designed as a DSS for tourism planning and development.

The GIS literature has predominantly been concerned with issues of mapping, while the operational research is concerned with inventing models based on data which may not always exist and solving these models. The DSS literature gives a point of critique of both of these subjects and provides necessary characteristics required to build a DSS. One of these characteristics is the need to match the technology very clearly with both user and problem.

It is argued in this research that most of the tourism models are not very helpful because these models are not particularly designed to fulfill the requirements of DSS.

A geographical DSS contains computer based tools that allow the user to organise, analyse, modify and evaluate information. These systems are usually data driven and incorporate GIS for spatial analysis. They can use established biological, economic, and/or engineering based numerical models within a graphical user interface (GUI) to produce comprehensive modeling systems and geographic representations of the results from the models. GUI is defined as a combination of window, menu, and icon selections designed to guide the user quickly and easily through a program. It provides the ability to set up a more realistic and effective modeling environment than those that have been possible through simple GIS application.

The uses of models in tourism are quite complex, therefore the decision support process must include visualisation of the computational results on custom built maps. For this reason the database functionality must support subset retrieval, constructing and maintaining maps, and database update.

Generally, every GIS aims to help the user make more informed decisions. The need for using spatial decision support systems comes from situations in which complex spatial problems are ill-structured or semi-structured and decision makers cannot define their problems or fully articulate their objectives [3]. The decision making process adopted to solve semi-structured spatial problems is often perceived as unsatisfactory by decision makers. It could be suggested that they require a flexible, problem-solving environment in which the problem can be explored, understood and redefined. Furthermore it is possible to offer trade-offs between conflicting objectives investigated and priority actions set.

5 Customised GIS Based DSS in Tourism Planning

GIS based DSS can be very effective and efficient for different types of users if the framework of its operation is established carefully. The research presented here includes the design, prototyping, evaluation, and implementation of a user-friendly tourism DSS for the use of public and private organisations. The main functions of this DSS are to display different kinds of geographic, socio-economic, environmental and tourism information of Cumbria that map onto a number of problems confronting the planner, in particular; to find potential sites for specific tourism attraction under the diverse criteria relevant to the users; and to demonstrate the possible impacts of tourism investments.

The system should assist planners in different stages of the tourism development and planning process, but the system is not aimed to provide decisions on behalf of the planners. By these means, the planners are being supported by the system but their mode of working is not undermined. It is a tool for improving the quality of planning and decision making. Planners and decision makers are predominantly seeking user friendly interface and key information.

The research has involved two main stages. The first stage of development of the system was concerned with the design and implementation of digital data integration and the spatial data browsing system. In the second stage of the research has been concerned with actual data analysis, especially the design and implementation of a user friendly DSS for analysing the data. The key features of the system are the easy and uncomplicated use of the system and its flexibility in representing a wide variety of complex tourism problems, in particular, that of analysing the impact of investment alternatives.

The system was designed according to framework outlined in Fig. 1. When designing GIS as a DSS for tourism planning and development (using the framework below) DSS principles were adapted [9, 4]. In this framework, database, models and GIS are all integrated. GIS can take data from the database system and use them. Models read data either from the database system or GIS, after which models use them for a specific modeling purpose. After modeling some output may return to the database system and some may go to the GIS for displaying. GIS may send output to the database or to the models for further evaluation. Data always circulates among these three systems. As a consequence a user gets final information from GIS using graphical user interface capabilities. A commercial GIS software package, ARC/INFO, was used to handle all three systems. However, each system might use three independent software, and information can be transferred to each other.

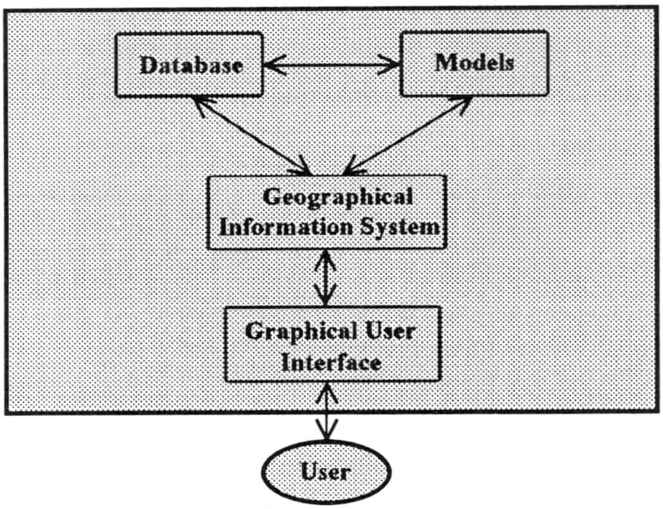

Fig. 1. General framework of Tourism DSS

Typical of DSS, the tourism DSS has four modules: (I) a data management system, (ii) analytical modeling capabilities and analysis procedures, (iii) display and report generators, and (iv) a user interface.

Within ARC/INFO, spatial data are stored as coverages for vector data and grids for cell or raster based data. All attribute data (including time series data) has been either created as an ARC/INFO file or converted to an ARC/INFO file.

Various interviews have been carried out to help identify user needs. Some of the important needs include the ability to support the user with timely and accurate information; ease of use; the ability to provides the casual non-expert user, and visualisation and reporting facilities. Users want a system which is easy to understand and use, provides quick and efficient information, and obtain assistance whenever they encounter problems. The system has been designed to be flexible and general enough to address a variety of planning projects.

The main menu and their related submenu options of Tourism DSS are shown in Fig. 2. *Cumbria Maps* menu has three options, draw, list attributes and query, to use basic Bartholomew Cumbria digital map data. In this computer system, every digital map data has draw, list attributes and query options to provide detailed information about the data coverage to the user. Draw options let a user draw different map coverages such as roads, lakes, accommodations. List attributes options list the attributes in chosen coverages. Query options display the attribute information in chosen single point, line or polygon. All menus and graphs are designed in colour output.

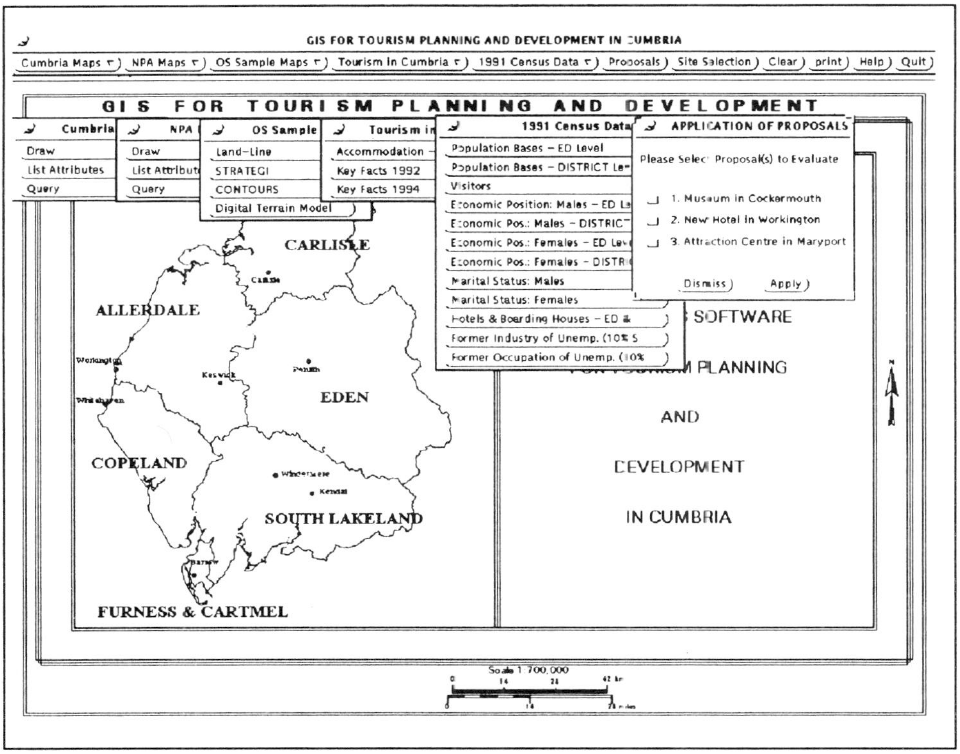

Fig. 2. Main menu of Tourism DSS and their related submenus

NPA Maps menu provides National Park Authority (NPA) owned digital map data, such as car parks, caravan sites, camping sites.

OS Sample Maps menu offers four different Ordnance Survey (OS) datasets options. These are Landline, Strategi, Contours and Digital Terrain Models. *Landline* menu provides 1:2500 and 1:1250 scale very detailed digital map data. Buildings, house numbers, electricity lines and many more important data are included in this option. *Strategi* menu provides 1:250000 scale map data in a bigger area and includes lakes, woodland and other information. *Contours* menu provides a contour data with other environmental data such as lakes and rivers. *Digital terrain models* (DTM) menu provides different DTM maps to help user to understand the ecological part of the area. A 3D model is also provided in this option.

Tourism in Cumbria menu offers three different options. The first option deals with Cumbria Tourist Board (CTB)'s accommodation data. Three searching methods, town name, post code and accommodation name, can be used to get detailed information about selected accommodation. Accommodation data is classified as serviced and non-serviced. The concentration of accommodation is also available in ED (Enumeration District) or district level, under the submenu option. Fig. 3 shows the distribution of CTB's total accommodation unit in each ED.

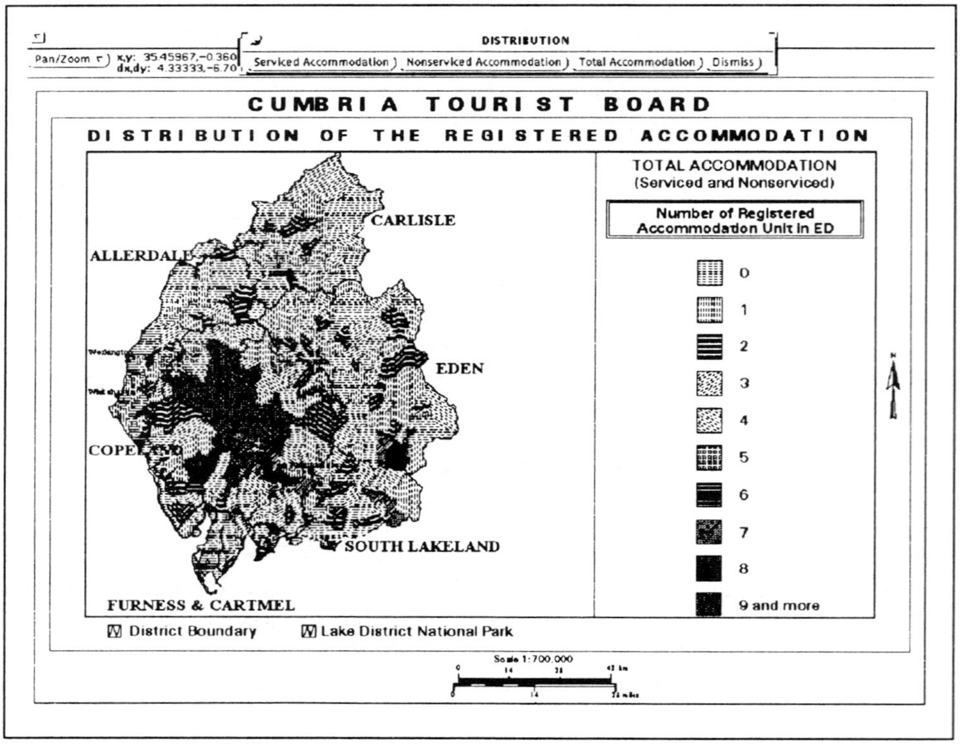

Fig. 3. The concentration of CTB's registered accommodation

As mentioned earlier, all graphs presented here are in colour and mainly designed to print in a A3 size paper. Each accommodation type (serviced and non-serviced) can also be drawn with other coverages (lakes, rivers, roads, etc.) and as always, list and query options also available in this option.

1991 Census Data menu provides several 1991 Census socio-economic data in ED or district level. The distribution of population in Cumbria, unemployment, economic activities, and marital status are some of the options provided in this menu.

Proposals menu lets a user evaluate different proposals (three proposals in this exercise, seen in Fig. 2) either alone or with a combination of proposals. Outcome of the evaluation process displays the possible economic and environmental impacts of the proposal(s). Demonstration of the cumulative impacts of several proposals gives user opportunity to see what would happen when these proposals were accepted.

The objective of the *Site Selection* menu of the Tourism DSS is to support a site selection process for tourism related businesses. The task of evaluating potential sites for industrial development is a multi-dimensional one, requiring knowledge of economic, social, political, environmental, and technical factors. In practice, site selection is usually performed by professionals who have a good understanding of current site specifications and availability. Professionals make their recommendation based on this knowledge. Since there are large amounts of data relevant to this problem, its analysis is time intensive and searches are often not extensive. Fig. 4 demonstrates the entrance of site selection criteria screen.

With regard to the site selection part of the tourism DSS, a well-designed system should help planners to evaluate quickly and accurately various alternative sites that satisfy several planning criteria or factors, for example, proximity to various transportation networks, access to energy, etc.

The success of any GIS system depends absolutely on the availability and the quality of data. In this research, different scenarios have been investigated to bring the necessary data together for the development of the DSS system. Potential sites are evaluated against a list of criteria. Finally, the sites are selected which meet all criteria and are ideal for the specific purpose (in this case tourism). Without the aid of this GIS, it would become more difficult to find a suitable site that satisfies all requirements when the number of criteria increases. The uses of ARC/INFO GIS to identify feasible sites for the tourist attraction in the North West part of Cumbria are mainly discussed in this research. The GIS operation therein constitutes little more than a sequence of map overlays derived from a Boolean search. The object of the overlay analysis is to determine which areas of Cumbria are potentially suitable for a tourist attraction using several siting factors (proximity to motorway, A road, B road, slope, aspect etc.).

After the result of the preliminary site selection, the multi-criteria evaluation method can be used to get a fuller evaluation of the potential sites that takes into account additional criteria. Planners or decision makers may immediately want to see the potential sites on the screen, but waiting for the result of the multi-criteria evaluation is time consuming. Here the use of tourism DSS without waiting for the result of the multi-criteria evaluation saves time, and this is very important for planners and decision makers.

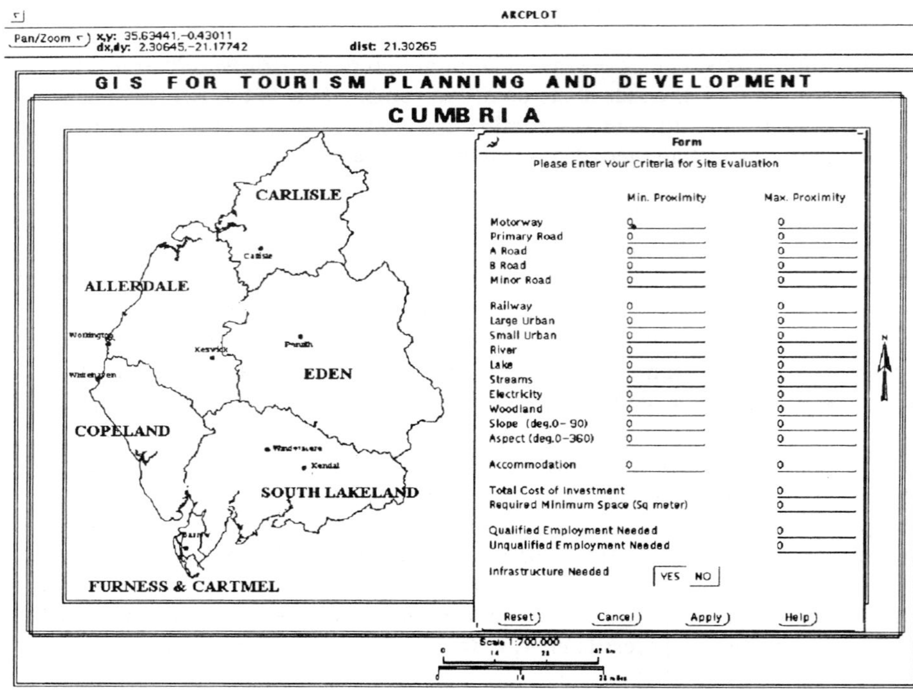

Fig. 4. Screen output of the Site Selection menu

6 Summary

GIS tools can provide planners and decision makers with valuable geographical based information through GIS. Decision makers also use analytical modeling with GIS using GIS as a part of DSS. In this research, one complex subject '*tourism*' has been taken as an application area to demonstrate how the features of a GIS can be integrated into an effective DSS. There is much evidence in this research that GIS can be a very effective and efficient DSS tool in tourism planning and development.

Concerning tourism related planning, application of Tourism DSS has shown that the system can be very useful for planners and decision makers in both public and private sector.

References

1. Archer, B. Demand Forecasting and Estimation. In: Ritchie J.R.B., Goeldner C.R. (eds.): Travel, Tourism, and Hospitality Research: A Handbook for Managers and Researchers. New York: John Wiley 1994 (pp. 105-114).

2. Beaman, J., Meis, S. Managing the Research Function for Effective Policy Formulation and Decision Making. In: Ritchie, J.R.B., Goeldner, C.R. (eds.): Travel, Tourism, and Hospitality Research: A Handbook for Managers and Researchers. New York: John Wiley 1994 (pp. 39-47).

3. Enache, M. Integrating GIS with DSS: A Research Agenda. Proceeding of Urban and Regional Information. 154-166 (1994).

4. Finlay, P. Introducing Decision Support Systems. Oxford: Blackwell 1994.

5. Fishwick, A., Clayson, J. GIS Development Project. Final Report to the Countryside Commission. Kendal: Lake District National Park Authority 1995.

6. Gamble, P.R. Strategic Issues for the Management of Information Services from the Tourism Industry to the Hospitality Industry: A Lesson for the Future. In: Progress in Tourism Recreation and Hospitality Management. Vol. 5. Cooper C.P., Lockwood A. (eds.): Chichester: John Wiley 1994 (pp. 273-288).

7. Green, H., Hunter, C. The Environmental Impact Assessment of Tourism Development. In: Johnson P., Thomas B. (eds.): Perspectives on Tourism Policy. London: Mansell 1992 (pp. 29-48).

8. Hjalager, A.-M. Dynamic Innovation in the Tourism Industry. In: Cooper C.P., Lockwood A. (eds.): Progress in Tourism Recreation and Hospitality Management. Vol. 6. Chichester: John Wiley 1994 (pp. 197-224).

9. Sprague, R.H., Carlson, E.D. Building Effective Decision Support Systems. Englewood Cliffs: Prentice-Hall 1982

10. Sussmann, S., Rashad, T. Geographic Information Systems in Tourism Marketing. In: Cooper C.P., Lockwood A. (eds.): Progress in Tourism Recreation and Hospitality Management. Vol. 6. Chichester: John Wiley 1994 (pp. 250-258).

11. Yzewyn, D., Brabander, G.D. () The Economic Impact of Tourism and Recreation in the Province of Antwerp, Belgium. In: Johnson P., Thomas B. (eds.): Perspectives on Tourism Policy. London: Mansell 1992 (pp. 157-174).

The Relevance of Meta Modeling and Data Warehouses for Executive Information Systems

R. Kirkgoeze[1], A. Kurz[1], H. Reiterer[2], A M. Tjoa[1]

Vienna University of Technology, Austria[1]
{remzi,kurz,tjoa}@ifs.tuwien.ac.at
University of Vienna, Austria[2]
reiterer@ifs.univie.ac.at

Abstract: Applications based on Data Warehouses especially decision support systems such as Executive Information Systems (EIS) are rapidly becoming a key to gain competitive advantage for businesses. Data Warehouse allows businesses to get data from operational databases and turn that data into useful information on which users can carry out their analysis. In this paper we will discuss the importance of meta modeling for EIS and Data Warehouses as analytical base in EIS for decision support.

1. Introduction

Data Warehouses collect information from several data sources and integrate it into single database where it can be analyzed and queried by users of the Data Warehouse. The relations stored at the Data Warehouse represent materialized views over the data at the sources [1]. The base data from which the views are derived is stored in one or more data sources, where these sources may be distributed, heterogeneous and/or remote from the Data Warehouse. Because queries at the Data Warehouse are often long and complex, a Data Warehouse may contain many materialized views in order to speed up query processing [2].

As shown in Figure 1, there are two important components in Data Warehouse: the Data Pump Component, allows extraction, cleansing and integration of data from operational sources, and the Data Analysis Component, responsible for fulfilling the information needs of specific end users. Most current commercial warehousing systems focus on the Data Analysis component, providing specialized index structures at the Data Warehouse and extensive querying facilities for the end user.

Data Warehousing has been used successfully by many organizations to provide a consistent view of data across an enterprise and grant end users to answer business questions without assistance from any specialist. It also builds an ideal base for DSS and EIS with its integrated central data store.

An EIS is designed to satisfy the unique information needs of executives by providing friendly user interfaces to access managerial information, including summarized business transaction data. In addition, an EIS also supports access to external news services, tools supporting decision analysis, and interpersonal communication software. One of the most important driving forces is that executives can use EIS to obtain relevant information on line to support their decision making in increasingly dynamic business environment [4].

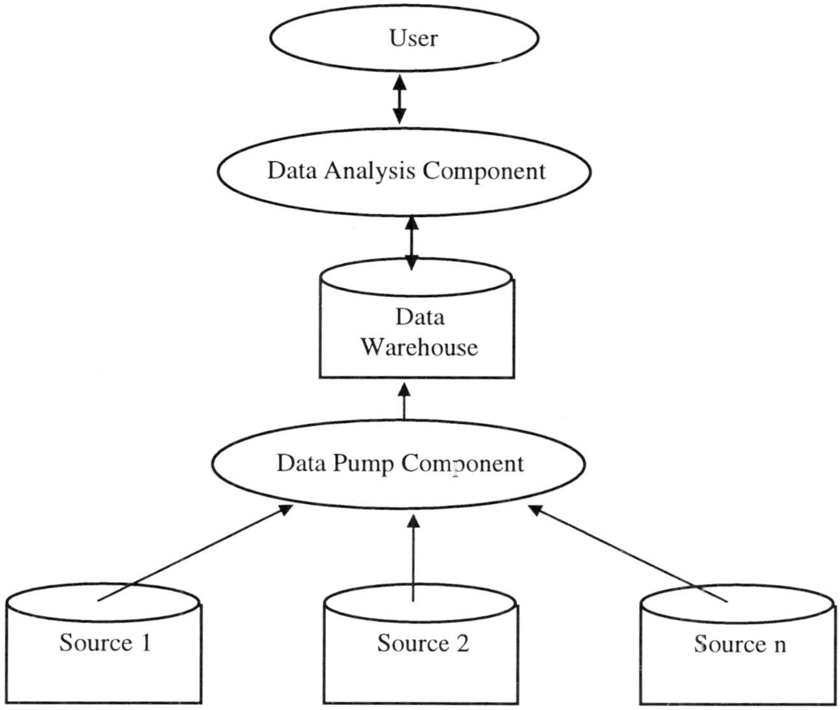

Figure 1. The basic architecture of the Data Warehouse

The emergence of EIS signals a shift of emphasis from development of information systems to the delivery of information produced by information systems. Rockard and De Long [5] identify four basic functions of EIS:

- Communication tools support executives' interaction with people using phone, voice mail, electronic mail, and fax
- Organizing tools, which include to-do lists, electronic calendars, group scheduling software, and project management tools, can be used to assists executives in organizing their tasks and managing their schedules.
- Data and status access tools allow executives to have access to external information via electronic news services and external databases. Executives can monitor critical performance measures of business operations stored in the information base.
- Decision analysis tools include spreadsheets, statistic and business graph packages, and data-driven and drill-down data analysis tools

Since executives rely heavily on both external information and the analysis of information by professional staff, it is important that staff members be included in the loop to provide soft data to the information bases. EIS are becoming a mechanism for enterprise-wide information delivery to all levels of organizations. Therefore, the EIS industry is changing the meaning of the EIS acronym from Executive Information Systems to Everybody's Information Systems or Enterprise Information Systems [6].

Technical components of existing EIS include: database system, graphical user interface, personal productivity tools, direct data import/export facilities, decision support tools, text and image management systems, and communication tools. Each of these address only a part of nowadays executives need. There is no direct support for the modeling of organizational models, an important feature that should be incorporated into future EIS [4]. Another aspect is more sophisticated visualization and navigation tools. Hypertext/hypermedia systems enable the capturing of unstructured nonlinear multimedia information and provide better user navigation [7]. There is no intuitive way to build a high-level user interface to integrate the diversified sources of applications and information required in EIS.

In this Paper we first describe EIS and DSS requirements to EIS systems. Secondly we focus on Data Warehouse and meta modeling for data warehousing. Finally we combine EIS and Data Warehouse and describe our model of Data Warehouse based EIS.

2. Executive Information Systems (EIS)

The definition of EIS varies according to the weight given to the three elements in the name itself. For some "an EIS is a set of tools designed to help an organization carefully monitor its current status, its progress toward achieving its goals, and the relationship of its mental model of the world to the best available signs about what is really happening". Some others state that "an EIS gives non-technical senior executives the ability to navigate through strategic information collected from company databases and on line information services". For our purpose, an EIS is a computerized system that provides end users with access to internal and external information that is relevant to decision making and other critical success factors.

The EIS has long promised to provide executives with the computing power and information access. The reality is that the EIS of today is moving down the organizational structure and the executive is still waiting for the benefits of the information technology. Strategic-planning capabilities, an external environment focus, ease use of learning and use, and custom tailoring to meet the unique needs of individual user are common properties of EIS.

EIS usually includes personal productivity, communication, analysis and coordination components that provide end users with a big range of services that increase productivity. It is agreed that an EIS has powerful analytical tools and the following characteristics:

- An easy to use and maintainable graphical user interface.
- Integrated capabilities for data access, security and control.
- On request drill down/roll up, slice/dice capability to lower and upper levels of detail.
- Behaving of organizational indicators.
- Functionality for decision support, ad-hoc queries and what-if analysis.
- Data analysis features
- Advanced report generation.
- Statistical analysis tools for summarizing and structuring data.
- Access to a variety of internal and external data sources.

Classical EIS tries with help of some querying and analyzing tools to satisfy above requirements by direct use of operational sources. Operational sources are designed to provide extremely efficient data access for large number of transactions involving very few records and to automate day-to-day operations. These on-line transaction processing (OLTP) systems insert and update data in databases at fast rates, but they are not very good to deliver meaningful analysis, answer ad-hoc queries and they do not keep historical data which is very important for EIS and DSS. All these affects the performance and acceptance of traditional EIS.

On-line Analytical Processing (OLAP) has emerged as a break through technology that can provide the foundation for fast, easy and affordable EIS solutions. OLAP, when integrated with a Data Warehouse enables very fast response, drag-and-drop navigation, point-and-click drill-down/roll-up among levels of details, exception reporting, what-if analysis [8]. In our model we suggest to use Data Warehouse/OLAP/EIS solution as an alternative to the classical , EIS to satisfy the needs of today's business users.

3. Data Warehousing and Metadata Modeling

The Data Warehouse is the foundation of DSS processing [3]. Because there is a single integrated source of data in the Data Warehouse, and because the data are accessible, the job of the DSS analyst in the Data Warehouse environment is immeasurably easier than in the classical environment.

The Data Warehouse is designed specially to meet the needs of decision-makers rather than the needs of OLTP systems. Unlike an OLTP system, a Data Warehouse is specialized for retrieving and analyzing information quickly and easily. Data Warehouse is specific database which consists of extracted and transformed operational data. Because a Data Warehouse is designed to support OLAP the data model of the Data Warehouse should be in the position to reflect needs of OLAP. OLAP needs can not be satisfied with classical ER modeling. Dimensional modeling has been represented to solve problems that have been faced in ER model to support OLAP.

Dimensional modeling is a technique developed to structure data around natural business concepts and to provide a foundation for sophisticated data analysis. In dimensional modeling, data structures are organized to describe business measures and dimensions. Business measures are the numerical data being tracked such as sales, production etc. Dimensions are natural business parameters such as product, time, region etc. that define each transaction.

The physical architecture of the dimensional model is described by many techniques. One of the popular technique, which has been used in our model too, is *start schema.* Star Schema, as the name suggests, the star schema is a modeling paradigm that has a single object in the middle connected to a number of object around. The star schema mirrors the end user's view of business query: a fact such as sales, compensation, payment and invoices are qualified by one or more dimensions such as time, product, region. The object in the center of the start is called *fact table* and the objects connected to it are called the *dimension tables.*

Before Data Warehouse data can be accessed efficiently, it is very important to understand what data is available in the Data Warehouse and where that data is located. Metadata or data about data provides DSS analyst with a catalog of data in the Data Warehouse. Before getting real data from Data Warehouse, user navigates through the Metadata of the Data Warehouse and only last operation is getting needed data from warehouse. Metadata is like the roadmap to data, it points to the location and meaning of various information within the Data Warehouse.

The design of Metadata for a Data Warehouse is a dramatic shift in paradigm for information analysts who have been analyzing and designing databases to support operational systems. The focus during the design of an operational database is to create normalized data models with atomic data. The design of Metadata for the Data Warehouse needs a complete change of these thinking. The focus of the design of

Metadata for the Data Warehouse is to present a rich range of relationships to the analysts, in most cases with a large number of redundancy, Because updates are not a primary issue within the Data Warehouse.

An important aspect of Metadata is the need to maintain mappings of data from the sources to the Data Warehouse through the process of extraction, transformation and integration. These mappings need to be maintained to verify the quality of data, synchronize and refresh data in Data Warehouse, and integrate the data.

Metadata for Data Warehouses can be grouped in three categories:

1. **Technical Metadata**: Contains Metadata that can be used by Data Warehouse administrators and designers. The following can be found in this category:
    - Information on the sources of the data
    - Information on the targets of the data.
    - Transformation information
    - Data mappings from source to target.
    - Cleanup rules.

2. **Business/Analyst Metadata**: Supports business analysts who do not have technical background. It contains the following kind of information:
    - Business terms.
    - Associations between technical terms/alias and the business terms.
    - Source data, derivation rules and currency data in common terms.
    - Description of the pre-defined queries in Data Warehouse which are described in business' terms.
    - Contact information for the owner of the data.
    - Authorization requirements for access.

3. **Enduser Metadata**: Contains Metadata for browsing and querying Data Warehouse. It contains the following kind of information:
    - Information about aggregation levels.
    - Drill-Down/Roll-Up information.
    - Information about previously created queries.
    - Request information to the Data Warehouse administration staff.

4. Executive Information Systems based on a Data Warehouse

Developing an EIS based on Data Warehouse must involve more than simply loading the available data. It also must be more searching than a response to a particular executive request. Instead, the approach must be entirely pragmatic. The Data Warehouse data model must address the conflicting goals of system flexibility and data delivery performance. Flexibility is required so that additional attributes, dimensions, and metrics can be added to the warehouse, complimenting and leveraging existing data. Quick query response is required in all effective interactive decision support systems, since it not only impacts end-user satisfaction but directly determines the number of analyses the DSS analyst can perform given limited time.

We suggest a three layer, *database / analysis / presentation*, model to satisfy the needs of the executives. This model has been successfully implemented in our WWW-DWH-EIS project. As it is shown in Figure 2, Data Warehouse is the bottom line of the model. It contains information from external and internal operational sources and presents them in a way that supports easy analyzes and querying. Data in Data Warehouse is organized into categories or subjects that are meaningful to business end user.

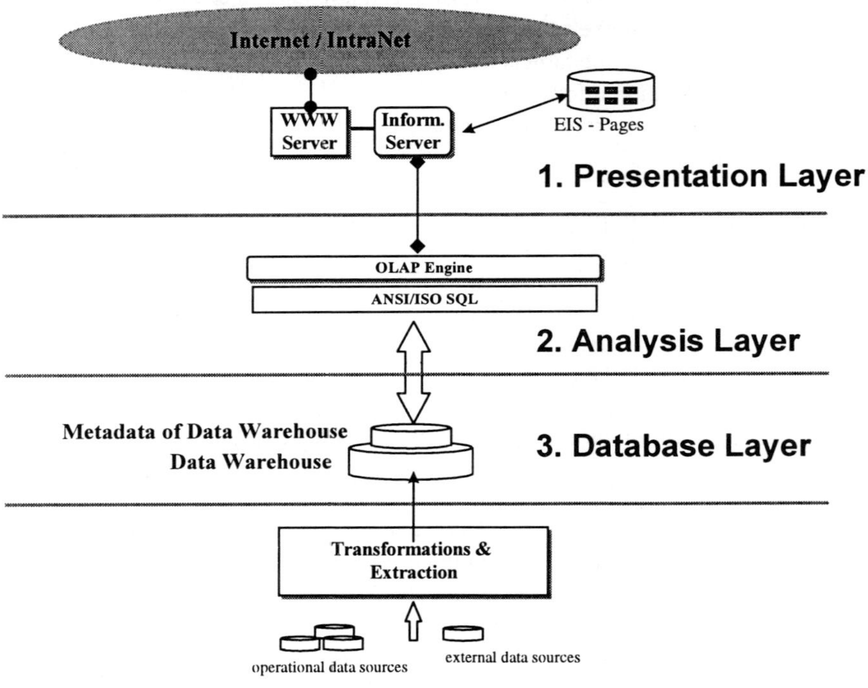

Figure 2. Three Layer EIS

The information that has been stored in the Data Warehouse is decoupled from operation sources. This issue makes loading new information into Data Warehouse very important. If data flow form operational sources to Data Warehouse is not satisfactory, the whole system will not give desired results. The Data Warehouse should be supported by tools to read, process and store significant data from operational sources into Data Warehouse quickly without being taken off-line. In our project we used Oracle as DBMS and *start-schema* for dimensional modeling to realize the Data Warehouse.

Analysis Layer is the connection between Data Warehouse and EIS-application. It merges data requests, Metadata and its own query building algorithms to create SQL queries to send to the Data Warehouse. Upon query completion, the Analysis Layer performs necessary data manipulations, such as cross-tabulations, before sending the data to the EIS-application. This layer requires a criteria set from the user, which specifies the attributes and metrics to be hold from Data Warehouse. It also uses a data rotation to specify how to orient Data Warehouse data on the X and Y axes for the EIS-application.

In our project we used a special meta script language to describe all user requirements to be performed by Analysis Layer. The meta script description of user requirement is processed by the layer and based on this process SQL-queries generated, data from Data Warehouse retrieved and all data manipulations done. Before sending data to the Presentation Layer again a meta script form of data is generated. The Presentation Layer interprets this script and generates appropriate screens for end user.

The Presentation Layer is the interface between business end user and the Analysis Layer. By using only standard input devices the user is in the position to formulate his wishes in this layer. This user input is transformed to a meta script language to send Analysis Layer. The prepared data by the Analysis Layer is interpreted and presented to the end user by using pre-defined user interfaces. This interfaces give the user the possibility to navigate through the data and get more detailed/global information. (drill-down/roll-up).

In order to have a common interface that can be used easily by any user we used Web browser as visualization tool for the Presentation Layer. The main task of the Presentation Layer in our model is to interpret and visualize pre-defined EIS-Screen, which are described by using meta script language. Meta script language of the EIS-Screen describes all screen components and data to be prepared by the Analysis Layer. This description is interpreted at run time and on fly HTML-pages are generated. The interpretation is done by Information Server(IS) which is the heart of the Presentation Layer.

During interpretation IS checks if data needed from Data Warehouse, if it the case, the description of the needed data is sent to the Analysis Layer and the received information is integrated into generated HTML page. We have implemented three ways of data visualization on the screen; regional map, cross-tables and graphical(bar charts). Each of these enables the user to navigate through the Data Warehouse. An important function of the Presentation Layer is the traffic lightening, which draws the attention of the business end user where to concentrate.

5. Summary

The data warehousing revolution has finally acknowledged the requirements of business users for easily accessible, accurate and timely information. Previous efforts have been worried with difficulties, both from a business politics and technology perspective. With the changes in business and managerial structure, coupled with technology advances, the realization of easy access to business oriented information has become easier.

The existence of a Data Warehouse allows organizations to retain multiple years of historical data. Standard query/report tools can access that data but performance and functionality of analysis will not be satisfactory for the today's executives. EIS supported by OLAP and Data Warehouse appear to be the best solution to enable executives to interact with one of their most valuable resources, information.

The Executive Information Systems (EIS) has long promised to provide executives with the computing power and information access. The reality is that the EIS of today is moving down the organizational structure and the executive is still waiting for the benefits of the information technology. By using Internet/Intranet in our project as Presentation Layer we tried to make use of current technology. This is a great step from Executive Information System to Everybody/Enterprise Information System.

6. References

[1] IEEE Data Engineering Bulletin, Special Issue on Materialized Views and Data Warehousing, 18(2), June 1995.

[2] V. Harinarayan, A. Rajaraman, J.D. Ullman. Implementing data cubes efficiently. In Proceedings of ACM SIGMUD 1996 International Conference on Management of Data, 1996.

[3] W.H. Immon. Building Data Warehouse. John & Sons, Inc. 1996.

[4] M. Chen. A Model-Driven Approach to Accessing Managerial Information. Journal of Management Information Systems, 1995, Vol. 11, No. 4, p. 33-63

[5] J.F. Rockart, D.W. De Long. Executive support system: The emergence of top management computer use. Homewood, IL: Dow Jones-Irwin, 1988

[6] P. Pinella. An EIS for the desktop. Datamation, May 1, 1991

[7] J. Conklin. Hypertext - An introduction and survey. IEEE Computer, 20, 9(1987), p. 17-41

[8] R. Kirkgoeze, A. Kurz. Using Internet as an EIS-Client. CON'96: Future Information Technologies, OCG Schriftenreiche Band 95, R. Oldenburg, 1996, p. 75-84.

SnowNet: An Agent-Based Internet Tourist Information Service

A. Erni, M. C. Norrie

Institute for Information Systems
Swiss Federal Institute of Technology (ETH)
ETH-Zentrum, CH-8092 Zurich, Switzerland.
email: {erni,norrie}@inf.ethz.ch

Abstract SnowNet is a project to provide Internet access to a variety of existing information services concerning snow and avalanche conditions in the Swiss Alps. To provide regular users, such as mountain guides, with access to the latest information without typical network delays, we use a combination of client and server agents. These agents cooperate to reduce network traffic and ensure that current pages are always available in the user's local cache.

1 Introduction

For tourist information services on the Internet to be effective, it is essential that the requirements of regular users can be met without the typical delays currently experienced using Internet browsers and the World Wide Web (WWW). The main reasons for such access delays are overloaded networks and servers and these problems are likely to remain with us in the foreseeable future. We have addressed these problems by developing an agent-based architecture for access to Internet databases. We are using this architecture to construct an integrated tourist information service (SnowNet) for snow and avalanche data in the Swiss Alps in conjunction with the Swiss Federal Institute for Snow and Avalanche Research (SLF) in Davos.

SLF currently provides a range of information services in support of the Swiss tourist industry and also the scientific community. These services include the generation of avalanche forecast warnings, snow cover maps and statistical summaries of snow and avalanche data. All of these services are of interest to some provider or customer of the Swiss tourist industry. For example, a tourist may be interested in general information on snow conditions, whereas mountain guides and those responsible for ski resorts would be interested in more detailed information on snow profiles and avalanche paths.

To put the current SLF services on the Internet would not in itself be a major problem. The goal of the SnowNet project is to provide an attractive, interactive and integrated service that meets the requirements and expectations of various user groups. This goal includes ensuring access to current information without delays and, further, controlling access to information and implementing a charging mechanism.

Our main concern is to provide regular users of the SnowNet information services with fast access to their desired information. Such users may subscribe with the information service and, by installing a special *internet agent* on their machine, in most cases, the required WWW pages will be prefetched into the user's local cache thereby avoiding network delays. Prefetching of data can be based on both user and application profiles which, together, enable user information requirements to be predicted and the currency of cached data to be ensured. We refer to this prefetching strategy as *active caching* since it means that data is cached not only as a direct result of user-initiated data transfer, but also on the basis of agent activities.

On the side of the information service, we have a *server agent* which controls access to the information services. The server agent registers users and cooperates with the internet agents to implement active caching and ensure that only new and updated information is transferred from the server. Our approach further reduces network traffic by decomposing pages in such a way that only updated parts of pages need to be transferred and using Java [Fla96] for application programs in order that graphics can be generated on the client rather than server side.

A major part of the development was to consider how to exploit and integrate existing Internet services and technologies. We have investigated several alternative products and approaches on a variety of platforms. Our current prototype system uses the Netscape 3.0 browser on the client side, the Oracle WebServer 2.0 [Ora96] on the server side, the Java programming language and Javascript, and HTML [Gra96] extended with specific tag extensions for agent communication.

In section 2, we describe the SnowNet information services in more detail to establish the basic requirements of such a system. Section 3 presents the overall architecture in terms of the client and server sides, the agents and the various information sets. In section 4, we provide some detail of our prototype and, specifically, the communication between the server and internet agents. Concluding remarks and an outline of future research is given in section 5.

2 SnowNet Services

SLF plays a very important role for the Swiss tourist industry and scientific community in providing a range of information services about the snow, avalanche and weather situation all over Switzerland. Data are collected from various sources in the Swiss Alps. Automatic sensor stations provide SLF with the newest data on an hourly basis and, during the winter, human observers provide the newest observations of the snow situation. All this data is stored and processed at SLF in Davos.

The information services include generation of avalanche forecast warnings, snow cover maps and statistical summaries of snow and avalanche data. Avalanche forecast warnings are generated daily, while snow cover maps are produced weekly with additional maps being produced in the event of heavy snowfalls. Some of these services, such as the basic avalanche forecasts, are provided as a public service free of charge, whereas others are charged to support the operation of SLF.

Currently users can access these information services in different ways. Textual avalanche forecasts and general information about SLF is already available on the Internet in normal HTML format. For more detailed information and graphical representations, the user has to download the desired data from a communication server in Davos in using special software called InfoBOX [SLF] which is sold by SLF. Data consists of measurements for 3-day periods for various locations in the Swiss Alps. InfoBOX comprises different programs for data presentation according to the kind of information required. The information can be displayed using different graphics or just as normal text.

Another method of information distribution is used for snow cover maps. Interested parties may subscribe to SLF and receive newly generated snow cover maps by fax. In certain newspapers, a snow cover map appears once per week. The various information services are being integrated such that they will be supported by a single database together with various application programs.

Many kinds of users access these information services. Examples are mountain guides or people responsible for ski resorts who need detailed information on the snow situation, such as snow profiles and avalanche paths, in order that they can predict avalanche danger. Other users are tourists just interested in the general snow situation in the Swiss Alps to decide, for example, where to go skiing. Since historical records are kept, it would also be possible in the future to provide extended services to help tourists plan holidays according to the likelihood of good snow conditions. Further, one could envisage the situation where such a service was provided as part of an integrated Swiss tourist information system.

As can be seen, SLF supports a range of multimedia tourist information services involving a large degree of data and user heterogeneity. It is therefore important to provide such services in an attractive and effective way that meets the requirements and expectations of the various user groups, while controlling access to certain information and implementing SLF's charging policy.

Nowadays, it is desirable to make such important tourist information service publically available by putting it on the Internet. As previously mentioned, some information about SLF is already available on the Internet and it would be no problem to add the data of the other services as simple HTML pages on the WWW. But the goal of the SnowNet project is to provide effective information services to users regularly accessing the data without having to deal with the usual performance problems of the Internet browsers and the WWW.

Because different kinds of people have different aims, we provide them with different access means to the information services. We distinguish between casual users and regular users. Casual users would include general Internet surfers, just curious as to what services are available, and individuals occasionally looking for some information about the snow and weather situation in the Swiss Alps. Such users are more interested in the available services than in meeting strict performance requirements in accessing them. For them, we integrate all existing SLF application services into a single, integrated tourist information service with a single entry point in terms of a WWW home page. This involves, not only design issues of producing effective interactive, multimedia presentations, but also mechanisms for access control and charging. An example screen shot showing this single entry point (in German) is given in figure 1.

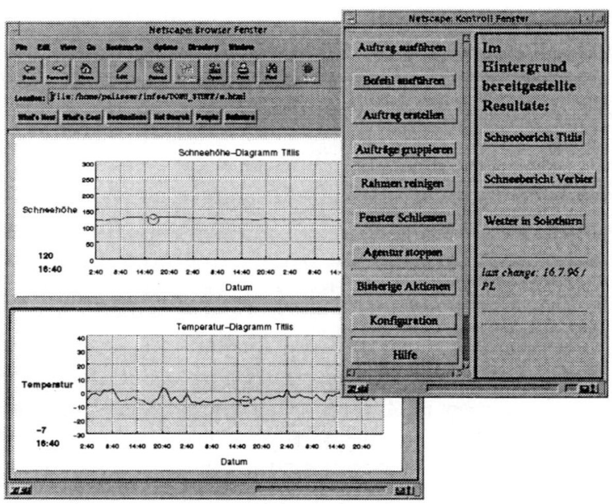

Fig. 1: SnowNet User Interface

As can be seen, the user interface uses the Netscape browser. The interface itself is implemented as HTML documents containing JavaScript code and Java applets. A control panel which controls the information services is shown on the right. The main panel contains two forms – one displaying a graph of snow depth and the other a temperature graph.

Our main concern, however, is the regular users, such as mountain guides, who frequently use the available tourist information services and are dependent on them. For them, our system should ensure fast access to the desired information. Other users who regularly access SLF information include ski resort managers, media services such as newspapers, government agencies and scientists.

Note that different user groups may have to pay different amounts for access to the same data. For example, government agencies, media services and tourists may be charged at different rates according to their expected information usage and frequency of access.

To address the needs of both casual and regular users, we developed an agent-based architecture which is presented in the following section.

3 SnowNet Architecture

We propose a general agent-based architecture for the development of integrated, multimedia information services for Internet databases, as shown in figure 2.

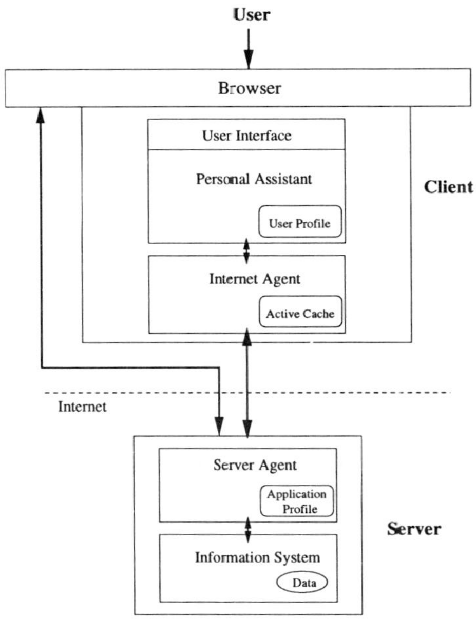

Fig. 2: SnowNet Architecture

A combination of server and client agents provides performance gains to users through semantic and active caching mechanisms based on user and application profiles. The client side actually consists of two cooperating agents referred to as the personal assistant and the internet agent.

Overviews of general agent models and technologies are given in [Mae94, WJ95]. In our architecture, agents have four properties: *autonomy*, *social ability*, *reactivity* and *adaptivity*. An agent is *autonomous* in that it knows its tasks and capabilities and may act without the direct intervention of users or other agents. An important aspect of an agent is that it must be able to communicate and cooperate with users and other agents and we refer to this as *social ability*. An agent also perceives its environment and reacts on changes that occur in it and we refer to this property as *reactivity*. An agent is *adaptive* if it has some learning capability and can adapt over time to improve its performance and adjust to new situations. Specific examples of agents which learn from user behaviour are reported in [MCaLMZ94, MK93].

The server agent controls all access to the database and application services. Casual users may access the information service directly via the usual WWW services. However, regular users may install local client agents and register with the server agent, thereby enabling them to optimise their working with the system and, possibly, gain access to additional application services.

For registered users, user response times are improved through a combination of techniques to reduce network traffic and perform *active caching*, i.e. prefetching of data into the local cache. The reduction of network traffic is achieved in two ways. First, cooperation between the server and client agents ensures that only new and updated information is transferred from the server, and other information is accessed via the local cache. Our approach goes beyond standard WWW caching as we can decompose pages and detect changes to only parts of a page. For example, a weather map could be decomposed such that the background, which remains the same, could be fetched from the local cache, while the updated information such as temperature readings are transferred from the server. Second, multimedia data such as graphics and animations are frequently generated by application programs rather than stored in the database. The use of Java for application programming means that rendering of raw data on the client side is possible.

The main task of the internet agent is to reduce access times for the user. This is done in cooperation with the personal assistant and also the server agent. Active caching is based on both user and application profiles. Based on the user profile, the personal assistant predicts when the user needs which pages, and the internet agent then loads these pages into the local cache. The server agent has an application profile at its disposal that describes the data types, the price of the data, the expiry date and access rights. Because the server agent keeps track of changing data, it can automatically send the newest version of the data to the internet agent without explicit request. The internet agent is also responsible for handling partial updates, when only parts of the data have changed.

A user profile consists of a set of event-condition-action (ECA) rules specifying actions to be executed after a certain event happens. For example, a time-triggered rule can be used to prefetch data shortly before the user normally requests that data. Other prefetch ECA rules may be used to specify access patterns in terms of browsing sequences. For example, if a user typically requests a certain page shortly after requesting some other page, we may use a rule to prefetch the second document immediately after the first one has been requested.

The personal assistant is responsible both for maintaining the user profile and for communicating with the internet agent to request services from the information server according to the information in the user profile. The ECA rules in the user profile may be specified directly by the user or be generated by a learning component of the personal assistant which monitors user behaviour. For example, in the SnowNet project, user monitoring by the personal assistant may reveal a usage pattern that a mountain guide requests the weather report between 5am and 6am every day. This information can then be used to generate a time-triggered rule that causes the personal assistant to request the relevant pages from the internet agent at 5am every day and store them in the local cache. The user may also request to have some special data always updated in the cache and, for this, the internet agent needs to communicate with the server agent.

In general, the personal assistant is a user-specific agent which controls the interface to the user and provides him with all of the available services. It is possible that such a personal assistant would also cooperate with other forms of domain specific agents such as an email agent.

In SnowNet, the server agent makes it possible to load just that part of an HTML page which has changed. It also provides the functions for charging. Because the server agent knows all the registered users accessing data from the information server, it is possible to keep track of the fees every time it sends data to a user.

4 Prototype

In section 3, we presented a general architecture for providing information services on the Internet that uses cooperation between client and server agents. In this section, we describe in detail how we developed our SnowNet prototype based on this architecture. Particularly, we show how a number of existing Internet services can be exploited in the development of such a system.

In SnowNet, we implemented the agents in the Java programming language.

Java is used because it is portable across different platforms and further supports object-oriented design and programming. The agents are standalone Java applications so they can run independently of any WWW browser. Communication between agents, both within a site and between sites, is over sockets. All output is formulated as HTML pages containing JavaScript elements for dynamic behaviour. These HTML pages are displayed by the Netscape Navigator Web browser. Communication between the JavaScript programs and an agent is achieved through use of LiveConnect, Netscape's internal communication interface. For the user interface, the agents rely on Netscape 3.0 since it supports frames and LiveConnect. The internet agent allows access to both standard WWW servers and to the server agent of SLF. The database management system on the server side is Oracle. The forms of communication between clients and the information server is shown in figure 3.

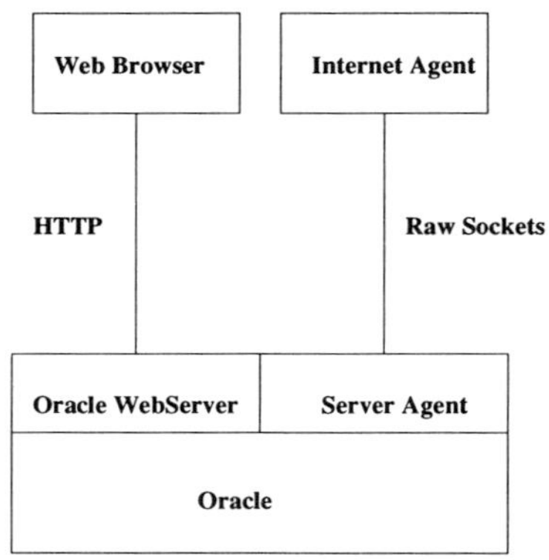

Fig. 3: Client-Server Communication

For communication with standard WWW browsers, the server agent relies on Oracle's WebServer 2.0. In the case of access by internet agents, the server agent uses Oracle's proprietary database interface. The content of the database is visible to the internet agents in a virtual hierarchical structure which is built by the server agent. The server agent itself performs all of the mapping between the hierarchical structure and the relational data in the database.

The primary goal of SnowNet is to provide high performance access to the the information server. In the best case, no single byte of data is transmitted unnecessarily and every page requested by the user is available, and guaranteed up-to-date, in the local cache *before* the user actually requested it. As discussed before, two methods – active caching and partial updates – are required to achieve this and we describe here in more detail how these were realised in the SnowNet system.

Active caching mainly takes place in the internet agent and means that documents are prefetched before the user actually requests them. Documents may be prefetched into the local cache as the result of a request from the personal assistant as it, in turn, predicts user requests. However, documents may also be downloaded into the cache as a result of old versions becoming outdated. This can occur in one of two ways. For documents that are produced at regular intervals, an expiry date can be included in the header which informs the internet agent when the document will cease to be valid. At this time, the internet agent can remove the old document from the cache and send a request for the document to the server agent. It is also possible for the server agent to initiate document transfers by notifying internet agents when documents have been updated. This is possible since interested parties are registered with the server agent. The internet agent informs the personal assistant that a new document has been cached and it decides whether any immediate action is to be taken such as notifying the user. Clearly, there is always a limit to the size of the local cache and it is therefore necessary for the internet agent to decide how best to use the cache resources. For a discussion of general caching strategies for WWW see [Wes95]. Detailed information about the SnowNet cache replacement strategy is given in [ENP96].

To improve efficiency, our system also supports *partial updates* in which only parts of an HTML page may be updated and it is not necessary to reload the entire page. Clearly, this feature is particularly useful where images, such as maps, are displayed. Support for this notion of partial updates is only possible in cooperation with the server agent and not for access to standard WWW servers.

Partial updates are implemented by extending HTML with a new tag. The tag is inserted by the server agent especially for interpretation by the personal assistant. The WWW browsers used by our users do not see these SnowNet-tags, as they are filtered out by the personal assistant and replaced by standard HTML commands. Note that all Web-browsers ignore unknown HTML tags. This makes such customised extensions of HTML feasible without causing compatibility problems.

The tag used for partial updates is as follows:

```
<SW7A_Agent COMMAND=command PARAM=parameters>
</SW7A_Agent>
```

The personal assistant executes the command and the result is placed between <SW7A_Agent> and </SW7A_Agent>. The command can, for example, be the instruction to fetch an image, much like the <IMG> tag in HTML.

Extending HTML in this way with specialised tags is very flexible and we used it as a general means of communication between the server and client agents.

5 Conclusion

The SnowNet system will provide effective Internet access in terms of performance requirements to a range of services in support of the Swiss tourist industry. Cooperative agents on the client and server sides enable an improved service for regular users, without affecting casual user access via the usual WWW services.

The general approach is applicable to many tourist information services on the WWW where, typically, there are a large number of casual users, but also many regular users for whom availability of information and acceptable performance is critical.

Currently, we have a prototype SnowNet system which supports a limited subset of the SnowNet information services. The system will be extended both in terms of the services provided and also the agent functionality. We plan to implement a second prototype using our own object-oriented database management system, OMS, which has a WWW interface.

References

[ENP96] A. Erni, M. C. Norrie, and L. Perrochon. Agent-Based Information Services for Internet Databases. Technical report, Swiss Federal Institute of Technology (ETH), 1996.

[Fla96] D. Flanagan. *Java in a Nutshell.* O'Reilly & Associates, 1996.

[Gra96] I. S. Graham. *HTML Sourcebook.* John Wiley & Sons, 2nd edition, 1996.

[Mae94] P. Maes. Modeling Adaptive Autonomous Agents. *Artificial Life Journal*, 1(1 & 2), 1994.

[MCaLMZ94] T. Mitchell, R. Caruana, D. Freitag abd L. McDermott, and D. Zabowski. Experience with a Learning Personal Assistant. *Communication of the ACM*, 37(7), 1994.

[MK93] P. Maes and R. Kozierok. Learning Interface Agents. In *Proc. of AAAI'93*, 1993.

[Ora96] Oracle Corporation, California 94065. U.S.A. *Oracle WebServer 2.0*, 1996.

[SLF] SLF, Swiss Federal Institute for Snow and Avalanche Research, Davos, Switzerland. *InfoBox.*

[Wes95] D. Wessels. Intelligent Caching for World-Wide Web Objects, 1995.

[WJ95] M. Wooldridge and N. R. Jennings. Agent Theories, Architectures, and Languages: A Survey. In M. J. Wooldridge and N. R. Jennings, editors, *Proc. ECAI-94 Workshop on Agent Theories, Architectures and Languages.* Springer-Verlag, 1995.

Expert systems as an important competitive advantage in tourism

N. Tödter[1], B. Brigl[2]

[1] Technical University of Dresden, Institute for Economics and Transport, Chair of Tourism Economics, D-01062 Dresden, toedter@rcs1.urz.tu-dresden.de

[2] University of Heidelberg, Department of Medical Informatics, Im Neuenheimer Feld 400, D-69120 Heidelberg

Introduction

Today, holiday-makers often are overtaxed in making the right vacation choice because they have to take into account a variety of information sources like operators' brochures, destination specific literature etc. In most cases, even the travel agency staff is not informed enough to assist the traveller in its decision process because the knowledge about destinations is very complex and is changing rapidly.

Nevertheless, decision support systems and expert systems are not used in the domain of travel counselling. Expert systems have the opportunity to derive problem solving proposals using formally represented knowledge ([1]). A special kind of expert systems are knowledge-based decision support systems which are able to solve complex decision problems. Successful examples are known from the domain of industrial services and manufacturing or from the medical domain ([2]). For the domain of tourism, HRUSCHKA and MAZANEC ([3]) discussed already in 1990 the problems of transforming information available in tour brochures and its evaluation in terms of motives into machine readable formats.

At the same time, mainly the bigger tourism companies are strategically preparing the introduction of new technologies, partly through the further development of computer reservation systems (CRS), to compensate for the budget deficits of closed systems and to evolve their advantages, partly through the development of strategic concepts about what products should be sold in future and what conditions have to be fulfilled for that. KRUPKA and TILEMANN mention as biggest challenge the Dynamic packaging and Trading ([4]).

Both aspects are important for the economic future of the tourism industry and the market position of travel agents and tour operators. The horizontal competition ([5]) among travel agents and tour operators and also other companies of accommodation and transport as the basic product elements is counselling in the best way to the customers. Therefore, quality, security and price are important criteria for competitive advantages.

The travel decision

The planning of a journey is a complex decision. Figure 1 shows a decision tree which describes exemplarily factors which influence a journey. HRUSCHKA and MAZANEC describe the process of a travel decision as the attempt of a diagnostical check to explore the benefits sought by the traveller, and the selected trip as the therapy to cure the wants and desires ([3]). The tourism sciences prefer a model of seven steps for the travel decision. First comes the decision between going on a holiday journey and staying at home.

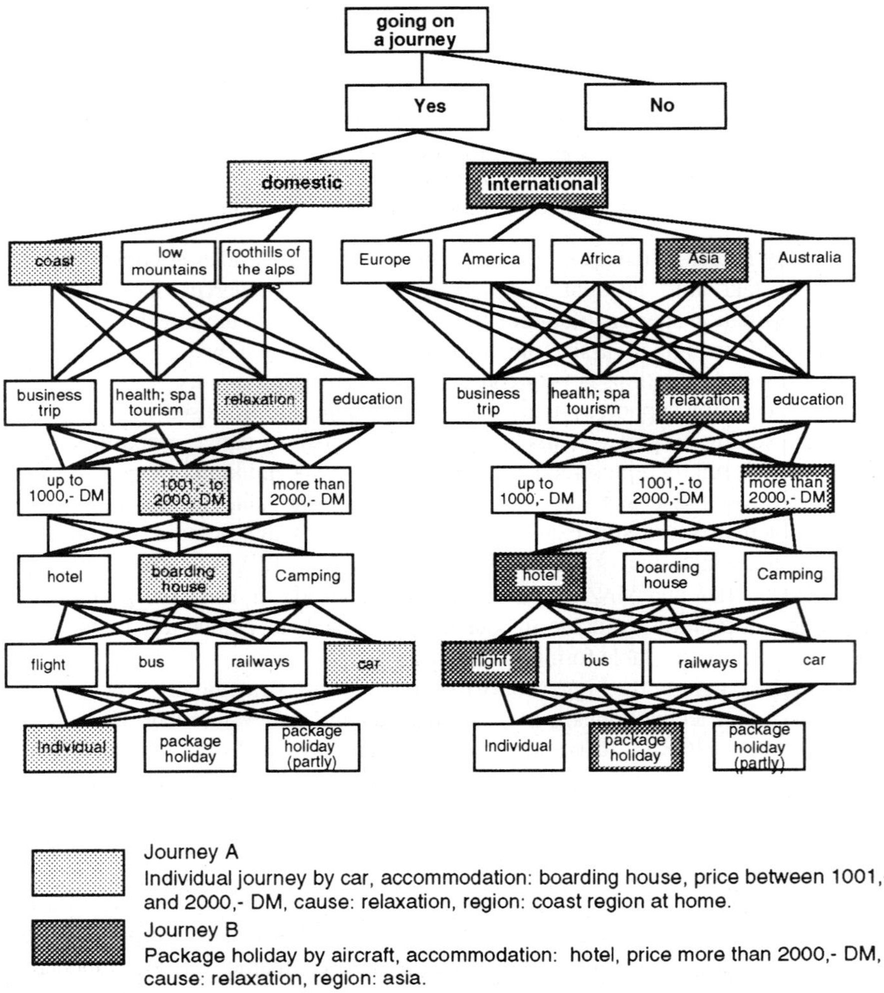

Fig. 1: Decision tree for the journey demand ([6]).

Nearly 70% of German people make one holiday trip per year which lasts more than four days. Those people have to decide in the next step about the destination. Nearly 30% of them 70% prefer a domestic destination and 70% are travelling to foreign countries. The next decision step asks for the cause/motivation of the journey.

Subsequently, the customer has to decide about the price, the accommodation, the transport and the organisation of the journey.

Most customers need much more information before booking their holiday trips "as the most beautiful days of the year". Normally, the clerk should assist the customer in his/her decision process. High quotes of civil cases in Germany concerned with 'missing holiday fun' and trips with defective services show clearly that many realised holiday trips had not been of best choice.
Tourism industry standardises the complex decision problem of travelling only in form of printed brochures. With few exceptions, destinations are in the focus of attention. There are special brochures only for travels like golf trips, spa tourism etc. Not taken into account are many other facets of customers motivations. Let us take an example: three different persons plan to take a holiday together and they each select their personally most important decision making criteria, e.g.

- island, sun, few other German tourists,
- golf, good medical support, no earthquakes,
- horse-riding and young people.

To find the right destination the customers had to study a multitude of offers and printed brochures. For the agency staff it is impossible to get the required information from the existing information and reservation systems. Often it is more a lucky chance if the clerk will find the right destination and product. Whereas decision support is only provided within the person-person-interaction, new technologies have the problem that they first of all have to care with criteria like price, destination or product.
To be able to place the customer in the centre of the communication process, new and comprehensive databases have to be developed, which offer different accesses and navigation possibilities. Depending on their demands, their knowledge and the time to invest, customers need corresponding information and communication possibilities. To a large extent this implies:

- if a customers knows date, destination, hotel and other services
  --> direct booking without delay by offering already known information,
- if a customer only knows that s/he demand a two week holiday in Spain for about 1000.- DM:
  --> direct offers comprising different journeys to Spain for about 1000.- DM and the possibility to get interactively more information about the different offers.
- if a customer wants to play golf, need a good medical care and is afraid of earth quakes:
  --> offer a golf holiday in an earthquake proof destination with good medical care([7]).

On the one hand, the tradition of closed on-line systems in tourism (computer reservation systems) is the optimising of selling an existing product and not to inform and assist the customers in booking the right vacation. On the other hand there exist more and more tourism information systems from tourism destinations without a product you can buy.

Quality of staff knowledge

Some important facts about die quality of staff knowledge are a result of an expert interrogation of the German institute of professional education (BIBB) in 1996 which was carried out by the Chair of Tourism Economics, Technical University of Dresden ([8]).

Permanent training concerning destination knowledge is necessary to enable agency staff to sell a destination. The knowledge necessary comprises not only product knowledge available in tour brochures or mediated in operators' courses but also knowledge about geography, politics, law, society etc. This kind of knowledge is mainly acquired through info and leisure tours, but also through the lecture of technical literature. Within the last five years, about 60% of the clerks of tour operators and travel agencies went on a info tour. The enormous demand for training and education is first of all a mean to refresh obsolete knowledge. The mediation of that knowledge is very time consuming. The enterprises invest considerably in training and education to offer a well-trained staff. Today as well the staff of travel agencies and travel operators tend to specialise concerning destination knowledge. The extent of necessary knowledge does not only increase because the number of destination increases but also because the customers are more experienced and are more informed through media like television, travel information events or others. The knowledge acquired through training is only valid for a small period of time and is related to the person trained. This leads to two problems:

(1) If a clerk is ill, occupied only half-time or leaves an enterprise, the enterprise also looses the clerks' expertise. The enterprise is often forced to specialise a new clerk.

(2) A clerk is not able to follow up changes about a destination continuously. S/he always counsels the customer using experience of the past.

To increase the quality of counselling, travel agencies need expert systems where the knowledge base is updated regularly, is up-to-date and can be called off individually at any time. Only if the counsellor knows more than the customer, s/he will pay for counselling. Otherwise new media will substitute travel agencies .

In future, the use of new technologies and the integration of expert system in the decision process will not only influence the competition within the branch of tourism, but especially will change the value added chains of tourism companies. Whether such processes will have revolutionary or evolutionary character will depend on the speed of the development of technical requirements and on the strategic planning of tourism companies. The communication and information systems that have to cope with the customers demands, have to care intensively about different motives of travelling. Active decision support which today needs personal advice can be completed by computer-supported decision support systems. The growing information sources intensify the problem of the travel decision process. Tourism information systems of today only insufficiently consider the customers desires and wants.

Expert systems in tourism

The problem of demand

The German holiday-maker have different motives for their holiday trip. Every year the German 'Studienkreis für Tourismus' (study circle of tourism) until 1992 and the German 'Forschungskreis Urlaub und Reisen' (research circle holidays and travel) from 1992 ask a representative selection of holiday makers about their different motivations. Recovery, recreation, time where you can do nothing, to do what you want to do, fitness, beauty, health, having fun, doing sports, meet people, cultural aspects are the main reasons of making holidays. These motivations give no answer to the question why e. g. more people spend their holidays in Spain than in Scotland. The image of a destination, experiences of the customer himself or of friends and promises in the brochures influence the potential customer in finding the right holiday trip which meets his/her wants and desires.

The problem of offering

Today travel operators need criteria to pack their packages before they get an answer about how many people will book their arrangement. Success and failure is not manageable at short notice.

The problem of counselling

The counselling problem had been discussed in the chapter staff quality. Firstly, it is a lack of up-to-date information about all aspects of a tourism product. Secondly, it is the problem to match the individual motives of a customer with a high standardised product arranged in the past.

These different kinds of problems concerning the travel decision process leads to the following considerations about the use of expert systems in tourism:

(1) Dynamic packaging as an expert system for tour operators

Dynamic packaging on the one hand means, that journeys are no longer composed and offered by the travelling organisations in advance, but that they are composed when the customer wants to buy a journey or when the distribution channel calls for it. Aim is, to link the different distribution channels together. Then, journeys can either be sold using catalogues or be distributed through kiosk-systems, CD-ROMs or Internet using the same database. On the other hand, the basis of tracing are strategic and tactic elements. A necessary prerequisite is the categorisation of tourism products using wholesale prices. With reference to dynamic packaging the prices are determined (just like an electronic market) when supply and demand come together . Therefore you need a relational data model, holds the whole spectrum of information about the enterprise inclusive facts of financial management, customer databases and databases of the suppliers of part performances. Next step is the integration of all relevant facts for the product to build categories and to index these facts. In the model of dynamic packaging the selection of the right product will be steered by an inference mechanism. Primary selection criteria concern the customer itself (e. g. accommodation, date, transport,

price). Secondary criteria describing the enterprise can be selected (e. g. capacity, modality of payment, prefer suppliers and so on) ([4]).
With dynamic packaging tour operators can make available 'just in time' products and need not keep ready the whole capacities a long time before booking. A further existing problem is, that it will not be interesting for tour operators to combine the motivations of travelling into these process. The motivation for dynamic packaging will be at first the re-engineering of the value added chain of the largest enterprises to get a competitive advantage.

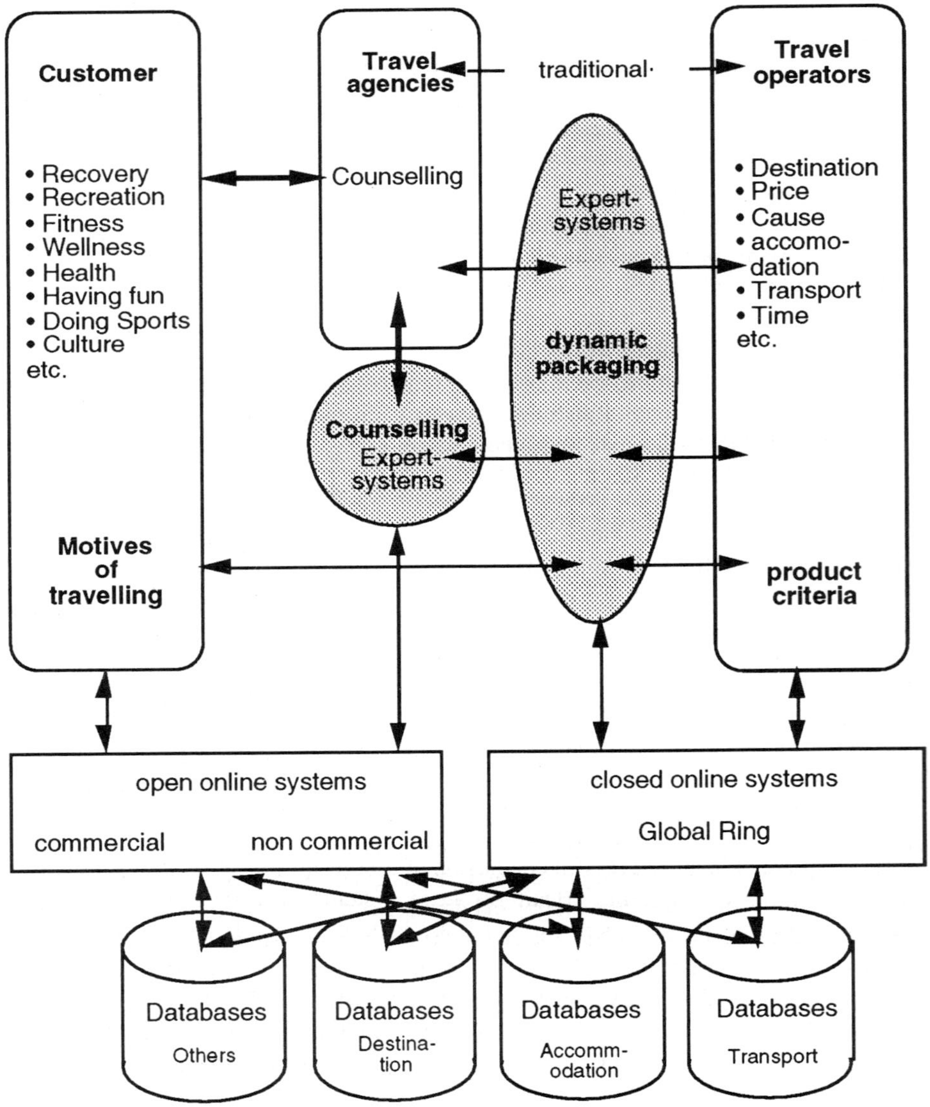

Fig. 2: Expert systems in tourism.

From the systems' development point of view dynamic packaging is a kind of construction problem, i. e. the problem solution is composed of a set of building blocks considering a set of additional conditions.

(2) Expert systems for travel agencies

Travel agencies will be able to use the expert system of dynamic packaging in the distribution process. Unfortunately they also will lose customers to the direct distribution channel. The dynamic packaging offered by tour operators will have a high attractivity for customers who know about their motives and perhaps use the fastest possibility to bock a journey.
Travel agencies need an expert system which focus on the counselling of the travel decision. An expert system for travel agencies can fulfil two different functions. First, it can provide up-to-date information about the climate, geography, culture, politics, different tour operators, accommodation-possibilities, security standard of transport-enterprises etc. For this, you can think about an hypermedia information system with comprehensive possibilities of navigation. Secondly, an expert system can derive journey offers from a customers' motive description. From the systems' development point of view, counselling is a diagnotic problem, i. e. a set of symptoms (in this case motives) is matched to one or more diagnoses (in this case journeys).
The integration of decision support functions into existing application software products like computer reservation problems demands for the integration of databases which contain customer-dependent data and knowledge bases which contain customer-independent knowledge. The quality of knowledge and the process of knowledge acquisition is decisive for the success of knowledge-based systems.

Conclusion

At the DRV-congress in 1996 in Orlando, USA, Romeiß-Stracke mentioned that in future the question 'Where do you want to go?' will be substituted by the question 'How do you want to feel?'. Instead of travel agencies there will be agencies of lifestyling ([9]). Without individual mass-tourism products as a result of dynamic packaging it will be impossible to answer questions like the one mentioned above. The implementation of dynamic packaging can be a revolution for the value added chain of tourism enterprises. But the introduction of dynamic packaging will only be one aspect of a more customer oriented market.
Travel agencies of today's standard will be the loser of the future. High standardised journeys will be sold more and more directly. For 'agencies of lifestyling' the act of selling a product is the successful conclusion of a high quality counselling task using modern multimedia systems. Heart of these systems will be a decision support system. Clerks need much more time to investigate the lifestyle of a customer. Only with decision support systems they can give up-to-date answers about all dimensions of a holiday trip. To diagnose wants and desires of customers correctly, agency staff need results of different sciences and the whole systematized information integrated in an expert system. Then it will be possible that they give an answer of the question:

Say, what kind of motives you have,
and I will tell you at what time and with whom you can go to what place.

References

[1] Puppe, F.: Einführung in Expertensysteme, Berlin. Springer,1991

[2] Feigenbaum, E, McCorduck, P, Nii, HP 1988: The Rise of the Expert Company: How Visionary Companies are Using Artificial Intelligence to Achive Higher Productivity and Profits. London: Macmillan, 1988.

[3] Hruschka, H. and Mazanec, J.: Computer-Assisted travel Counseling, in ANNALS of Tourism Research, 1990, pp 208-227

[4] Krupka, B. and Tilemann; B.: Touristik: Surfing on the Tourism-Net, in: Little, A.: Management im vernetzten Unternehmen, 1996, pp 202-228

[5] Schulz, A. et al: Tourismus und EDV, 1996

[6] Freyer, W.: Tourismus-Marketing, Manuskriptfassung, 1996

[7] Tödter, N.: Multimedia und Online-Systeme im Tourismus; Einsatz neuer Technologien als Wettbewerbsvorteil in vernetzten Tourismusunternehmen in: Freyer, W.; Scherhag, K. Zukunftswerkstatt Tourismus, 1996, pp 281-303

[8] Freyer, W. and Tödter, N.: Karrierewege von Beschäftigten in der Reisebranche, Forschungsbericht des BIBB, 1996

[9] Romeiß-Stracke, F. Statt Reisebüro eine Agentur für Lebensgefühl. FVW international, Nr. 25/96, p 13

Distributed software Agents for WWW-based Tourism Information Systems (DATIS)

Thomas Steiner
HEC / INFORGE, University of Lausanne, Switzerland
thomas.steiner@hec.unil.ch
http://www.hec.unil.ch/people/tsteiner

1 Introduction

With the Internet becoming more and more popular, an increasing number of online systems allow travelers today to access *interactive destination information systems*. Such systems on the WWW do not only offer comprehensive descriptions of tourist destinations. They also provide access to sophisticated reservation systems, where the user can book flights, pass hotel reservations, rent cars, buy concert tickets, etc.
In contrast to this, the actual *information retrieval* approach on the WWW (browsing) might unfortunately become inappropriate for efficient information gathering during the customer's destination selection process. The difficulty level of the systems and the time required for the information retrieval process make online destination information systems unpopular.[Ng96] Moreover, from an information processing perspective, there is a limit to the number of information resources and browsing steps a human being can deal with.
In our paper we suggest a meta-architecture of *distributed software agents* that aims to significantly enhance the information retrieval process for *WWW-based Tourism Information Systems* (TISs). The DATIS project innovates with a society of distributed software agents that keep themselves informed about the availability of tourism specific information sources within the context of the WWW. As a result, they build a meta-architecture that is masking the underlying, hierarchical structure of the Internet and the corresponding information retrieval complexity.
The use of a meta-architecture in our concept makes information retrieval highly flexible, much less formal than traditional access methods, and supports heterogeneous information sources and their eventual change. Moreover, the suggested meta-architecture maintains *physical data independence* for the destination specific design of WWW-based information systems. As a result, we also believe to improve the design, communication and execution *autonomy* for tourism information providers on the Internet.

2 Software Agents

The word agent - in a sense other than 007 or travel agent - is becoming more and more popular in press, in education, as well as in industry. The question 'What is an agent ?' is an embarrassing one. The problem is that *the term agent is widely used* - even so widely, that it seems to become popular to sell an important number of concepts in some

'agentified' form. There are nearly as many definitions of the term agent as there are papers, willingly relating to a common understanding of what it *could* be.
For most of the authors, an agent is a *physical or virtual entity* that is able to act in an environment, to perceive it, and to directly communicate with other agents. Moreover, agents are generally guided by a set of *trends*, have their own *resources*, a partial *representation* of the environment, as well as some competence to offer *services*.
Hence, following such very general definitions, a human being - with eyes, ears, and other *sensors*, hands, legs, mouth, and other *effectors* - and a light switch both fall under the term agent. For this reason, we believe that a stronger notation of agency is required in order to describe the software agents we are referring to in the following.
The best *definition* to describe our type of agents seems to be the one provided by [Maes95]: Software agents are 'computational systems that inhabit some complex environment, sense and act [...] in this environment, and by doing so realize a set of goals or tasks for which they are designed'.
In particular, the agents we speak of are computer programs that *are distributed over the WWW*, contain a representation of this environment, are skilled to communicate with other agents by the Internet, and are able to reason in some way about the actions to take:

DATIS agent = actions + representation + communication + reasoning

From an architectural point of view, DATIS provides a framework to coordinate the behavior of several specific agent types that will be explained below. Different approaches may be followed when *coordinating agents* to form an agent society. The architectures most frequently use either shared memory or message passing for inter-agent communication.
In the DATIS project, the software agents use *message passing* in order to establish a communication with other agents : They all speak *KQML* (Knowledge Query and Manipulation Language - a DARPA standard) as an ACL (Agent Communication Language) and exchange messages through the *Internet* as the underlying communication medium.

3 The DATIS project

At the current stage of the project, DATIS describes the *theoretical framework* for the prototype implementation of an agent society. It has been developed preliminary to future implementations which will encompass both human computer interaction contexts, and distributed artificial intelligence issues in the domain of WWW-based TISs.
The DATIS agent society consists of four *agent types* : *Information Watchers* (IW) survey WWW-servers for tourism related information. They periodically scan these resources for appearance of tourism specific keywords which are stored in a global dictionary. On every change, they report to their corresponding *Information Broker* (IB). IBs have a double function : on one hand they redirect change notifications to *Directory*

Agents (DAs), which each keep a dynamic reference list with URLs related to their specialization. On another hand, IBs publish their dictionary to *GUI agents* (GAs). GAs are the visible part for the user : they assist the user in the browsing process and offer the new service provided by the underlying agent society. With respect to the browser profile, they consult DAs and check their dynamic entries for appearance on the HTML-page that is actually displayed in the browser.

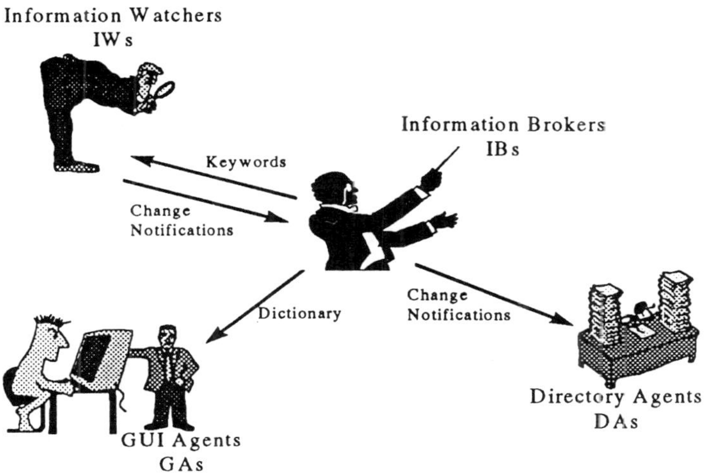

fig. 1 : the DATIS agent society

The four types of agents correspond to *four predominant tasks* a potential guest is confronted with, when browsing the Web for destination specific information: Searching the Internet for tourism related information sources by means of a browser (GAs), keep a list of interesting information sources (DAs), periodically review these resources for eventual change (IWs) or new information, and finally put all this together in some kind of browsing behavior (IBs).
Following the *agentification principle*, we have first determined the human activities that are best replaced by a software agent when searching the WWW for tourism related information. The second stage consists in physically creating the processes - writing the software agent programs in our case. In the third stage of the agentification process, the software agents are tied together through inter-process communication and build an agent society.
From *the tourist's perspective*, a DATIS session contains the following steps - please open `http://pcbfsh1-128-2a.unil.ch/DATIS/` with your browser:

1. The potential guest opens a DATIS-GA with a WWW browser.
2. A DATIS-GA basically consists of a menu and a sub-browser, in which the user can click on hyperlinks in the traditional way. Until here nothing spectacular, the application simply offers a browser within the browser.

3. The user personalizes the browser by specifying the research criteria (first option in the GA menu). Research criteria are chosen from a list of keywords that the GUI agent uploads from the Information Broker.
4. The sub-browser processes the page and marks all links of potential interest with a particular sign - actually an edelweiss:

5. The user is free to follow the links suggested by the system or not. If he or she follows hyperlinks marked with an edelweiss, he or she will browse towards WWW-pages that correspond to the specified research criteria.

The power of DATIS for potential tourists lies in the fact that GAs *completely mask the technical infrastructure* of the system. DATIS users do a priori not realize that their enhanced browser implies a whole society of agents devotedly working behind the curtain.

From *the destination's perspective*, a DATIS session contains the following steps:

1. The tourist destination designs its WWW pages and publishes them either on its own WWW server or with an Internet Service Provider.
2. The tourist destination makes sure that an IW is created and configured - by opening it anywhere on the WWW in a browser - which will watch the pages for eventual change. This can either be done by the site manager on a PC at the destination, or be delegated to any person with an Internet connection and a WWW-browser.

And that's all to be done at the destination! The power of DATIS for tourist destinations lies in the fact that the *design autonomy* for destination related WWW information sources is guaranteed. Besides traditional WWW publishing recommendations (HTML, netiquette, etc.), there are no supplementary presentation standards to be respected - you are free to make use of the full representational abilities of the WWW!

From *the system administrator's perspective*, a DATIS session contains the following steps:

1. The system administrator keeps a WWW-server running, from which the four agent types are served in the traditional way.
2. The system administrator opens and configures an IB in a WWW-browser. DATIS updates the other agent types, and the next time an agent is distributed to a browser, it already knows that the IB is available.
3. Additionally, the system administrator might also open IWs for tourist destinations that are not willing to run an IW on their own resources - as IWs, DAs, IBs and GAs may sit on any node of the WWW.

The power of DATIS for system administrators lies in the *distribution* of the agents: Each agent type is designed to be opened as a an application in a WWW-browser. The only restriction is that the agents need a valid e-mail account in order to communicate

with other agents.
This 'weakness' can in the same time be seen as a strength: The use of Internet mail as the underlying inter-agent communication medium allows *off-line functionality* and improves *fault tolerance*. The messages are left on the mail server until the agents retrieve them. This implies e.g. that IW agents even don't need to be activated permanently.
We believe that DATIS will be a very flexible system for two reasons: the use of more or less platform-independent WWW programming tools - or at least of tools that seem to evolve towards such portability - significantly increases *environment flexibility*.
The fact of using Internet mail as a an *inter-agent communication medium* is all the more an elegant solution, because it is already available in the environment for which the agents are implemented. We consider the - moreover global - availability of a message passing medium to be a lucky punch, as by far not every Multi-Agent System (MAS) designer can make use of such a powerful tool...
In a first time, the DATIS prototype is planned to *focus on tourist destinations in Switzerland* - with a `.ch` in the domain name. We will begin to direct our IWs to such Swiss tourist destinations and give a certain number of IBs and DAs to the system. On another hand, this doesn't necessarily imply that the DATIS prototype will not be globally available: from the beginning, IWs, IBs, DAs and i.e. GAs are supposed to be functional on any node of the WWW.

4 Logic Programming for the WWW

A lot of the current work on WWW concentrates on the improvement of document presentation. On another hand, there is also a increasing need to improve information gathering. In the following we describe the use of *logic programming* to improve information gathering in the context of the WWW.
The WWW environment imposes very *special requirements on the information gathering problem* - i.e. the need to deal with variety, change, and autonomy of both clients and information sources.[Fikes95] The available retrieving technologies (browsers etc.), often return information that is incomplete or irrelevant because of an inappropriate mapping to heterogeneous information sources.
Before diving deeper into WWW issues, it might be opportune to very briefly sketch out the most important characteristics of the system [Dobson96]: The WWW uses *hypertext*, allowing information to be inter-connected with hyperlinks, which users may traverse to navigate through the information space.
The WWW uses a standardized HyperText Markup Language (*HTML*) which allows text, graphics and hyperlinks to be combined quickly and simply. On the WWW *publication is decentralized* : every organization or person with an Internet connection can make information available to the world.
Finally, the WWW *subsumes other services,* so that within a single framework, one can access hypertext, File Transfer Protocol (*FTP*) compliant files, news, mail, and all other Internet services through the mechanism of Uniform Resource Locators (*URLs*).
A good example of such subsuming of other information services is the attempt of some

researchers to access relational databases via the WWW. These approaches use a gateway to generate queries and convert the results back into HTML.
More recently there have been suggestions to use a server-side logic base, which can be accessed through a query implemented by a CGI-script, and which converts the results to HTML before it delivers them back to the client. We call this centralized approach the *integrated logic base*.
In the DATIS architecture we use a second - and still less deployed - way to implement WWW-based logic bases : the *distributed logic base*. Here we consider the logic base to reside *within* the HTML-page that is distributed to the browser.
The distributed logic base is inspired by the ability of societal information gathering approaches to *communicate knowledge queries*. It implements the *downsizing and distribution* of knowledge, and can therefore lead to a reduction of the information overload, when combined with a pertinent source and domain model.
From an implementation point of view, a distributed WWW-based logic base consists of a HTML-page body which *embeds* a *KQML-interpreter* (ontology-independent outer shell), as well as a *Prolog-interpreter* (ontology-dependent implementation of KQM services).

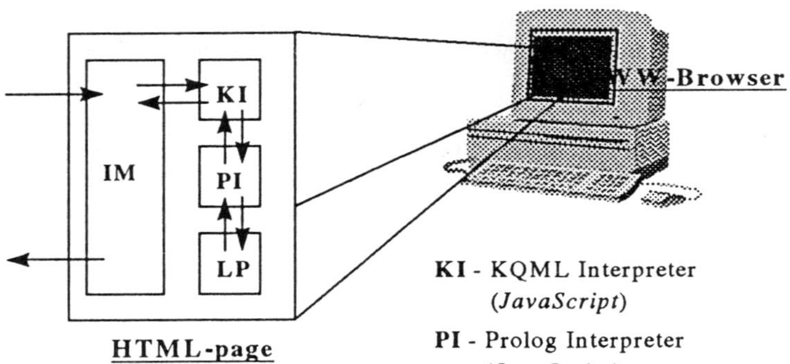

KI - KQML Interpreter (*JavaScript*)

PI - Prolog Interpreter (*JavaScript*)

LP - Logic program (*Prolog*)

IM - Internet Mail (*ActiveX*)

fig. 2 : the 'inside life' of a WWW-logic base agent

KQML-performatives are sent between agents using the Internet's Simple Mail Transfer Protocol (*SMTP*) service. Once the performative has been extracted from the mail body, it is interpreted by the KQML-interpreter which maps the ontology-independent KQM requests to implementation-specific Prolog predicates.
The *Prolog predicates* are declared by a logic (Prolog) program that is executed by the HTML-embedded Prolog-interpreter. In the DATIS architecture, each of the four agent types is characterized by a specific logic agent program.
The agents are linked together by the underlying SMTP-infrastructure. The *third stage of*

the agentification process - the binding of the individual agents into the society - is satisfied by the execution of the agent application in the browser, and the availability of inter-agent communication through mail exchange. This tight integration of the communication support and the logic (Prolog) program ensures a semantically sound treatment of the meta-terms and facilitates distributed logic programming. [Bonnet96]
The interpreters are programmed in *JavaScript* to support a maximum of platform independence. The HTML-page - which physically represents the body of an agent, embeds two *ActiveX* components for exchanging mails with other agents over the Internet. All the internal parts of the agent are scripted together to form the agent entity. This corresponds to the *second stage of the agentification* model, the creation of the agent process.
A very important requirement for Web-distributed logic bases is that the rules they consist of have to be *maintained properly*. An out-of-date Web-logic base will be of no use for agents accessing KQM services. We have chosen the *model of reified behavior* explained in [Bonzon93] and [Kuokka96] to define our logic agent programs in a convenient way.
The reified behavior model offers - in a sound theoretical framework - a loose distinction between *theories* (programs and data) and *meta-theories* (theories about theories). These are outstanding assets for the support of ontology-independent KQM services and the programming of societal agent based applications.
Document-retrieval use of hypertext-embedded knowledge via societal information gathering approaches is in contrast to traditional *fact-retrieval* usage of Web databases. The suggested distributed logic bases can be seen as a category between relational databases and traditional HTML documents.

5 RAD of WWW-based Software Agent societies

With the paradigm shift to highly inter-connected computers and programming tools, programming languages have a new duty: whenever possible, they should contribute to practical *distributed application development*. In such distributed, heterogeneous environments though, traditional programming languages often fall short in exploiting the potential of the new trend - it becomes i.e. much harder to bring *logic* programming closer to the new mainstream technology.
This is all the more a sad story that many of the *requirements* for the new tools and languages of WWW applications *are already met* by existing logic programming environments, i.e. Prolog: rapid declarative prototyping, automatic memory management, database interfaces, symbolic computation for decision support and information management, etc. [Bonnet96]
Quite often, far reaching *mathematical modeling* has lead to prove the correctness of intuitive assumptions in agent theory, and from a purely theoretical point of view a lot of them have made their marks for their coherence and formal soundness.
Unfortunately a lot of these models specify very narrow theories that cannot directly be achieved in practice. We believe it to be the agent programmer's work to check the very abundant theoretical body of agent-related theories with a healthy portion of *pragmatic*

criticism.
To keep in touch with implementation issues when flying through the theory cloud, *Rapid Application Development* (RAD) tools gain very high importance for software agent developers. Finally, what's the value of any brilliant agent theory, when there's no instantiation possible in the world of everyday problem solving?
RAD tools such as Delphi, Visual Basic, IntraBuilder, etc. can assist multiagent system designers in testing different information states of agent systems. *Prototyping* with these RAD tools can therefore be seen as a bridge between theory and practice. We believe that RAD of software agent societies increases our understanding of existing DAI systems, that it will aid the design of future systems, and that it will encourage the reuse of models for multiagent problem solving.
In the following, we will briefly mention the RAD tools we use to implement the prototypes of the DATIS architecture. We do not enter the subject of presentation details here - i.e. the use of custom WWW publishing tools and Internet server technology will not be mentioned in detail.
For rapid high-level prototyping, we use the *Amzi!Prolog* IDE with extended predicates written in and imported from a Delphi DLL. In contrast to traditional procedural environments, Prolog separates the control structure and the *declarative* structure of a program. The innovative potential of Amzi!Prolog lies in the *Logic Server* which allows RA Developers to *extend* core Prolog functionality with extended predicates written in Host Languages.
The other way round, Prolog logic bases can easily be *embedded* in a Host Language application - Amzi!Prolog comes with interfaces for C++, Delphi, Visual Basic, PowerBuilder, and many other RAD tools. More recently socket-support, CGI-extensions and a Java class of the Logic Server have been made available.
We write our *logic agent programs* with this RAD tool. Once the Prolog code has been tested, it can be embedded into the HTML pages - which correspond to the body of our agents -, and then be processed by the Prolog interpreter.
The Prolog and KQML interpreters are written in *JavaScript* with Borland's *IntraBuilder*. For portability reasons, everything of DATIS that will be distributed to a tourist's or destination's browser needs to be environment-independent. Over the past few years, Java/JavaScript has evolved as a standard that addresses such portability issues.
We believe IntraBuilder to be a valuable RAD tool for lightweight agent society prototyping. Offering HTML-integration of custom databases, Java applets, Java scripts and ActiveX technology, IntraBuilder *potentially supports all three inter-process communication styles* of distributed programming: live data structures, message passing and distributed data structures.

6 Conclusion

In [Bloch96] we have illustrated *some trends shaping modern tourist destination strategies*, such as e.g. the mass customization of services, the interactive design of products with customers and the increasing information intensity of products. It is clear

that with tourists shifting from service to *self-service* consumers, innovative destinations will try to offer original micro-services, from which the guest will form a personalized service package.
More recently, tourist destinations have begun to offer such micro-services *with new Information and Communication Technologies* (ICTs). The number of tourist service providers actually dispatching information on the Internet is impressive. On another hand, the actual information overload on the Internet does unfortunately not really speak in favor of new ICTs as a distribution channel for tourist micro-services.
In order to guarantee at least a minimal structure of WWW-based TISs and a convenient way for potential tourists to access them, *sophisticated information gathering techniques* will be needed, as stated above.
The DATIS project attempts to use *a societal approach* to information gathering in the context of WWW-based TISs. With a MAS that is distributed over the Internet, we try to provide browsing assistance to potential guests surfing on the Internet.
There is no doubt that such assistance is crucial for the tourist's decision making process. We are convinced that in the future only advanced information gathering techniques will allow new ICTs to fully play their role as a leverage for modern business trends in the tourism industry.

Acknowledgements

The author would like to thank the Inforge department at the University of Lausanne, i.e. Prof. P. E. Bonzon, for providing the technical and intellectual framework for the project. Furthermore, DATIS is a PhD project supported by the Swiss National Science Foundation (Project 12-47141.96).

References

[Bloch96] Bloch, M., Pigneur, Y., Steiner, T.: *The IT-enabled extended enterprise, applications in the tourism industry*; in S. Klein & al., 'Information and Communication Technologies in Tourism', Proceedings of the ENTER96 conference, Springer 1996.

[Bonnet96] Bonnet, Ph., Bressan, S., Leth, L., Thomsen, B.: *Towards ECLiPSe Agents on the Internet*, Proceedings of the 1st Workshop on Logic Programming Tools for INTERNET Applications, http://www.cs.mu.oz.au/~ad/lp-internet/eclipse/ea.html.

[Bonzon93] Bonzon, P. E., Kuokka, D. R.: *Decision support systems from an AI perspective*, 5th Annual Conf. on Intelligent Systems in Accounting, Finance and Management, Stanford University, 1993.

[Dobson96] Dobson, S.; Burill, V.: *Towards improving automation in the WWW*, invited talk for the British Computer Society seminar on New Directions in Software Development, University of Wolverhampton (March 1995), http://www.cis.rl.ac.uk/proj/www/docs/bcsndsd.ps.

[Fikes95] Fikes, R., Engelmore, R., Farquhar, A., Pratt, W.: *Network-based information brokers*, in Knoblock, C., Levy, A.:'Information gathering from heterogeneous, distributed environments', Papers from the 1995 AAAI symposium, AAAI Press.

[Kuokka96] Kuokka, D.R., Bonzon, P.E.: *Reifying Behavior in Decision Support Systems*, International Journal of Intelligent Systems in Accounting, Finance & Management, Vol. 5 Number 2, June 1996.

[Maes95] Maes, P.: *Artificial life meets entertainment: Interacting with lifelike autonomous agents*, Communications of the ACM, 38 (11), 1995. Special Issue on New Horizons of Commercial and Industrial AI.

[Ng96] Ng, F. Y. Y., Sussmann, S.: *A personal travel assistant for holiday selection - a learning interface agent approach*, in S. Klein & al., 'Information and Communication Technologies in Tourism', Proceedings of the ENTER96 conference, Springer 1996.

[Steiner95] Steiner,T.: *Information technologies and destination management in tourism: a conceptual framework*; Master thesis, Ecole des HEC, Université de Lausanne, October 1995;
`http://www.hec.unil.ch/people/tsteiner/office/tourism`

Agent Oriented On-line Bibliography

The author maintains an agent oriented bibliography on-line. At the time of writing, the bibliography contained more than 300 agent related references. We invite interested readers to have a look at
`http://pcbfsh1-128-2a.unil.ch/svr/intrasrv.isv?apps/AgentBiblio/agents.jrp`
for further references in the context of this project.

Realizing Travel Malls: A Logic Programming Based Approach

Subhasish Mazumdar

Department of Computer Science
New Mexico Institute of Mining and Technology
Socorro, NM 87801. USA.

mazumdar@cs.nmt.edu

Abstract

The vision of a Travel Mall, if fully realized, would allow the tourism industry to come to grips with the potential of the advances in information and communication technologies as well as the needs and demands of customers in this new age. In order to realize this vision, certain issues need to be addressed; that of information integration is crucial. We have presented an architecture for a travel mall that tackles this problem and addresses all the important issues. We outline how our approach exploits an extension of logic programming.

1 Introduction

The personal computer revolution has recently been joined by rapid changes in communications technology; we are left trying to fathom the ramifications of this synergy. First, relentless improvements in hardware and software have provided workstation capabilities to PCs. Second, advances in networking coupled with Internet services such as the World Wide Web (the fastest growing service on the Internet and still growing at a phenomenal rate) have brought planetary dimensions to the information access capability of ordinary individuals; the ease of use and low price of Web browsers have translated this capability into power. Every service industry must take cognizance of this phenomenon; the tourism industry in particular, with its heavy dependence on the world of information, must both exploit and cope with the consequences of these changes.

We concentrate on two of the most crucial agents: the customers and the service providers. The former includes both individual and corporate entities; the latter includes the travel providers, the hotels, the owners and managers of tourist attractions, including even small ones, e.g., a Ma-and-Pa family establishment with a special-interest collection located near a more widely known tourist location. The main question is how will customers obtain information relevant to their interests and what can be done to enhance the possibility that the providers' service information reach the customers.

The Internet is largely responsible for promoting the feeling among customers that travel and tourism can be organized differently from the past; their needs are changing along with their information access capabilities. With a self-service mentality, they are desiring customized holiday packages; they wish for more frequent, but shorter travel, last-minute reservations, global advice, service quality, and market transparency [1]. Existing medium and large service providers are finding traditional travel agencies inadequate for addressing these changing customer trends. Tiny and small service providers feel that in the increasingly smaller world, they too should be able to reach out to customers, but cannot. It should be noted that all providers, large and small, can post on the Web information regarding their service available; the problem is that there are few *systematic* channels for that information to reach the customer. At the same time, while the Internet provides access to very large amounts of data, customers feel frustrated when they cannot distill relevant information from the bewildering mass of data.

Bloch *et al.* have studied the Information and Communication Technology's enabling role in the new tourism industry [1]. Their envisaged solution is the construction of a *travel mall*, a specialized information mall, i.e., an on-line place where large number of online buyers and sellers can congregate and where commerce can be centered. The hope is that a customer using the Internet would access the travel mall directly and cut down on search time and effort, while enjoying increased success. The following needs must be fulfilled:

- connection of small and large suppliers to on-line facilities (such as the Web);
- integration of information from all suppliers;
- capability to search for product and services that match customer needs (not always known a-priori);
- customization of information for a specific customer;
- service that is flexible and dynamic;
- provision of value-added services, e.g., weather information, visa requirements, and experiences of past travelers.

The following would be useful:

- synthesis of composite information from data on separate trip modules (as increasingly self-reliant customers prefer to obtain) after checks on their mutual consistency.
- proactive targeting of prospective customers for a particular service and/or special packages.

The main problem with realizing this vision is that information integration is very difficult. Note that the Web was designed to facilitate *access* to information, not the *integration* of that information. *Lack of centralized control*, a crucial factor behind the success of the Web, is also its weakness. The sources of data exhibit no uniformity in their systems and platforms; some use old *legacy* systems while others others can partially migrate to open systems. Some use sophisticated DBMSs while others use flat files. We therefore contend that Bloch *et al.*'s suggestion that we should demand common formats for data sources through industry standardization is impractical: it goes against the very nature of the Internet and the truly global dimensions of its full potential.

Moreover, a truly diverse (nationally, ethnically, culturally, geographically) set of providers would use different terms for the same information and have subtle differences in *meaning* even when there is agreement at the symbolic level. Various components of a holiday package may well have their prices published, but they may include all, some, or none of local tax, entertainment tax, value-added tax, etc. The term *attached bath* carries different meanings across the globe; the simple query *find the cheapest single room with attached bath* may be impossible to answer *precisely* even if every provider went on-line with identically structured data on identical computer hardware running identical operating system software. Thus, we must rule out any approach that requires a central organization to comprehend and merge all pre-existing and new sources of data; the autonomy and ephemeral nature of such sources only exacerbate the problem. Neither common data structure nor common data semantics being feasible, any travel mall must cope with heterogeneity in the structure and semantics of their information without demanding a standardization of data.

In this paper, we outline our approach which, we believe, can address the difficult problem of information integration and can fulfill the needs enumerated above. This paper is structured as follows. In the next section, we outline our approach and remark on our current prototype. Subsequently, we summarize related work. Finally, we make concluding remarks while enumerating the advantages of our approach.

2 Our Approach

In this section, we summarize our approach: we list various dimensions of the problem to be solved, and for each, we outline the technical aspects of the solution. We refer to the *Travel Mall Information System* as the *TMIS*.

1. We believe that the World Wide Web is currently the best hope for bringing various hardware and software platforms communicating online. Old and new systems can publish their information using the Web either by putting it on their own PC with Internet connection or via an Internet Service Provider. Further, customers with their PCs can access the Web via widely-available browsers such as Mosaic, Netscape, and Internet Explorer.

   In response to appropriate invocations from the TMIS, the provider's system must be able to send data — this can be done using CGI scripts. Those providers using Database Management Systems (DBMSs) can make use of products (e.g., *web.sql* from Sybase) offered by the major DBMS vendors for achieving similar ends.

2. The TMIS offers the customer an initial Web page whose design admits of infinite possibilities. While this presentation is left to the TMIS designer, it must allow a help facility with which its vocabulary is clearly explained. The power of hypertext can be exploited to facilitate in following relationships among terms. For example, with regard to lodging facilities, the notions of different kinds of room sizes, bed sizes, attached baths/showers, etc, need to be clearly explained. It is also necessary to delineate the relationships such as between the specific and the general, e.g., how transportation is related to land travel, cruises, airport hubs, etc. Terms for value added services such as weather reports etc would be integrated.

   To achieve this need to access definitions and query relationships among terms, we use a declarative logic-based approach: specifically, *Datalog* rules. Such rules capture the implicit structure of the knowledge domain of tourism by encoding the relationship of various terms while allowing for flexibility and dynamic changes. Using a Logic Programming methodology, these rules can be used to answer queries and to guide the customer from general terms to more specific ones.

3. After browsing the TMIS web page, customers can specify their detailed queries using Web-standard HTML forms. These forms are screens that can be viewed and filled with standard World Wide Web browsers. The customer can fill in the blank fields with values (e.g., *Los Angeles* in the field marked "destination") or inequality predicates (e.g., trip cost should be *less than 1000* dollars). Such specifications joined together is a query. Thus, the customers do not have to learn the underlying information model (e.g., the Entity-Relationship model) or the syntax of a new query language.

4. The TMIS must also interact with the providers using the same Web tools. The provider is given the TMIS' initial vocabulary, a description of the relations among the terms, and constraints that hold among them. By default, any contributor of this TMIS implicitly agrees to the vocabulary and meanings. However, in the extremely likely case where a provider disagrees with the meaning of certain terms, it is free to *link* its own meanings with that of the TMIS.

   For example, the TMIS may allow the customer to obtain, for any participating hotel, a list of the various types of rooms and their costs. A provider hotel may have a list called *roomlist* of costs for each room type, but these costs may include taxes while the TMIS' costs do *not* include taxes. Therefore, the costs of our hotel provider would have to be transformed by deducting an appropriate percentage. Linking consists of connecting the symbol *hotel* of the TMIS with the symbol *roomlist* of the provider while specifying the adjusting transformations. We show below that a Datalog rule serves this purpose.

The above example would be handled by the Datalog rule (where capitalized names stand for variables and lower case names are constants):

```
hotel(Rmtyp,Cost) :- send_data(d1,'rmlist',Rmtyp,Netcost),
                     Cost = Netcost*0.9 .
```

which says that the data about "hotel" (in the TMIS) which may be conceptualized as a table with columns named "Rmtyp" (i.e., room type) and "Cost", can be obtained from the source d_1, by asking it to send data on "rmlist" (i.e., room list) which again looks like a table with columns named "Rmtyp" and "Netcost". However a value of "Netcost" obtained from d_1 must be multiplied by 0.9 to make it equivalent to the TMIS's notion of Cost.

This rule would be contributed by the provider hotel (data source d_1); installation of this rule in the TMIS "links" the symbol "rmlist" of the provider with "hotel" of the TMIS. Similar rules from other providers who have differences in meanings would provide similar links. If a customer queries the TMIS about the cost of a certain type of room in hotels, the query would be answered by following every link from the TMIS to individual data sources. The link from the TMIS's symbol "hotel" to our provider (source d_1) would be followed by posing to it the query on "rmlist" and transforming the values obtained (through multiplication by 0.9). The customer would be given the merged results from all such providers on all such links.

Logic programming associates procedural semantics with predicates [12]; we extend the paradigm [4] by associating *remote procedure calls* (RPC) with *distributed* predicates. This is what we mean by extending logic programming.

5. While we require the providers to install such *linking* rules, it is impractical to demand a knowledge of Datalog from the providers. Instead, Web based forms can be used with which providers examine the various symbols used by the TMIS and indicate what relevant data they have and what has to be done to make their data fit the TMIS's semantics.

   For example, consider a TMIS with information on "hotel" with three columns in a tabular representation: chain (the national or international chain a hotel belongs to), room type, and cost (the last two as in the previous example). A provider would indicate for each of these three whether or not it is covered, and if so, whether or not there are any transformation rules required. A data source d_2 may indicate that it does not cover the first (because it is not a member of any chain), that it can supply the other two columns with the exception that cost must be corrected for tax (with a 10% reduction as in the previous example), and this transformation is specified in another module called "CostConvert". Using this information from the provider, using our prototype [8], the TMIS would synthesize the Datalog rule:

```
hotel(Chain,Rmtyp,Cost) :-
    not_covered_filter(hotel,Chain,nochain),
    pre_convert(hotel,Cost,d2CostConvert),
    send_data(d2,rmlist,Rmtyp,Cost),
    post_convert(hotel,Cost,d2CostConvert),
    not_covered_message(hotel,Chain,nochain) .
```

This rule is similar to the previous rule except for the *not_covered_filter* and *not_covered_message* predicates. These would make sure that a customer who is interested in hotel chains will get the message that this provider does not have any chain information to offer. The pre- and post-convert predicates take care of conversion of TMIS costs to the provider's costs and vice-versa.

Periodically, the administrative controller of the TMIS reviews the peculiarities of contributors and decides if they are interesting and/or general enough to be incorporated into the TMIS.

6. From the customer's point of view, it is not only important that they be able to understand clearly the semantics of the terms used by the TMIS but how the relevant contributors agree or differ with them. For example, a customer who queries costs of hotels and receives information provided by data source d_1 as part of the answer should be informed that this provider has a non-standard notion of costs (i.e., different from that used by the TMIS).

   This is easily done by the query answering system by remembering the Datalog rules used to process the query and checking them to see if they contain a *send_data* predicate along with *pre_convert* or *post_convert.*

7. Owing to the great variation in meanings of terms unadjustable through any transformation, certain queries cannot be answered precisely. The query asking for single rooms with attached baths within a set price may be hard to answer if some providers claim to have attached baths that are only *semi-attached* or attached but *not quite full baths.* Ignoring these differences may mislead the customer, while ignoring such providers will not serve the customer.

   We propose to tackle such inevitability of imprecise answers by using query annotation as suggested by Motro [5], [6], [7] but within the logic programming framework. An answer that includes data sources who have a non-standard interpretation of the query would be annotated with this discrepancy (similar to a list with footnotes containing caveats).

8. Constraints on terms used by the TMIS can also be very useful in detecting errors. These constraints express certain invariants that hold in the real world, e.g., the arrival time of a flight must precede the departure time of a connecting flight. Many errors in data entry can be detected by checking that these constraints actually hold on the data. This checking can be done periodically by the TMIS by accessing the provider's data relevant to these constraints. If an inconsistency is detected in a data source, then answers to future relevant queries from that source can receive a warning annotation.

   The constraints can also be used during query evaluation: remote calls can be saved if the answer is known to be empty and the customer is better off with the knowledge of the existence of a conflict between the query and a constraint rather than an empty answer.

9. Information about completeness of subsets of answers can also be communicated through annotation [5]. For example, if museum has *all* the works of Picasso, then that is a useful piece of information for queries on paintings by Picasso or the holdings of that museum.

   This is also very useful in answering negative queries. Normally, answers to negative queries are based on the absence of positive facts. For example, the query "find all modes of transport that do not involve air flights" can be answered by all such modes of transport *known to the TMIS*. However, in case all such modes are completely known by the TMIS, it would be useful if the query answering system could assert that the answer to this query is indeed complete, i.e., that all such alternative modes of transport have been covered in the answer.

10. Providers may have special information available for certain kinds of customers. This means that the query answers would depend on the customer.

    To achieve such information customization, a customer profile can be stored. Then parameters from this profile can be sent to the providers as part of the request to send data.

11. As customers increasingly adopt the self-service mentality, they want to obtain pieces of trip information and then stitch them together into their own travel plan. This trend can be supported by the TMIS by allowing the storage of results obtained from queries for parts of a trip in the form of tables. These results can be joined to form information about a whole trip. This needs a rudimentary DBMS capability in the TMIS but not with the providers.

2.1 Current Prototype

We looked for a Datalog/Prolog interpreter that would support remote calls implied by the *send_data* predicate, but we were unable to find one. While some Prolog interpreters such as XSB allow foreign predicates, they could not fulfill our requirements. For our prototype, Randi Pettersen implemented [8] a restricted version of the required interpreter. It is not a true *interpreter*: the rules are translated into *C* functions and compiled. It behaves like the interpreter in that it can handle any query against the set of rules, but is unable to add rules without recompiling. We are yet to implement annotation of answers; it will be implemented by adding a parameter to send_data. Future work on this prototype will also address issues regarding optimization and communication.

3 Related Work

It has been long recognized that even partial integration of information sources presents a number of research problems (Litwin [3]). The basic approach of researchers is to make the semantic connections among related terms explicit (Sheth [10], Spaccapietra and Parent [11]). Wiederhold [15] postulates the need for intelligent interfaces among databases that are called *mediators*. Such mediators are active knowledge-based systems that transform, access, and merge appropriate data from multiple databases. The logical approach has been pursued by Qian [9], Whang *et al.* [13, 14], and Boman and Johannesson [2]. The first uses intelligent communication and compilation modules that translate the symbols and the results; however, the data sources are transparent to the customers and flexibility and dynamic requirements are not explored. The third uses a very powerful non-standard logic that appears to be too complex for practical realization. The second has the closest approach to ours; but it does not use distributed predicates and cannot handle imprecise answers.

4 Conclusion

We have outlined an approach for fully realizing the vision of a travel mall. There are several advantages of our approach.

- Our approach is scalable: the expanding number of users of the network will not directly influence the number of such TMISs.
- The most ordinary provider will be able to be participate with the greatest ease.
- It does not matter whether a provider uses flat files, or a relational DBMS or a deductive object-oriented one. All that is required is that the remote source respond to a message asking for data on a symbol and some parameters. As indicated earlier, CGI scripts can be used to respond to such messages.
- The data sources will *not* be transparent to the customers. Thus, the customers can restrict the sources to be considered and also use their

knowledge of the source of each data item in the integrated answers they get for their own evaluation, judgment, and analysis in order to make individual preferences.

- The TMIS allows dynamic expansion of its vocabulary.
- Using query annotation, we can tackle the problem of inevitable imprecision in query answering.
- Logic and its standard inference processes provide clear semantics for queries.

We have indicated how our approach addresses each of the needs enumerated in the Introduction while outlining how our extension of logic programming takes care of the issues. Among the useful features of a TMIS, the only one we have not not addressed is that of proactive targeting of customers but feel that it can exploit the technique we have used for customization of information. Using rules and heuristics about classes of customers, various packages can be brought to the attention of a given customer.

Among the needs, the problem of information integration was the one that we have emphasized because it is vital for the realization of a TMIS that can tap the full potential of the Internet in favor of the tourism industry connecting tiny through large providers to customers the world over.

Acknowledgments The author is supported in part by a grant from the National Science Foundation under contract IRI-9509789 and a grant from Sandia National Laboratories.

References

[1] M. Bloch, Y. Pigneur, and T. Steiner. The IT-enabled Extended Enterprise: Applications in the Tourism Industry. In *ENTER 96: Proceedings of the International Conference on Information and Communication Technology in Tourism.*, 1996.

[2] M. Boman and P. Johannesson. Epistemic Logic as a Framework for Federated Information Systems. In S. Deen, editor, *CKBS 90: Proceedings of the International Working Conference on Cooperating Knowledge Based Systems*, pages 255–270. Springer Verlag, 1991.

[3] W. Litwin. From Database Systems to Multidatabase Systems: Why and How. In *Proceedings of the Sixth British National Conference on Databases*, pages 161–188, 1988.

[4] S. Mazumdar. A Deductive Approach to Semantic Heterogeneity in Multidatabase Systems. In *Proceedings of the Third International Workshop on the Deductive Approach to Information Systems and Databases (DAISD-93). Costa Brava, Catalonia, Spain*, pages 99–121, 1993.

[5] A. Motro. Integrity = Validity + Completeness. *ACM Transactions on Database Systems*, 14(4):480–502, December 1989.

[6] A. Motro. Relational Meta-Answers. In Z. Ras and M. Zemankova, editors, *Intelligent Systems: State of the Art and Future Directions*, pages 371–386. Ellis Horwood, 1990.

[7] A. Motro. Annotating Answers with Their Properties. *ACM SIGMOD RECORD*, 21(1):54–58, March 1992.

[8] R. Pettersen. A Federation of Semantically Heterogeneous Databases. Master's thesis, New Mexico Institute of Mining and Technology, 1996.

[9] X. Qian. Semantic Interoperation Via Intelligent Mediation. In *Proceedings of the Third International Workshop on Interoperability in Multidatabase Systems (IMS 93)*, 1993.

[10] A. Sheth. Semantic Issues in Multidatabase Systems. *ACM SIGMOD RECORD*, 20(4):5–9, December 1991.

[11] S. Spaccapietra and C. Parent. Conflicts and Correspondence Assertions in Interoperable Databases. *ACM SIGMOD RECORD*, 20(4):49–54, December 1991.

[12] M. Van Emden and R. Kowalski. The Semantics of Predicate Logic as a Programming Language. *Journal of the ACM*, 23(4):733–742, 1976.

[13] W. Whang, S. Chakravarthy, and S. Navathe. Relational Schema Integration: Dealing with Inter-Relation Correspondences and Querying over Component Relations. *Computing Systems*, 6(3):319–352, 1993.

[14] W. Whang, S. Navathe, and S. Chakravarthy. Logic-Based Approach for Realizing a Federated Information System. In *Proceedings of the First International Workshop on Interoperability in Multidatabase Systems (IMS 91)*, pages 92–100, 1991.

[15] G. Wiederhold. Mediators in the Architecture of Future Information Systems. *Computer*, 25(3):38–49, March 1992.

Electronic Marketcoordination in the Travel Industry: The Role of Global Computer Reservation Systems

Dr. Axel Schulz

Lufthansa Systems GmbH, Am Weiher 24,
D-65451 Kelsterbach, Germany
E-Mail: 100566.355@compuserve.com

1 Introduction

One of the few examples for the global realization of electronic market coordination are the Computer Reservation System (CRS) or Global Distribution Systems (GDS) used in the travel industry. Because of the ever increasing competition, new technologies and changed distribution channels, a discussion has started about the role of CRSs. Before going into detail about current criticism and the systems´ future role, we will briefly describe what CRSs are all about:

2 Overview

Ideally, CRSs are systems for purely (information transfer) logistical functions. They store current information about all available service providers and have the necessary infrastructure to transfer such data. This means that the systems also perform additional tasks related to service distribution, which in the area of goods are typically carried out by freight forwarders. They support the transport of goods (or the right to a service) by eliminating the physical distance between the producer and the sales mediator or the consumer respectively. Therefore, CRSs can be characterized as supporting distribution process since the current systems do in no way support negotiations between the producer and the consumer. CRSs are a combination of infrastructure measures offered to interested providers in the tourist industry. Similar to an empty supermarket, the system provides a sales area in the form of memory and communications media. The providers can then fill the empty shelves by providing their data to the CRS.

The development of such systems was already started at the beginning of the sixties the objective being at the time to automate *seat reservations procedures* of airlines. Today, the largest civil data centers and the most powerful communications networks are being used exclusively for this purpose. After several changes in the marketplace, the best known global CRSs are AMADEUS, GALILEO, SABRE, and WORLDSPAN with AMADEUS being the only one which is largely owned by European *airlines*. Together with GALILEO, it is the leading system in the European market. GALILEO, however, also has a strong position in the US market after it merged with an American system. SABRE and WORLDSPAN have been developed in the US. So far, SABRE is the only CRS which is wholly owned by one single airline.

	AMADEUS	GALILEO	SABRE	WORLD SPAN
USA/Mexico/ Canada	U: 7.839 T: 29.313	U: 13.347 T: 59.980	U: 17.496 T: 82.441	U: 9.200 T: 41.106
Europe/ Russia	U: 21.729 T: 59.804	U: 12.561 T: 39.487	U: 4.028 T: 11.477	U: 5.474 T: 10.208
Latin America/ Caribbean	U: 4.631 T: 8.247	U: 1.100 T: 1.918	U: 1.985 T: 6.328	U: 53 T: 268
Far East/ Australia	U: 920 T: 2.034	U: 3.686 T: 12.871	U: 4.113 T: 14.246	U: 0 T: 0
Africa/ Middle East	U: 374 T: 768	U: 1.632 T: 5.144	U: 490 T: 1.452	U: 600 T: 1.368
Total:	U: 35.493 T:100.166	U: 32.326 T:119.400	U: 28.112 T:117.914	U: 15.327 T: 47.950

Fig. 1. Comparison between the number of users and the number of terminals of global systems (as at the beginning of 1996)

3 Basic Functions

Using the global CRSs, *travel agencies* can make reservations directly from their terminal with any airline and on all continents without having to go through a coordination process or contract negotiations. All global CRSs provide the basic functions for the reservation process such as product presentation, reservation, fare quote & ticketing and additional services.

3.1 Product Presentation

For a CRS, the most important source of information is the presentation of the products and services offered by providers in all areas related to the travel industry. Each group of service providers has individual screen categories the contents of which particularly represent the complexity of their offerings and specific features of their serivces. The product "flight" for example does not require complex descriptions since it is sufficient for a neutral product presentation to state the departure and arrival times, the route, availability of particular reservation categories and possibly the fare. It may, however, be difficult to describe the products of other service providers appropriately with only a limited amount of information. Hotels for example have so far only been able to provide information about the price, the size of the bed and its approximate location which alone is not a meaningful description for potential customers. For this reason, it is intended to link the particular offers to visual multimedia technology to be able to provide more detailed information to the customer.

3.2 Reservation

The core function of reservation systems and the main reason for developing them is the reservation of offered services in the travel industry. To this end, a so called Passenger Name Record (PNR) or Guest Name Record (GNR) is created for each passenger or each group of passengers. These records contain all services-related customer information. At the same time, this information is transferred to the internal inventory system of all service providers who thus have the latest information about availability at any point in time and can use it as a basis for a new offer. In addition, the system can also store customer related information such as e.g. all services provided to a certain customer, type of payment, service information etc.

3.3 Fare Quote & Ticketing

Just like the product praesentation, the fare quote, ticketing and voucher generating process depend on the type and the complexity of the services offered. For flights, multiple fares are being offered which differ according to the reservation category, the date of the journey, the day when the reservation was made, the route and the length of the stay. That means that practically each fare needs to be calculated individually. In addition, fare quotes can change daily. The prices of other service providers, however, are relatively fixed so that in most cases, they stay an unchanged part of the offer. For the ticketing, travel agencies receive a fixed number of ticket forms which may only be used after confirmation has been received from the respective service provider. Typically, the print-out of any additional information material is not supported by most systems as it is not always necessary, like for example a print-out of a hotel or rental car reservation.

3.4 Additional Services

Because of the increasing competition in the market, system operators were forced to offer not only the three essential components of an information and booking system but also additional services. Today, the user has direct access to essential travel information and can find further information in all CRSs about trade shows, visa regulations, particular events etc. Furthermore, programs and interfaces have been developed which facilitate the internal administration for each travel agent. Also invoicing, accounting, customer and quota management increasingly depend on the respective CRS. User prompting has significantly improved so that even inexperienced users can easily learn how to work with the reservation procedure. Since more and more PCs are being used in this area modern user interfaces have been introduced.

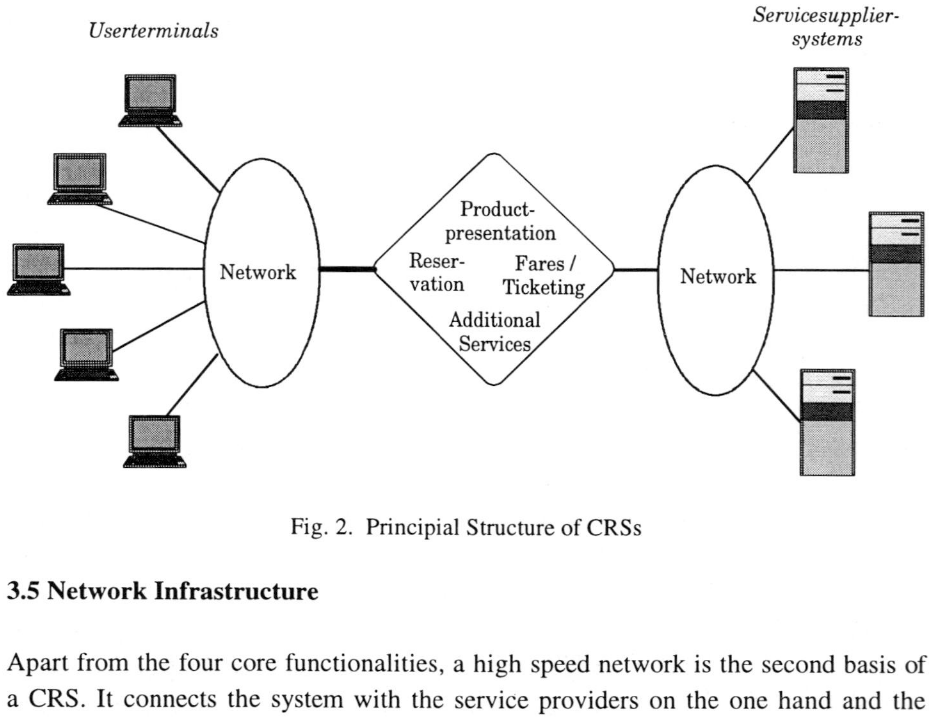

Fig. 2. Principial Structure of CRSs

3.5 Network Infrastructure

Apart from the four core functionalities, a high speed network is the second basis of a CRS. It connects the system with the service providers on the one hand and the travel agents on the other. In Europe, travel agents are mainly linked via interfaces to the cooperating national systems in which case the local network infrastructure is being used. Travel agents using the CRS AMADEUS are linked via various networks. In most cases, they are integrated in the national START network. In other countries, they are linked either via the SITA network or via Datex-P X.25.

The city and airport offices of airlines are mostly integrated in their respective internal network.

The network of the global CRS AMADEUS has additional 14 network nodes distributed throughout the whole world. They are connected to each other by 13 lines with 1.92 Mbyte/s and 11 lines with a lower transmission rate. These lines are utilized 60% at most. By automatically establishing alternative links, an availability of 99.7% is achieved since, in case of necessity, any message can be transferred via other nodes.

4 Criticism of the CRS

Recently, there have been controversial discussions about CRSs. They have been called inflexible dinosaurs not being able to fulfill current or future market needs. Criticism mainly relates to the insufficient scope of services offered and the high distribution costs.

4.1 Insufficient Scope of Services

With most airlines it is possible today to book a flight easily and without problems merely by pressing the respective keys on the computer keyboard. However, there are still numerous needs for services which the current CRSs cannot yet fulfill sufficiently. With regard to product presentation, the offers are almost exclusively limited to the category of airlines. Other service providers (such as hotels, rental cars, tour operators, charter airlines etc.) only have a very limited reservation rate or are not displayed at all. In Europe, such tasks are mainly performed by national distribution systems. However, even today it is still easier to book a flight to a far off holiday resort than simply reserve a room in a small hotel nearby.
While the functionality "reservation" is working without problems on the basis of modern access technology "fare quoting" is still complicated, incomprehensible, and confusing process. Price transparency or even price negotiations are not supported by the systems. Therefore, it always requires a lot of time to find the best fare. One of the reasons for these shortcomings is that the CRSs are owned by airlines who are not interested in making their products comparable. Most recently however, travel agents have increasingly demanded "Best Buy"-functionalities and information about negotiated fares.

Another weak point in the CRSs have always been additional functions since such functions would not increase the number of sold flight tickets. Only now modern graphical user interfaces are being developed in the various systems for the front office. Sufficient back office functions have been offered only very rarely.

4.2 Distribution Costs

Apart from the criticism about the internal capacities of the systems, one further point of discontent have been high distribution costs. Airlines have an enormous need for streamlining their organization in order to stay competitive. In many sub-areas, some optimization and efficiency measures have already been taken. Now the only costs which can still be changed and reduced are personnel and fuel costs as well as costs for sales. Generally, sales costs within this industry amount to 20 - 25% of the total revenue which is perceived as far too high. So it is the commission for travel agents and reservation fees paid to the CRSs which are now being checked and reviewed very closely.

There are two approaches which can help to reduce distribution costs. On the one hand, the on-line services provide additional possibilities of cost-effective direct distribution to the end customer. Many airlines now offer access to their internal service provider systems via on-line services and in particular via Internet. This also helps reducing costs incurred by travel agents and fees to the CRSs. One of the industry leaders in this area is Lufthansa: Its interesting concept "Info Flyway" will provide statistical information on CD-Rom and will make it possible to make direct reservations via an on-line interface. Moreover, some airlines now offer last minute tickets at particularly good prices in auctions via Internet.
By using on-line services it will be possible to reduce costs for travel agent commissions and CRS fees and increase the load factor at the same time. However, this will only attract a small part of the overall clientele since in Europe for example only 3% of the population at most have access to on-line services.

The so called "no frill carriers", one of them being Southwest Airlines, have taken this one step further. In all areas, they avoid unnecessary costs (service, change of bookings, meals, etc.) and merely offer basic services, i.e. to transport a passenger from A to B at the best possible price. They also do not use or offer via the CRSs any of the normal distribution channels but process all incoming bookings with the help of powerful internal Client/Server systems. As in the past, customers now go back to calling the airline directly and pay there and then with their credit card. This aggressive procedure has forced many other airlines to also make direct reservations and by-pass the CRSs. Some carriers, such as Swissair, have come to an agreement with major travel agent chains which then can make their reservations directly in the system of the respective airline. Some of the major travel agents like American Express offer the possibility to various airlines and hotels to display their services directly in the travel agents´ internal systems. This could sufficiently substitute the scope of services provided by CRSs and the cost of communication between the service provider and the travel agent can be reduced. However, this would increasingly counteract the idea of creating a global electronic market.

5 Conclusion

In short, it can be stated that on the one hand the global computer reservation systems have reached a significantly high economic and technological level and travel agents would hardly be able to do their work without these tools. On the other hand, however, because of the high costs incurred many providers avoid using this consistent concept with its complicated functional requirements and worldwide networks. Therefore, the CRSs are now at a tuning point. Either the systems adapt to the new market conditions or they will become an early example for the destruction of the worldwide electronic markets.

References

Berninger, H.: Opas Carrier ist tot, in: Touristik REPORT, 6(1996), pp. 6 - 8 .

Caldwell, B.: Ticketing is up in the air, in: InformationWeek, 23.5.1994, pp. 80 - 82.

Feldman, J.: Biting the hands that fees them, in: Air Transport World, 6(1996), pp. 141 - 146.

Humphreys, B.: New Developments in CRSs, Paris 1995.

Jegminat, G.: Die Reisebüros müssen für sich selbst sorgen, Fremdenverkehrs-wirtschaft International, 15.3.1996, pp. 24 - 25.

Maglitta, J.: Lean, mean flying machines, in: Computerworld, 28(1994)28, pp. 81 - 84.

Schmid, B.: Elektronische Märkte, in: Wirtschaftsinformatik, 35(1993)5, pp. 465 - 480.

Schulz, A., Frank, K. und Seitz E.: Tourismus und EDV, München 1996.

Schulz, A.: Sind CRS überflüssig ?, veröffentliches Manuskript, ITB Medientag, Berlin 1996.

Schulz, A.: Vertriebskoordination und Angebotsoptimierung in elektronischen Märkten am Beispiel computergestützter Reisevertriebssysteme, Dissertation, Nürnberg 1995.

Informed Decision Making in Tourism Management Closing the Information Circuit

K.A. Froeschl, H. Werthner

Dept. of Statistics, Operations Research and Computer Methods
University of Vienna, Universitätstr. 5/9, A-1010 Vienna, Austria
[froeschl,werthner]@smc.univie.ac.at

Abstract. Tourist information systems (TIS) play an increasingly important role in the promotion of tourism destinations. Currently, this vivid discussion about TIS puts heavy emphasis on the distribution and selling aspect of tourism management. However, the importance and opportunities of effectively utilizing the back-flow information – which, in fact, can also be seen as an immediate indicator of a destination's business performance – obtained from TIS as a by-product for business planning and decision making processes has not yet been recognized properly. Our contribution outlines an operative information management framework aiming at a deep integration of tourism marketing and distribution activities with performance monitoring and decision making processes, thus closing the current gap in the – electronic – tourism information management loop by adding an *upstream* information flow from markets to decision makers.

1 Introduction

Electronic media assume a more and more prominent role in the tourism business. In particular, the convergence of tele-communication and information processing by computers interconnected in large-scale networks, like the Internet, opens powerful new avenues in product promotion, marketing, and even sales in virtually any information-intensive business branch; in tourism, this strategy is now pursued quite vigorously with so-called tourist information system (TIS). While there is nothing wrong with this tenet, the current emphasis on information distribution by new media and their broadcast and transaction capabilities (for instance, direct selling) overshadows yet another unprecedented opportunity of exploiting information technology (IT) in tourism management: shifting the focus from distribution and marketing – *downstream* information flows – more to product development. We claim that by establishing tourism management information systems at destination level fed with *upstream* information flows a promising potential for improving overall economic destination performance is offered. Given that the economic actors are interconnected electronically, these complementary upstream information flows are arranged rather easily in technical terms. Not yet resolved, however, is the data integration and data linking logic indispensable for setting up a formal framework supporting effectively the managerial decision making and planning processes. This paper investigates the economic and strategic preconditions for introducing destination management information systems (DMIS) of the type proposed and pinpoints the operative targets prospectively aimed at.

2 Three Premises About a Business in Change

Premise 1: Tourism is an information business for structural reasons [3]. Tourism is based on contractual agreements about services, the consumption of which necessarily will take place in the future. Moreover, the tourist has to travel to the product; in advance, *information* about the product can be gathered only (i.e., in economic terms, it is a "confidence good"). This specific circumstance induces an urgent need for information on both, the consumer's and supplier's sides entailing high information seeking costs and causing informational market imperfections [6]. These, in turn, lead to the establishment of specific product distribution and – comparably long – information and value-adding chains as well as specific approaches to information management (including the information brokering role of travel agents). Apparently, also tourism service suppliers respond "structurally", as competitive information management (gathering, distribution, providing access) gains economic advantages and, thus, becomes a "core" competence. As a matter of fact, the design of TIS must be considered *strategic* in organizational change and business process re-engineering because information influences crucially the driving factors of competition in pivotal respects such as product development and adaptation, product differentiation and branding, pricing and calculation, as well as distribution and sales modalities.

Now, while the output of business processes is communicated to the market (in terms of product information flowing to the potential customer/purchaser), running a business and making decisions themselves are based necessarily on market information, mostly represented in terms of statistical aggregates (summaries, trends, forecasts) reporting on market behavior and competitor performance. Both, downstream and upstream information flows, create a *tourist information network* tying together all market participators and, apparently, reflecting the economic relationships between them. Technically, the information entities, or contents, conveyed through this network are as divergent as products, contracts, transactions (bookings), or market statistics.

Premise 2: Tourism undergoes a thorough structural change. The observable changes in consumer behavior (tourists ask for "better" service, are becoming more mobile and critical but less loyal; tendency of making more and shorter vacations, decreasing time span between booking and consumption) provoke corresponding changes in the tourist offer: while the number of products increases these products become more specific and take into consideration individual preferences, leading to "mass customization" with a broad range of "component" products to be combined flexibly.

This general trend can be met by, inter alia, tighter integration of a destination's service suppliers and shortening distribution channels in both regards, viz. *time* (faster disposition with respect to price and product gestalt, topicality of offer) and *space* (by-passing of intermediaries and seeking direct access to the market and new consumer segments). All of this, however, comes at the cost of increasing complexity of strategic planning; as the tourist infrastructure becomes more flexible and markets more competitive, response cycles to market stimuli decrease calling, in turn, for faster amortization of investments as well as primary suppliers seeking to allocate higher shares of added value (especially by yielding transaction cost reductions). Quite naturally, evading fierce price competition also leads to a further segmentation of markets and supplier co-operation strategies within destinations.

In coping with structural change in tourism, modern IT plays an important role. By facilitating the nearly unrestricted dissemination of tourist information (world-wide networks, multi-media, etc.), market participants and especially end consumers get better access to product information, giving rise to both, global visibility of destinations and a global merge of market segments. This increase in distribution competition accelerates business processes and spurs the dynamics of innovation, calling for improved utilization of IT at the management level. Moreover, we argue that IT provides a managerial framework for integrating a destination's tourist offer either.

Premise 3: Tourism business goes electronic. For obvious reasons, this development is part of the general structural change in tourism (and elsewhere) stated in premise 2; on the other hand, it also provides a new and prospective basis to tackle offensively structural change. For one thing, IT is increasingly used for destination marketing and, recently, direct selling and booking, thus establishing *electronic tourism markets* [4]. As a matter of fact, TIS – like the Tyrolean Information System [5] – help to improve downstream information flows with respect to quality, content presentation, speed, and accessibility and – as has been recognized generally – offer a vast potential for improving efficacy and efficiency of virtually all transaction phases. However, taking a destination-oriented point of view of a "TIS", only few initiatives to implement such systems have been undertaken, and several of these approaches have failed. To date, inter-organizational information systems are still used predominantly by major distributors and carriers (for instance, GDS and CRS) to support their marketing and sales activities although this is subject to change with spreading utilization of digital tele-communication including the home consumer market.

Given that ever larger portions of tourism sales (via Internet, WorldWideWeb, etc.) and business operations (for instance, between service suppliers interconnected by electronic data interchange such as EDIFACT, or the like) take place through electronic media, a cardinal prerequisite for setting up comprehensive electronic markets encompassing *all* market participators and transactions is fulfilled. This, in turn, opens up unprecedented opportunities in tapping tourism transaction data such that electronic markets, in principle, report on their actual status in a timely and accurate as possible manner. We consider a specific arrangement of information systems fed directly from the emerging electronic tourism business network such that, contrary to hitherto emphasis, the main beneficiaries – by attaining market power – of these tourism management information systems (MIS) are the tourism destinations.

3 The Strategic Position of a DMIS in the Tourism Information Network

From the premises stated it is apparent that the economic position of tourism destinations as a whole can improve only if the crucial resource "information" is handled more efficiently. Hence, we suggest to enforce direct marketing efforts by establishing (electronic) "tourist information and booking desks" accompanied by a destination management's back-office MIS/EIS (executive information system) tapping market data from the network as exhibited in **Fig. 1**.

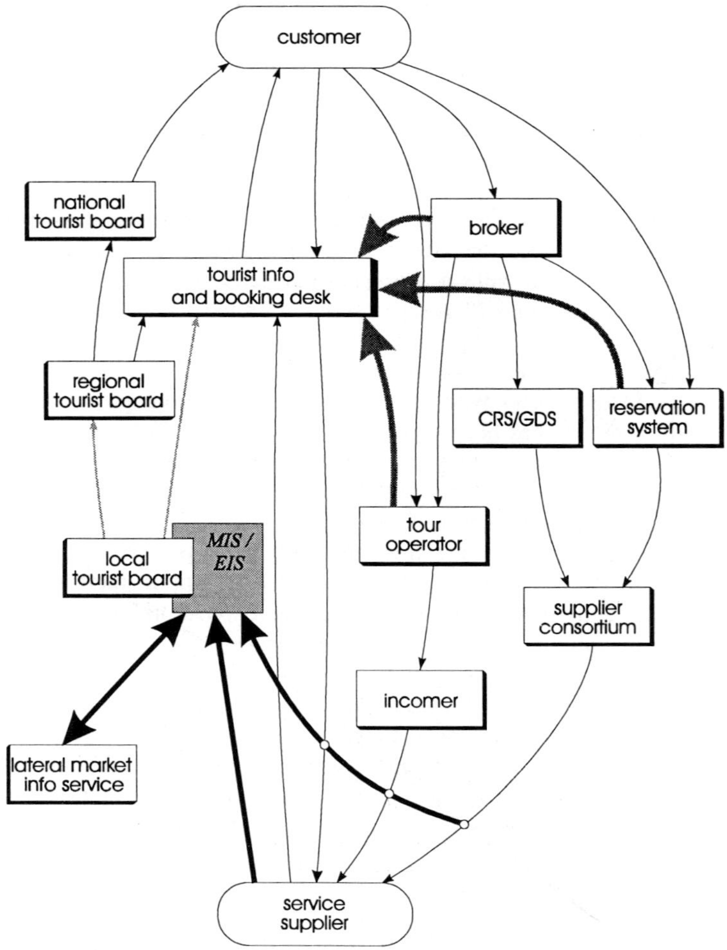

Figure 1: Tourism Information Network including Destination MIS/EIS

The introduction of MIS/EIS at destination level can be interpreted as a strategic measure to facilitate organizational change and increasing economic professionalism of basic service suppliers. In particular, we argue that the MIS/EIS is indeed placed most appropriately at the destination level because (i) marketable tourist offers are composed typically out of the services of several suppliers of a destination, and (ii) at higher than destination level neither product offerings nor economic interests are homogeneous to a degree rendering effective business policy making prospective or even feasible any longer.

In **Fig. 1**, arcs show the flow of information as well as their direction, between the two "poles" customer/end consumer and basic service supplier; in this context, the latter is to be understood, essentially, as a *local* service provider in that the economic sphere of operation is typically limited to the destination considered.

In order to couple both aspects, dissemination of product information and gathering of market (performance) information, the boxes "tourist info and booking desk" and "MIS/EIS" should, ideally, be combined to an *integrated* DMIS implying that this would enable a direct back-flow of operative business data to the MIS/EIS component. This appears all the more desirable as, in functional terms, this component builds the informational backbone of destination policy making and, for this reason, ought to fit into the general framework of planning and decision making as shown in **Fig. 2** [1].

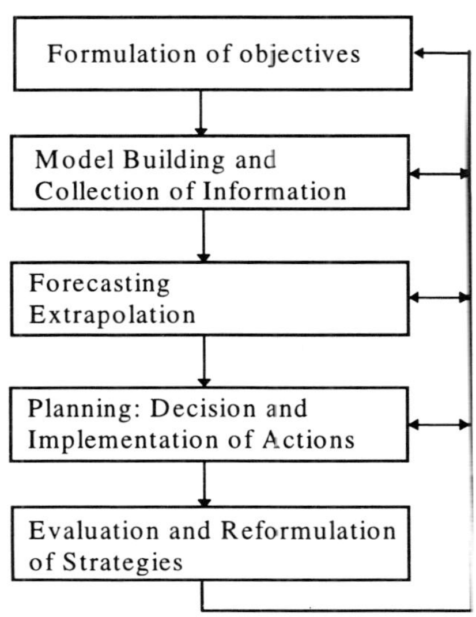

Figure 2: General Model of Planning and Decision Making

Basically, the incoming edges of the MIS/EIS box in **Fig. 1** correspond to the information collection stage in **Fig. 2**, while the "Implementation of Actions" stage shown in **Fig. 2** delivers policy back to the information network or, topologically, feeds it into the downstream channels the strategically most important of which is represented by the "tourist info and booking desk"-box of **Fig. 1**. "Model Building", in the present context, refers to the arrangement of a formalized business model in terms of information structures capturing faithfully "destination reality" and integrating all economic parameters of relevance in destination tourism management while "Forecasting/Extrapolation" concerns a set of tools for assessing destination performance against varying economic assumptions and business strategies; apparently, these assessments provide the rational foundation for decision making in the subsequent "Planning" stage.

4 Operative Targets of DMIS in the Tourism Information Network

After having "isolated" the tourism destination as the economic unit of main concern, the decision maker and his operative target proper must be pinpointed. In the sequel, by 'destination' is understood (i) a – usually contiguous – geographic area (ii) comprising all – local (sometimes regional) – tourism service suppliers (iii) creating jointly a more or less homogenous touristic offer. At destination level, the primary economic agent and decision maker is, in most cases (for Europe, at least), the *local/regional tourist board*; this board operates as economic *proactor and facilitator* on behalf of the community of tourism service providers located in a destination. In general, tourist boards are non-profit institutions responsible for all extramural tourism promotion, marketing, and local point-of-offer mediation activities of a destination; occasionally, regular business activity may also encompass sales.

In all this, the board aggregates both, services and information, from basic providers to the destination's genuine tourism offer. Conversely, however, the board often has rather limited directive power as basic service providers typically are, in legal terms, independent businesses not bound to follow anyone's suggestions; in reality, of course, boards can and, in fact, do reinforce the implementation of their destination policy and accompanying economic recommendations by a variety of measures while, at the same time, aiming to give individual service suppliers as much room to move as possible.

Simply speaking, the ultimate operative target of tourism destination management consists in optimizing the destination's *overall* economic performance. Improving economic performance in increasingly competitive markets is, of course, a many-faced endeavor; for one thing, decision making is tied deeply to a mixture of micro-economic and macro-economic as well as social and cultural objectives (such as development of local infrastructure, ecological sustainability, preservation of cultural heritage, etc.). Moreover, economic tourism objectives in the narrower, business-oriented sense comprise both monetary and non-monetary aspects such as destination *image* and maintenance of customer *goodwill.* Additionally, destination policies should foster coherence of service standards, strengthen joint commitment to agreed upon operative goals, and establish a firm "corporate spirit" shared by all service providers; for the sake of the latter, destination policies also have to have an eye on keeping internal interests of (locally competing) service providers in balance in a short- to long-term range.

Against this background, in our view a DMIS focuses most appropriately on the development of a tourism offer fitting both, the very economic and touristic potential of the destination and the (self-) defined operational economic – in the widest sense – (multiple) objectives in an integral fashion. More specifically, the proposed DMIS is conceived as a platform for *product development* such that product aggregation and design is facilitated more quickly, in a more flexible way, and based on much a broader decision frame – all of which are salient criteria for responding better to the challenges of structural change in tourism. Additionally, as product distribution and marketing is forced more and more to conform to technology-driven presentation styles and standards – inducing a "homogenization" in the way tourism products are conveyed along downstream channels to wholesalers and end consumers – which leave but little room for the transmission of individual product variations and "customized small-scale offers", product development must react accordingly. Thus, broadly speaking, the DMIS can be considered practically a destination manager's *workbench* used for transforming a destination's basic, and hence usually rather heterogeneous, service offer into a well-targeted presentation-ready product line of homogeneous service packages as well as component products destined for integration into yet more aggregated tourism products beyond destination level.

In defining the functionality of the envisaged DMIS, input and output interfaces need to be clarified first. Basically, DMIS input comprises

- a comprehensive and faithful as possible representation of the basic touristic offer of all service providers operating in the destination ("local suppliers"),
- the back-flow (from local suppliers) of information about the operative business performance, and,
- to a limited extent, at least, information about the tourism market(s) and its (their) development – trends in tourism – in general.

DMIS output, on the other hand, is a tuned set of product offers hopefully in line with tourism markets and advertised target populations. In particular, DMIS output should (i) be prepared properly for electronic (direct) selling and distribution and (ii) enable performance evaluation of the destination efficiently.

Given these major determinants, there are several internal DMIS tasks resulting immediately. First of all, the different data sources need to be combined effectively. This intramural **data integration** entails a rather intricate task, though, because of the inherent heterogeneity of source data: different types of service providers (such as accommodaters, funicular companies, bus carriers, etc.) deal with different sub-markets, are confronted with different service requirements and customer behavior, and have different views on their operative business even if they create jointly the same, or comparable, products. Hence, it comes as no surprise that internal business management and system solutions are divergent, in particular with respect to formal product/service descriptions. Apparently, the situation is the more heterogeneous the greater the differences in product types or services provided. Yet, in spite of this inevitable heterogeneity, data integration is bound to start with data and information structures available at source. Thus, in addition to the technical fusion of source data (to be resolved by format conversions and the like), this *a posteriori* data integration implies the much more critical task of putting different heterogeneous business models into context; in other words, source data need to be *harmonized*, or set in relation *semantically*.

Basically, the requirements of semantic data integration [2] can be met by arranging a conceptual "tourism super-business model" (TSBM) of an all-encompassing (but usually fictitious) tourism enterprise delivering the destination's *entire* tourism offer by putting together all the local suppliers' services treated as internal production components, or departments. To feed the TSBM appropriately, the DMIS must be equipped with a very wide "input gate" allowing it to embrace virtually any kind of business data of (possible) relevance in later decision making stages. In particular, the capability to obtain different conceptual TSBMs (that is, competing approaches to harmonization) on top of the real business models used by contributing local suppliers is of crucial importance as the structure of the most suitable TSBM may not be known beforehand. In terms of information transmission from local suppliers to the common DMIS platform, a kind of "broadband" interface has to be envisaged comprising two main components, viz.

- *pre-sales* information conveying a description of the local suppliers' basic offer (accommodation description, lodging capacity, kind of infrastructure/facilities, etc.), and
- *post-sales* information conveying timely summaries (that is, statistical aggregates) about the operative business (customer frequencies, effective demand structure, utilization of capacities, etc.).

Pre-sales information covers both, a long-term component related to the local supplier's production structure (determined economically mainly by bound capital and fixed costs resulting thereof) and a short-term component of service supply (variable components of service such as in-person services offered, days/hours of operation, price lists, contractual conditions, etc.).

Technically, the management of pre-sales information is quite different from post-sales information. While pre-sales information is probably the hardest to unify semantically (for instance, think of the close to insurmountable task of harmonizing the price information delivered by different local suppliers), it is, at the same time, a rather *static* type of information and, hence, mostly independent of the operative business. However, despite its heterogeneity, pre-sales information presumably is available in the desired granularity all the time. Thus, in this respect, the very creation of a reasonable TSBM remains the outstanding challenge.

As to post-sales information, the situation is almost the other way round. By its nature, this type of information establishes the primary feed-back loop transmitting operative business data to the DMIS; hence, post-sales data is highly *dynamic* and must be reported/recorded on a continuous basis. Fortunately, compared to other, e.g.

industrial, settings where even real-time instruments for performance evaluation may be called for, weekly reporting cycles will be acceptable in most cases (though, occasionally, half-week periods may be more appropriate). Anyway, utilizing post-sales information at DMIS level implies the provision of a means for **dynamic linking** of reported performance data which, therefore is yet another cardinal DMIS task. Contrary to pre-sales information providing mainly resource descriptions, post-sales information refers, basically, to resource *utilization* but, in addition to mere benchmarking, reported data should fulfill several marketing-relevant criteria (e.g., guest mix and consumption structure), too. For evident reasons, destination performance evaluation gets the more useful the better it fits the structure of the destination's actual tourism offer. Admittedly, this data reporting prerequisite is likely to introduce a new demand on information delivery (as well as, by the way, costs) for local suppliers and could be seen as an element of business process re-engineering emanating from the introduction of a DMIS. But, as it is advantageous in performance evaluation if benchmark data are subdivided according to product structure (or broad product groups, at least), local suppliers should be encouraged to capture market response conforming to product structure as far as possible. This way, by linking post-sales data at TSBM level, match/mismatch between offer and actual market response can be measured and, based thereupon, information about possible future policy adjustments or product redesigns gained.

Since, within the TSBM, post-sales data can be related to both short-term and long-term components of pre-sales data, the emerging decision frame covers a wide range of policies from long-term infrastructure development (investment policies) down to very short-time yield management (pricing policies). Specifically, DMIS-based product development could be augmented – at the disposition level of day-to-day management – with an undelayed adaptation of variable product components to market behavior; in other words, if covered by the TSBM, timely business performance data may improve current destination operation assuming that service providers actually join in such a shared business scheme.

5 Conclusion, With a Suite of Open Problems

We have described a conceptual framework for a destination management information system for *planning and decision processes* based on a so called tourism super business model. This framework features two, in a sense orthogonal, structural components, viz.

- an *informational structure* integrating the informational collation of the complete potential tourism offer of a destination with monitoring data about business performance (guests portfolios, product performance, etc.), reports on competitor performance as well as general indicators of development in tourism;
- an *operational structure* providing a repository of models for product development and offer composition, including the derivation of consistent and homogeneous product groups, model calculations (simulation) vis-a-vis (extramural) tourism

market data, strategies for decision making (forecasting) and product marketing and distribution simultaneous to the operative business monitoring.

A further ancillary benefit of such a DMIS which – due to the technical integration of local suppliers – emanates from the intensified exchange of information, is a more effective information sharing.

However, such a DMIS raises also several (and not simple) problems such as legal aspects, or the large volume of data which has to be stored and managed. Additionally, the overall accessibility and visibility of product data will lead to a further increased competition with a stock market of tourist products, with – most probably – decreasing prices (since these products represent contracts about future services completely devoid of value after expiration). A similar problem arises with the definition of an appropriate level of sharing – and thus visibility – of business performance data, because there is an obvious trade-off between disclosing one's own business data and obtaining the competitors' data. While there is no direct answer to the first problem, a possible solution to the latter case might be that (i) performance micro-data are stored at the "lowest" geographical level and should be statistically classified and aggregated in a variable manner (conforming to the own marketing needs, but without jeopardizing a competitor's identity) at another site or at a higher geographical level, and (ii) access to data won't be free of charge. Drawing such a scenario it can be foreseen that there will emerge a set of *tourist information brokers* providing such services, similar to the Internet Service Providers of today.

References

1. **Dörner, D.** (1989) *The logic of failure* (in German). Rowohlt, Reinbek/Hamburg.
2. **Froeschl, K.A.** (1997) Metadata Management in Statistical Information Processing. Springer, Vienna (forthcoming).
3. **Schertler, W.** (ed.) (1995) *Tourism as Information Business* (in German). Ueberreuter, Vienna.
4. **Schmid, B.** (1994) Electronic Markets in Tourism. In: Schertler, W., Schmid, B., Tjoa, A., Werthner, H. (eds) Proc. of the 1st Intern. ENTER Conference, Innsbruck, 1994. Springer, Vienna, pp. 1-9.
5. **Werthner, H.** (1993) TIS - Tyrolean Tourist Information System (in German). *Wirtschaftsinformatik* **35**, No. 1, pp. 43-50.
6. **Williamson O.E.** (1985) *The Economic Institutions of Capitalism.* Macmillan, New York..

The value of information technology: A case study and a framework

G.J. van der Pijl, H.T.M. van der Zee, P.M.A. Ribbers
Tilburg University, Nolan Norton
e-mail: G.J.vdrPijl@kub.nl

1. Introduction

Over the past years, many attempts have been made to measure the value of IT according to a variety of criteria. In 1993 Brynjolfsson summarized the principal studies of IT and productivity[1]. He concluded that: "The relationship between information technology and productivity - the fundamental economic measure of a technology's contribution - is widely discussed but little understood. The general unease and the blurred discussion about the determination of benefits of IT confirm the need for better measurement, frameworks, and tools, to assess and monitor its value. In this paper we describe how the value of information technology was measured at ANWB, The Royal Dutch Touring Club. The case described demonstrates that the value of IT for an enterprise can not be expressed in a single measure. Measurements at different levels have to take place in order to get a clear picture. In the second part of the paper we demonstrate how measurement on different levels can be applied systematically by using the BtripleE framework. The paper is based on the doctoral dissertation of van der Zee[2] that was written under the guidance of Ribbers and van der Pijl.

2. A case of IT value measurement

2.1 ANWB

ANWB is a large European, commercially run membership association in the travel and tourism industry. The core range of services of the organization are managed by a number of Business Units, and include the supply of access to many kinds of information, e.g. about road and trip conditions, hotels, restaurants, camping sites, and cars, over the telephone (voice-response), by computer (Videotex and Internet), in magazines, brochures, maps, guidebooks, CD-I, etc. Services also include car breakdown assistance, diverse assistance in case of emergency, and a broad array of private insurance policies. The company is a truly information-based organization: IT is strategic in and of itself. In addition to the IT that supports managerial business processes, such as general ledger systems, accounts payable and accounts receivable systems, personnel information systems and payroll processing, inventory and logistics systems, etc., the company employs a host of IT systems that directly support its primary processes. IT also *enables* the company to develop products and services effectively, disseminate products and services efficiently, and provide customer service through data bases and multimedia distribution systems.

At the moment of measuring the value of IT, the company was involved in at least five large, time- and money-consuming but strategic IT projects. Having experienced a large growth of IT costs over the past three years, two recently appointed board members decided to have the value of IT for the organization measured. Discussing the meaning of measuring the value of IT they concluded that it meant to have at least the following serious questions investigated and answered:

- Do we spend our money on the right IT and does IT offer us any strategic value?
- Are we able to effectively use our IT?
- Are we capable of managing and implementing our projects?

We will now discuss how we measured each of these aspects of the value of IT in the case of ANWB.

2.2. Measuring the spending on IT and its returns

To be begin with we had to investigate the amount of money that was spent on IT and the distribution of those costs. Next IT costs had to be correlated with the organization's performance on two related dimensions: business and strategic performance. We briefly describe the measurement of each of these factors now.

Actual spending on IT and the distribution of IT cost

Total annual IT costs were collected for a number of years to perform trend analysis, and to benchmark IT spending against other companies. The centralized IT budget was easy to retrieve, since it was managed by the (centralized) IT department, but all business unit controllers had to be visited frequently to collect, check, and double-check information on decentralized IT spending.

It appeared that over a period of three years, total IT spending had more than doubled in absolute terms and almost doubled as a percentage of revenue. However, when compared with data of other information-intensive organizations such as banks and insurance companies, the total annual amount of money spent on IT as a percentage of revenue was not alarming. It seemed, based on these comparisons, that the organization did not necessarily spend too much money on IT, but the *growth* of overall IT costs needed management attention.

IT development costs appeared to have grown much faster than any other category of IT costs. In fact, IT development had grown tremendously, while maintenance of existing IT systems had been neglected.

The questions "How much do we actually spend ?" and "On what is that money spent ?" could be answered by analyzing IT expenditure by IT activities and resources.

Business performance through IT

To evaluate the contribution of IT to improving business performance, we suggested to evaluate the relation between IT costs and multiple (including *non-financial*) business measures for different categories of business performance before. As many non-financial performance indicators are specific to sectors we didn't make comparisons with other organizations, but rather looked at historical trends. This approach, however, proved to be somewhat troublesome. IT costs per non-financial business measure had more or less doubled over the past three years. For example, IT costs per member nearly doubled because the limited growth of number of members was 2,6 percent per year while IT costs had increased more than 100 percent over the past three years.

This misalignment of numbers was mainly caused by the fact that IT investments do not have an immediate impact; the benefits of IT can take several years to show results. An econometric study of Brynjolfsson[1] found lags of two to three years before the strongest impacts of IT were felt. Especially the benefits of infrastructure, while potentially large, are indirect and often not immediate.

In conclusion, the valuation of the contribution of IT to improving the organization's business performance could not be effectively performed because of this time lag. Although always the case, the time lag problem was extreme at the case organization, since the enormous growth of IT expenditure troubled the analysis of relationships between IT expenditure and its benefits.

Strategic performance through IT

The second dimension of business value of IT deals with the question to what extent IT costs are distributed over the areas in which results are expected to generate strategic success.

We interviewed several members of the organization's management. They told us that those areas that contribute most to the organization's success are those that increase its ability to create more value for more members, to launch new products and services rapidly, and to improve operating efficiencies.

It appeared that allocating annual IT costs of each of these critical business areas would be too complex and alien, mainly because of the unusual high number of shared, infrastructural IT systems. Instead, as a practical yet viable alternative, we assessed which groups of business functions were of highest strategic importance, relative to others, and in which groups of business functions IT was concentrated most..

For each business area(is group of business functions), we took a snapshot of the current degree of automation, measured by the level of actual automation as a percentage of potential for automation. We also constructed an imaginary picture - a projection of the near future - consisting of the current IT concentration complemented with the impact of the systems under development. This picture would provide a basis to answer the question whether the organization would be gaining strategic business value through IT, after completion and implementation of the IT systems under development.

To create this picture, the current degree and the perceived quality of IT support for each of the seven defined business areas was estimated during a series of workshops with middle management. Then we included the potential effect of the systems under development (new initiatives), on the basis of discussions with the business area managers, and arrived at a future "IT concentration score" combining the IT in place and under development. If the IT concentration score of each business area is the same as the strategic importance of the business area itself, IT would be strategically aligned with business objectives[3]. Figure 1 shows the results of this exercise. It tells us that the organization's IT was only partly strategically aligned with its objectives.

Business Area	Strategic Importance	Degree of Automation	Quality	New Initiatives	IT Concentration Score
Membership Administration and Support	●●	●●	●	Yes	●●
Membership Representation	●●●	●●●	●●●	No	●●●
Sales of Products	●●●	●●	●●	Yes	●●●
Sales of Services	●●●	●	●	Yes	●●
Road Assistance and Emergency Center	●●●	●●	●●	Yes	●●●
Signposting	●	●●●	●●	No	●●
Facilities & support	●●	●●●	●●	Yes	●●●

Figure 1: Strategic Impact of IT at ANWB: ● = low; ●●= medium; ●●●= high

Reviewing Figure 1, it can be argued that the organization had invested and was investing in the right strategic areas, with the exception of over-investments in Membership administration and support, Signposting and Facilities & Support.

2.3. Measuring the effective use of IT

The second question to be answered, "Are we able to effectively use our IT ?" needs focusing on the three perspectives of:

- Business processes and business activities,
- Users of IT,
- IT supply.

At this level, we measured the availability of IT to business processes and activities (IT coverage), user satisfaction with IT, and effectiveness of IT from technical perspectives.

IT effectiveness from a business process and business activities perspective

To determine the effectiveness of IT, first the availability of IT to support the 29 operational processes and 19 management processes that we collectively defined, was evaluated through the measurement of the level of current IT support and the exploration of remaining opportunities for further IT support. In fact, this was a more detailed analysis of the IT support in the business areas mentioned before, using the same approach, but in greater detail. As said, we also investigated the level of IT support in the near future, supposing that the systems under development were implemented and used.

The resulting picture of actual IT support as a percentage of potential IT support showed a relative high degree of coverage of existing IT in many of the management processes compared to operational processes. This was surprising, since most cases of IT effectiveness measurement that we experienced show that operational processes are further automated than management processes. When we completed the picture with the additional, projected level of IT support of the systems under development, and supposing that these were implemented and used, this difference of coverage disappeared. This analysis demonstrated to management that a balanced approach to IT coverage in both operational and managerial processes had been followed. Maybe unconsciously, but nevertheless successfully.

IT effectiveness from a user perspective

IT effectiveness from a user perspective was measured in two ways:

- the perceptions of employees of the overall effectiveness of IT, employees were surveyed anonymously.
- the satisfaction of users of 21 identified IT applications was measured in more detail by asking them to rate the performance of the IT application on a list of user satisfaction attributes.

The results of the first survey revealed a lack of interest in IT and IT applications in general, and a shortfall of commitment to take on responsibility for successful IT usage. The organization's management felt that these findings confirmed their expectations.

The results of the second survey successfully revealed the satisfaction levels of users of the 21 IT applications. The effectiveness of each application could be expressed successfully, and an extrapolation could be made to convey an overall effectiveness satisfaction rating for the whole. Two important applications in particular showed severe shortcomings. Also, some satisfaction attributes were rated consistently poor for all applications

Based on these findings, a number of practical actions to be taken were recommended.

IT effectiveness from a technical perspective

We also looked at more technical effectiveness attributes of the 21 identified IT applications. Structured questionnaires filled in by and later discussed with maintenance professionals, system managers and operations management, asked them to rate several specific and relevant effectiveness aspects of maintainability, architecture, and operability. The answers successfully led to the determination of the more technical-related quality aspects of the individual IT applications, as well as to the common an therefore overall technical quality of the application portfolio.

2.4. Measuring the management of IT

The third question, "Are we capable of managing and implementing our projects ?" was answered through the use of the concept of the Balanced Scorecards[4] for IT supply and associated measures of IT supply effectiveness and efficiency[5]. The effectiveness and efficiency of two important IT supply processes of the organization's internal IT supply organization were measured: IT Development Management and IT Infrastructure Management. For both IT supply processes a number of measures were applied, reflecting the distinct perspectives of the Balanced Scorecard concept.

Effectiveness and efficiency of IT Development Management

IT development in general was measured using indicators of three perspectives of the IT Development Management Scorecard:

- *Customer perspective:* for seven recently developed IT applications, the costs per Function Point delivered were measured and benchmarked against a data base containing IT development measures of other organizations.

- *Internal perspective:* a further break-down of quantified performance indicators by development phase demonstrated some interesting deviations from benchmarks, e.g. the feasibility study phase of an average project took twice as long as at other companies. Also, performance to budget was measured and discussed. Finally, the existence and actual application of methods, procedures, techniques, rules and other management controls were verified, using a series of different measures.

- *Innovation and learning perspective:* the number of training days per employee per year was measured and benchmarked, showing under-investment in that area.

The effectiveness of the projects under development was further and more specifically investigated by reviewing these projects. In workshops attended by both users and IT professionals, a series of effectiveness factors were discussed for each project. Measures were selected for effectiveness factors such as risk management, project management, availability of facilities and support, allocation of required skills, commitment and involvement of users, etc.

The application of the Balanced Scorecard measures for IT development, in conjunction with the benchmarking of IT development performance against other organizations, appeared to be a successful approach for demonstrating the effectiveness and efficiency levels of IT development. We selected a set of measures in such a way that management's questions could be answered and that the selected measures could be used practically within the time frame available.

Effectiveness and efficiency of IT Infrastructure Management

The measurement of IT Infrastructure Management focused on three main aspects related to the central operation of data centers and corporate data communication networks:

- Level of service to users (the customer perspective),
- Technology resources employed in relation to the work load (internal and financial perspectives),
- Human resources employed in relation to the work load (internal and financial perspectives).

The *level of service to users* was measured using the indicators of response time for on-line services (overall and during peak times) and availability of IT services. Measures were compared with benchmarks, as usual.

The measurement of technology *resources employed* focused on available and used CPU MIPS, disk space, tape capacity, printing capacity, etc., in relation to the work load (measured in used capacity of the mainframe computers).

The measurement of *human resources employed* focused on the number of people performing data center-related tasks, in relation to the work load.

Finally, as with IT development effectiveness and efficiency, the existence and actual application of methods, procedures, techniques, rules and other management controls were verified, during workshop discussions with Data Center Operations' management and employees, using a series of measures on interval scales. Specific measures of the innovation and learning perspective were not applied in this case.

3. To a general approach of IT value measurement.

The general unease and the blurred discussion about the determination of benefits of IT confirm the need for better measurement, frameworks, and tools, to assess and monitor its value. In this paragraph we demonstrate how the approach taken at ANWB can be generalized into a framework for IT value measurement: the BTRIPLEE framework.

Systematic and consistent measurement of IT value must be based on two key, but currently deficient, attributes:

- *An overall management framework:* because the realities of IT application in organizations are so complex, a conceptual scheme for simplifying and ordering them is desirable.
- *A set of key measures for value:* these allow the planning and control of IT, varying according to the objectives of the organization and the level of the framework for which the measures are constructed.

3.1 The BTRIPLEE framework

Stoner and Wanke[6] define management as *a systematic process (a systematic way of doing things) of planning, organizing, leading, and controlling the efforts of organization members and of using all other organizational resources (finances, equipment and information) to achieve stated organizational goals.* This continuous cycle of planning and control requires standards, measures and norms. Before managers can organize, lead, and control, they must make the plans that give purpose and direction to the organization, deciding what needs to be done, when and how it needs to be done, and who is to do it. The traditional planning and control cycle is illustrated in Figure 2.

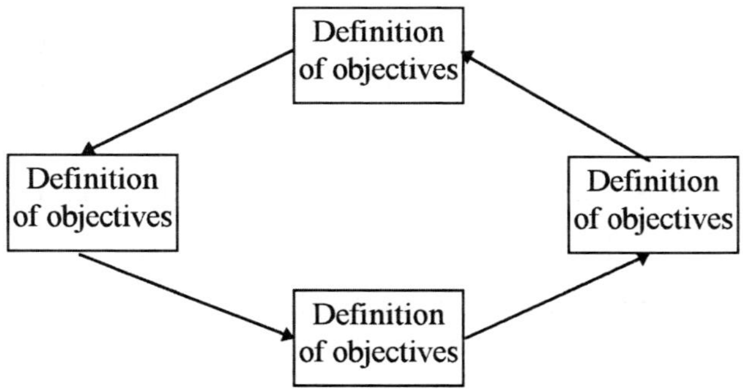

Figure 2: The Planning and Control Cycle

Planning and valuation of IT should not be different from any other planning and control cycle, however requires its specific approach. The framework we propose is based on a concept of demand and supply. The business organization is exerting a demand for information to support its management processes, its operational production processes, and its transactions. Through information technology a part of that information need will be supplied. An internal or outside IT supply organization will be responsible for delivering information services. Looking from a demand perspective we introduce two levels of analysis: business planning and IT planning. From a supply perspective we introduce the level of IT supply planning (see figure 3).

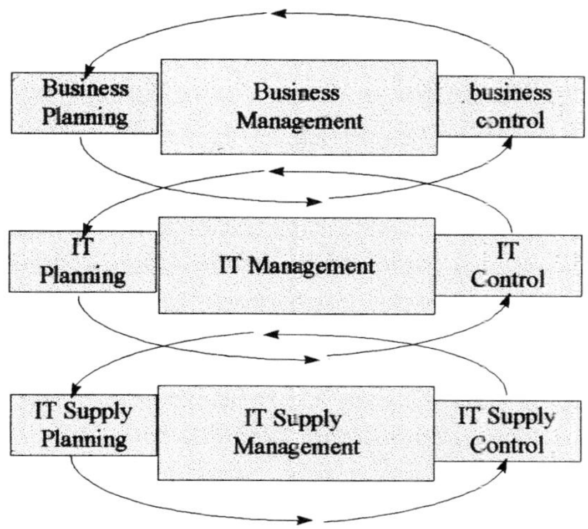

Figure 3: planning and control at different levels

It is our assertion that by assigning value to IT at *each* level, and in its full context, the overall IT value question can be approached. The different *planning levels* of business planning, IT planning, and IT supply planning, are shown in Figure 4. Although the levels of planning are interdependent and iterative, allowing for top-down and bottom-up initiatives, IT can be planned and controlled at each *distinct* level by applying *distinct* sets of measures.

Business plans are derived from business objectives and business goals. Business plans determine how the things that have to be done will be done through the configuration of business processes and business activities, along with the allocation of resources. The linkage of business objectives with its processes, activities, and resources, called business management, is illustrated in Figure 4.

IT strategies and IT architectures are not hierarchically derived from business plans; IT planning processes rather *align* IT strategies and IT architectures dynamically with business objectives and plans, and business processes, activities, and resources. This alignment, called IT management, is also illustrated in Figure 4.

Equally, the IT strategy and IT architecture are used to guide IT supply planning processes, which result in the construction of IT delivery processes and the definition of IT development projects. In Figure 4, the link between the IT strategy and IT architecture at one hand, and IT delivery processes and IT development projects at the other, is called IT supply management.

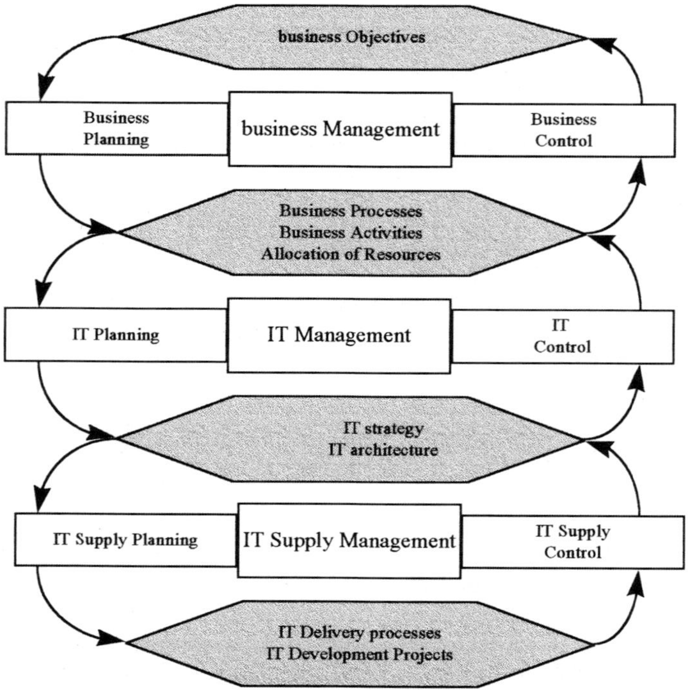

Figure 4: Planning, Deliverables and Control at Different Levels

Replacing "control" by "value measurement"

"Control" at the three levels in Figure 4 requires measurement and standards at the levels of business management, IT management, and IT supply management. At each distinct level different questions must be answered and distinct sets of measures applied.

- The measurement of aggregate cost and ultimate effectiveness of *all* IT relate to the business management level of our framework. The extent to which all IT enables and contributes to meeting business objectives effectively and efficiently, has been coined the **B**<u>*usiness value of IT*</u>. The label "Business Control" is therefore replaced by the label "Business value of IT", as shown in Figure 5. The <u>*Business value of IT*</u> can be defined as *the worth of IT for an organization as a whole, expressed in terms of* *organizational performance improvement at minimum costs.* Looking at the business value of IT answers the first question put by ANWB.

- The IT management level includes the measurement of the results of effective IT planning and hence the determination of the effectiveness of IT in supporting business processes, activities, and employees, regardless of associated costs, has therefore been called the **E***ffectiveness of IT*. This is also reflected in Figure 5. The *Effectiveness of IT* is defined as *the extent to which IT satisfactorily supports business processes, business activities, and business employees, regardless of associated costs.* Measuring the effectiveness of IT answers part of ANWB's second question.

- The third layer, IT supply management, includes the measurement of the results of IT supply planning. At this level, the effectiveness and efficiency of the supply of IT products and services are measured, therefore called **E***ffectiveness and* **E***fficiency of IT supply*. An appropriate definition of *Effectiveness of IT supply* is *the extent to which the supply of IT products and services aligns with the business's requirements as defined by an IT strategy and an IT architecture, regardless of associated costs.* By measuring the effectiveness of IT supply the other half of ANWB question 2 is answered. The definition of *Efficiency of IT supply* is *the extent to which effective IT is supplied at minimum costs.* Considering this level answers ANWB's last question.

The framework that links IT planning with the valuation of IT at the defined levels is called the **BTRIPLEE** framework, after the first letter of the name of each level of value measurement. Since the levels of planning and value are interrelated, both downwards and upwards, any IT planning and valuation framework must support both top-down alignment as well as bottom-up impact planning initiatives. The bi-directional arrows in Figure 5 reflect these aspects.

In conclusion, to assign value to IT is to ask questions at the different levels together and to place answers in context. Although the different planning levels are interdependent and iterative, the value of IT must be measured at each distinct level, by applying distinct sets of appropriate measures.

Reading the BTRIPLEE framework from bottom to top, value of IT is realized, if:

- Required IT products and services are satisfactorily developed, maintained, and operated (IT supply effectiveness), consuming a minimal amount of resources (IT supply efficiency),
- IT has successfully contributed to the performance of business processes, activities, and employees (IT effectiveness),
- IT is used to its full potential in terms of its possible contribution to organizational performance, at minimal cost (Business value).

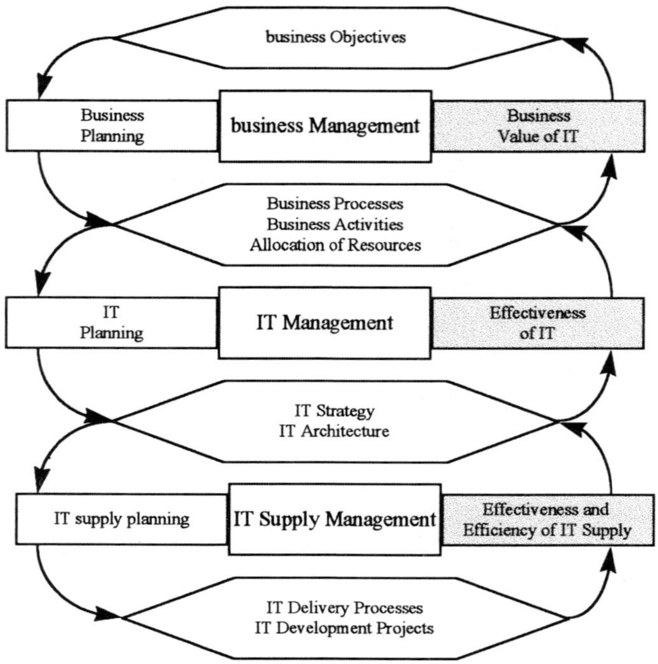

Figure 5: The BTRIPLEE Framework for IT planning and Validation

3.2 Measures of the business value of IT

An organization can improve its short and longer term performance in different ways. The measurement of the business value of IT is concerned with the relationship between the costs of IT and its contribution to the improvement of an organization's performance, measurable in three related dimensions:

Improving *financial* performance (reducing or avoiding operational and labor costs, increasing business productivity and revenue) through the traditional application of IT to improve the effectiveness and efficiency of the organization.

Improving *business* performance (enlarging market share, improving customer satisfaction, shortening order fulfillment times, etc.) through the innovative application of IT.

Enhancing *strategic* performance by reconfiguring the network of businesses involved in the creation and delivery of products and services, and by incorporating IT in products and services, thus reconfiguring the business scope.

A closer look at the three relevant value dimensions is taken, leading to an aggregated approach to correlating IT costs with business value.

IT costs and financial performance

Financial performance is measured by financial indicators such as profitability, productivity, earnings, etc. Profitability can be increased when operating costs are decreased, e.g., by means of improving productivity and efficiency through IT, and by increasing revenue. The business value of IT at the organizational level in that case is clear. It is often assumed that the total level of IT costs, measured as a percentage of the revenue of the organization, may be used as an indicator of the "right" spending level. This ratio is benchmarked (compared) with that of other organizations to judge whether it is right. If the organization's IT costs are about in line with those of others in the same industry, management might stop worrying because the numbers show they are at industry average. Although the percentage of revenue spent on IT is not a valid and reliable value-for-money indicator on its own, an analysis of IT costs for comparative purposes within the same industry at a gross level and in combination with other indicators can be worthwhile to perform. As is illustrated by the case study below, such an analysis may reveal abnormal patterns in IT spending and may lead to further, more focused investigations into the value question.

IT costs and business performance

The second valuation dimension is the relation between IT costs and business performance, measured by using financial and non-financial performance indicators such as competitiveness, new product sales, product development lead times, manufacturing lead times, distribution lead times, customer satisfaction, etc. To relate IT costs to improvement in business performance, a focus on the *results* of the organization's activities in the market is necessary, so that external-oriented measures of business performance are needed.

When linking IT costs to non-financial business performance indicators, e.g., IT costs per customer served, such ratios help in determining the business value of IT from an historical perspective (trends) or an external comparative perspective (benchmarking).

IT costs and strategic performance

The third dimension of measuring the business value of IT is to relate IT costs with the strategic performance of the organization. Strategic performance can be measured by the extent to which an organization realizes specific management objectives - its critical success factors: the most critical activities of the organization that contribute most to its succes[7]. It is possible to determine whether the best value of IT is obtained by relating IT spending to these critical success factors. In other words, the level of "strategic alignment of IT can be revealed by determining whether IT costs are aligned with the business strategy and distributed over the CSFs. This approach is based on the notion that IT costs should be focused on areas that give the greatest return: areas of greatest relative importance to the organization's business.

3.3 Measures of effectiveness of IT

The next level of the BTRIPLEE framework deals with the value of IT measured by its contribution to improvement in the performance of business processes, activities and employees. Three dimensions must be taken into account when determining the effectiveness of IT[8]. These dimensions are derived from the objectives and requirements of business processes and business activities, users of IT, but also from the objectives of IT supplying functions in relation to different types of IT.

Measurement of the effectiveness of IT is thus related to the:

- Support and enabling of business processes and activities, and its availability to business employees,
- Effectiveness as perceived by the people who use it,
- Technical aspects that stem from architectural and infrastructural requirements expressed by IT supplying functions.

The first measure to determine the effectiveness of IT is the degree to which IT capabilities exist to support the effective and efficient execution of business processes and activities. In fact, it is the degree of automation of business processes and business activities that is assessed (called coverage), indicating the extent to which human labor has been eliminated from business processes. Also, the availability from the users' perspective of IT applications and, in particular, customized user interfaces and workstations is determined.

Secondly, users must be satisfied with the context and content of IT to be effective, which are measured in ease-of-use, accessibility, flexibility, reliability, and security. Measuring the users' satisfaction with available IT capabilities is a way to measure their requirements and needs for effective IT; at the same time the need for education, training and coaching for users to employ available IT to the maximum is revealed.

Thirdly, requirements that stem from the IT supply functions (maintainability and operability of IT applications, adherence to architectural standards, etc.) are important to include in effectiveness measurements. Although they don't have a direct impact in the business sense, they *are* important for the effective and efficient supply of IT services and its ongoing maintenance.

3.4 Measures of the effectiveness and the efficiency of IT supply

According to the BTRIPLEE framework, IT supply performance is considered the result of two elements: IT supply effectiveness and efficiency. Efficiency-orientation is concerned with technical issues and optimization of production capability, while effectiveness-orientation is concerned with issues of superb support of use of systems, their impact on users, and organizational performance. In fact, it is the combination and balance of effectiveness of IT supply and the minimization of costs that determine IT supply success.

These concepts can be applied to order performance measures of each of the main IT supply processes:

- IT Supply Management, also known as Management of the IT organization,
- IT Development Management, including System Maintenance,
- IT Infrastructure Management, including Data Center Management and Data Communications Management,
- Account Management,
- Client Support, including End User Computing Support and Help Desk Support.

4. Conclusions

Most companies find it hard to asses the value of IT for their activities. Nevertheless one needs to know how to value IT when taking decisions on management of and investment in IT. There is therefor a need for good value measurement tools. Measuring the value of IT is a multidimensional task. Good measurement asks for a framework that connects the different levels to be considered. This paper demonstrates that value measurement can actually be done in a larger tourism oriented organization. Moreover it generalizes the principles used at ANWB into the BTRIPLEE framework for IT value measurement.

---

[1] Brynjolfsson, E. (1993), "The Productivity Paradox of Information Technology," *Communications of the ACM*, December, vol 36, nbr 12, 67-77

[2] Zee, H.T.M. van der (1996), In search of the value of information technology, doctoral dissertation, Tilburg University Press, Tilburg

[3] Henderson, J.C. and N. Venkatraman(1993), "Strategic Alignment: Leveraging information technology for transforming organizations", *IBM Systems Journal,* vol 32, nbr 1, 4-15

[4] Kaplan, R.S and D.P. Norton (1992), "The Balanced Scorecard - Measures that Drive Performance," *Harvard Business Review*, January-February, 71-79.

[5] See van der Zee (note 1) for a more elaborate discussion of the measures used

[6] .Stoner, J.A.F. and C. Wankel (1986), *Management.* Englewood Cliffs, N.J.: Prentice-Hall.

[7] Rockart, J.F. (1979), Chief executives determine their own data needs, *Harvard Business review, March-April*, 81-93

[8] Van der Pijl, G.J. (1993), *Kwaliteit van Informatie in Theorie en Praktijk*, Doctoral dissertation Catholic University of Brabant, Tilburg.

National Co-operation and Strategic Alliances
- The tourism business in Norway enters the Net

Kari Aanonsen
Senior Research Scientist
NR, Norwegian Computing Centre
Gaustadalleen 23, P.O Box 114, Blindern, 0314 Oslo, Norway

The rapid growth of the Internet as a distribution channel for tourism products has intensified the interest for electronic information systems and networks within the tourism business. New entrants on the Internet are choosing tourism as an area for new business, and traditional tourism companies are rethinking strategy and IT-investments to incorporate Internet. A number of actors are looking for ways to create business through this new medium. This paper focuses on how the organised part of the tourism business in Norway has met this challenge by choosing co-operation and strategic alliances as the main elements of their strategy. The paper is based on a project led by NORTRA, the Norwegian Tourist Board, and run by the Norwegian Computing Centre. Experiences from the first phase of the project were presented at ENTER96 (Aanonsen, 96). The current phase leads up to launching an application on the Net by January 1. 1997. In this paper we discuss the main challenges of this phase, which are building the network for national co-operation within the tourism industry and establishing strategic alliances within the IT and Internet world.

1. Background - A national framework

In 1994 Norway launched an initiative for building a National Information Network, NIN (NIN,95). The tourism business was chosen as one of 10-12 application areas. The goal of the project is to introduce and extend the use of computer networks within the tourism industry, particularly for connecting the numerous small and medium sized businesses around the country. We have chosen Internet as our main network, but are also following the development of other international network projects such as the EU-project TIM[1]/SAM (TIM, 93) which are using point-to-point connections through ISDN or ATM-networks (TIM, 94).

The focus of our project is twofold: using the Net as a tool for co-operation and communication within the industry, and using the Net as a marketing and sales tool. We have started with the last focus and are now building a national framework on World Wide Web (WWW) for marketing and selling Norwegian tourism products.

---

[1] Tourism Information and Marketing, RACE-project R2078

NORTRA took a leading position in the project during the first phase. The main activities in this phase were to develop demonstrators and define concepts and strategies. The concept consists of three elements (Aanonsen, 96); the organisational model for co-operation within the industry, the model for financing the system and the technical solution. We have chosen distributed models for all of the three elements, to encourage participation of tourism business at all levels and all over the country.

The organisational model is based on existing models for co-operation in the industry with regional and local tourist boards. The technical solution consists of a Web-application built as a framework for HTML-based Web-presentations connected through URLs, and basic product information from a network of regional and local databases.

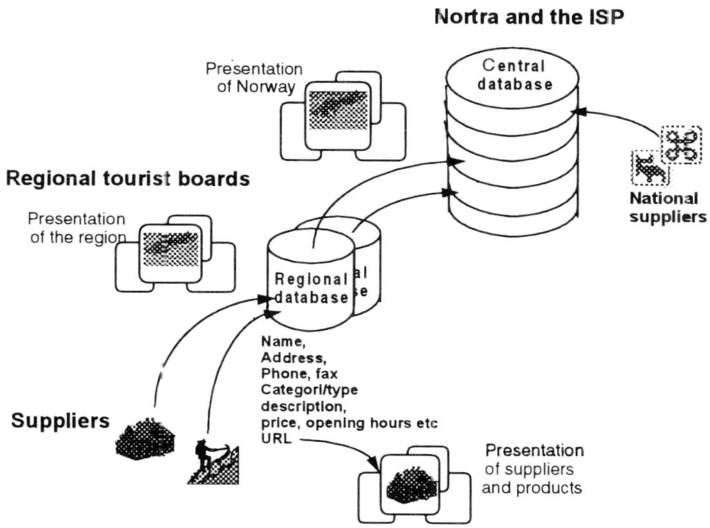

Figure 1 "The technical solution"

The financial model is based on distribution of costs and income. The costs for developing the framework is carried by the Internet Service Provider (ISP) with contributions from Nortra and the Norwegian Research Council. The regional tourist boards pay an annual fee to the ISP and Nortra to be connected to the framework. The regional tourist boards can take similar fees from local tourist boards and suppliers for linking them up. Tourist boards and suppliers at all levels may sell advertisements on their own pages, with a commission to the ISP.

To implement these models three main strategic decisions where made in the first phase of the project. The strategic decisions were to distribute responsibility for information to the regional level, to base the technical solution on regional Destination Information Systems and to establish an alliance with a national Internet Service Provider. The reasons for these decisions and our experiences from implementing them during the current phase, will be addressed in the next chapters of the paper.

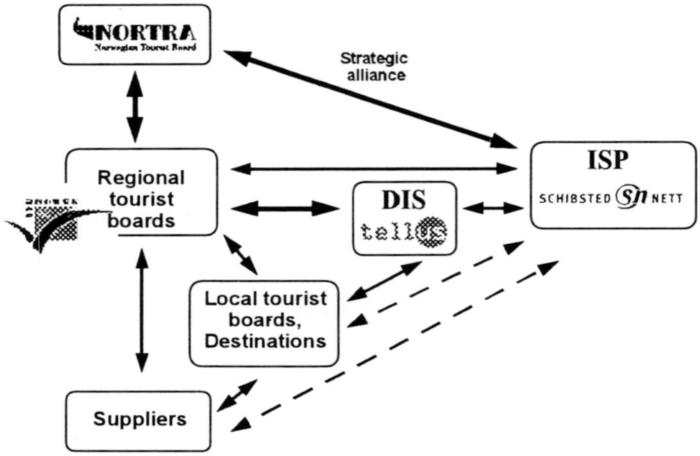

Figure 2 "Co-operation Network"

2. Nation wide co-operation between Tourist Boards

The Norwegian tourism business is organised as a hierarchy with Nortra on top, five regional tourist boards covering most of the country, a number of local tourist boards or "destination companies" on the next level and the suppliers at the bottom

2.1 The role of the regional and local tourist boards

The regional tourist boards play a central role in the model for co-operation in this project. The responsibility for information about each region is distributed to these partners. The tourist boards produce presentations of their regions, provides general practical information and basic information about the suppliers in the region. The basic information may include links to presentations from each business.

The regional tourist board is also responsible for organising the participation of local tourist boards and destinations within their region. These partners will be responsible for producing the local presentations, with practical information about their destination. It is also possible to give these partners responsibility for the basic information about local suppliers and for linking their presentations to the framework. The first kind of information can be created with any HTML-editor, but the basic supplier information requires a database and routines for transferring the information to the central database.

2.2 Defining the role of the regional tourist boards

There were a number of reasons for choosing a model based on co-operation with these partners; from the view point of the NIN programme, Nortra, the customers and the local tourism industry.

One of the goals set by the National Information Network program (NIN,95) was to build a nation wide network within the industry. To manage this we needed regional partners who could reach out to the numerous suppliers scattered around the country. The regional tourist boards have a well established organisational network for this.

A central element in the system is a product database, which the consumer can use to search for offers from all over the country. Nortra and the Norwegian tourism industry does not have a central database with tourism information, such as for instance the national tourist boards in Denmark, Finland, Holland and Ireland. One of the first strategic decisions Nortra made in this project was to *not* build and maintain a central national database. It would require more resources than Nortra wanted to put into it, and require co-operation with the large number of local suppliers around the country. The regional tourist boards are Nortra's main partner for this, and Nortra decided to establish a distributed database in co-operation with them.

During the design of the system we found that a crucial point was to be able to present the full spectre of products for the customers. A number of existing systems on the Net contain only information from a limited number of businesses. We designed a system where at least basic information from all businesses should be presented. To implement this we needed co-operation with partners who could provide such information. The regional tourist boards produce brochures with "yellow pages" containing lists of all the business in the region.

For the local tourism industry it is important to present their products in a context where the other regional products are presented. The context should also include general presentations which profiles the region. The regional tourist boards were the natural partner to archive this. We did also choose a financial model where potential income from the system could be directed back to the tourism business. Since the regional tourist boards in Norway are partly owned by the local suppliers, they were again the natural partner to choose.

2.3 Co-operation with the regional tourist boards

Most of the companies we have termed "tourist boards" are actually independent LTD-companies responsible to their boards and owners. Their main business is to market their region and its products on behalf of the suppliers. This means that co-operation between these companies has to be based on professional business agreements where each part can justify investments to their boards. In the end each participant has to decide on which IT-solution and which partners to use.

An agreement for co-operation between NORTRA and these partners is already established for brochure production, participation on fairs and campaigns etc. Co-operation on the Internet follows the same model, and the existing agreement will be extended to include this.

At this moment the five main regional tourist boards have agreed to participate in the national initiative. Each of them will produce information about their region and make it available on the Net according to the requirements based on the technical solution. Their first step has been to negotiate with vendors of Destination Information System (DIS), to find a common system for providing information about products and offers from all the regions.

3. DIS connects traditional tasks with new medium

For the tourism business Internet is just one of many distribution channels. Brochure production, handling of daily information requests, participation in fairs and campaigns, assistance to tour operators abroad, providing information for visitors etc. are some of the main tasks for the tourist boards. An Internet solution which is separated from these, will not be efficient, economic or get an acceptable quality.

One of the main challenges for an Internet system, is to maintain and keep the information updated. With a large number of partners who are not regular users of the Net, it is important to find solutions where the information is checked regularly as part of other routines in their daily work. The project's approach to this was to choose an Internet solution based on information from Destination Information Systems, which were used for a number of tasks.

3.1 Choosing a Destination Information System

The regional tourist boards had different IT-systems for their main tasks. Hardly any of them could be described as complete Destination

Information Systems. One of the systems was mainly used to store information for the brochures. The information was updated once or twice a year, before each new production of brochures. An other system was used at the counter in the main tourist information office in Oslo and at the tourist boards in one part of the country. The system was also used at hotel counters to provide tourist information for hotel guests. The information was regularly updated at the tourist information office, the tourist boards, and at the system vendor's office. The third system was used to register data for Information kiosks.

Our main reason for choosing these systems as potential partners where the fact that they were relatively well established in the industry. The tourist boards and the small and medium sized tourism businesses we wanted to include in this Internet system, have fairly simple technical equipment and competence. They need technical partners who can relate to this situation and give the necessary support. We considered other systems that could provide better technical solutions and international partners, but found that their support routines and commercial standing did not meet our requirements.

One basic technical requirement for the regional system was to provide structured databases. By establishing such databases around the country, the tourism business is taking an important first step towards more advanced systems for electronic marketing in the future.

3.2 Main requirements

The first step in choosing a Destination Information System was to identify a set of requirements. The system should satisfy both requirements from the Internet solutions and from analysis of routines and tasks at the tourist boards. The main requirement from the Internet solution was a structured database which could deliver data to the national database.

The other main requirement was routines for quality assurance of the data. The routines for updating information at the regional board has mainly been connected to brochure production, which means that they do a major updating job once or twice a year, except for events. An Internet solution requires continual maintenance as the data has to be correct at all times. One of the systems had extended routines for this maintenance, with expire date on all the information and automatic faxes to the owners of the information. This system also had routines for distribution of a simple version of the database to local tourist offices. This solution could help maintaining the data, by including this job in the routines at the tourist office, where most of the data is collected anyway.

4. National ISP[2] for Internet presentation and marketing

The third strategic decision from the project, which we will discuss in this paper, is to establish an alliance with one of the national Internet Service Providers (ISP). There are to main reasons for this choice. The first is related to marketing and Internet expertise, the other is financial.

4.1 Marketing on the Net

Marketing on the Internet requires continuos effort and attention. Since the main way of navigating on the Net is through links, it is important to create as many links as possible to a presentation. This requires knowledge of which Web-sites to be linked to, and how to get links there. The Internet Service Provider we chose has partners in a number of countries and does this kind of marketing also for other customers.

The Internet Service provider will also contribute to the marketing by including our presentation in their own main font page. This front page is marketed continually in a number of ways, for instance by being updated with news every day. The front page is the default home page for all their access customers. Including our tourism application in their front page will add value for them by drawing more visitors to their site. This was one of their reasons for establishing this alliance.

4.2 A professional Web-application

The rapid development of Web-functionality through new features in the HTML-standard and new technologies such as Java and ActiveX, requires co-operation with professional Web-application developers. One of the means to create traffic on a Web-site is keep it "alive" and up to date with the new technology. The national Norwegian tourist board decided not to have this kind of expertise in house, and chose in stead to establish the alliance with the ISP.

One of the main next steps in the development of Internet functionality is services for handling money transactions. In this version of the application the suppliers will have to include their own services for booking and selling products. The ISP will include a simple email to fax-service at this stage, but they are working on general services for payment over the Net, which will be made available through our application as soon as possible.

4.3 Development costs

The other main reason for establishing this alliance is financial. As part of the strong competition we now see between national Internet Service

---

[2] Internet Service Provider

Providers, they are investing in advanced solutions for different kinds of content. Since tourism information is one of the most popular, an alliance that combines a good technical solutions with high quality content was regarded as interesting for both parties.

The Internet Service Provider will develop the Web-application, based on the specifications we developed in the first phase of the project, in co-operation with main partners form the tourism industry. We are also participating in the design activities during this development phase. To create the user interface based on Nortra's graphical design, we have involved the marketing company responsible for the Nortra brochures, in the development process.

5. Challenges in the next steps

The main challenge with the solution we have chosen in Norway is to establish relationships and agreements which are based on business concepts for all the participants. Doing business on the Internet is a new challenge and an area with new opportunities and threats for everybody. New roles and positions has to be established.

5.1 Competion from other actors on the Net

Internet creates new competition for the traditional tourism organisations. A number of actors see potentials for new business on Internet in connection with travel and tourism. The main actors in Norway are now local and national newspapers, the Norwegian telecom company, marketing companies and new Internet and multimedia companies. The main services these companies are offering the tourism industry are creating Web-presentation and linking them to their Web-sites. Some of the companies also connect suppliers and tourist boards to the Net. The main source of income on the Net today is advertising. This makes the tourism industry attractive for two reasons. Web-sites with tourism information attracts net users and the number of "hits"/visitors on the site grows. This number is the basis for selling advertisements on the site. The other reason why tourism is a popular business for Web actors, is the chance to get the advertisements from the tourism business itself.

It is now a challenge for the tourism organisations to take an active role on Internet and create a positions for themselves as the main mediator for the tourism industry, also on this medium. The tourist organisations must use the advantage they have as the supplier's partners on brochures, and as their professional partners on strategy and marketing. Their expertise on the tourism industry, which is based on market analyses and a wide range of experience from the business, are competitive advantages the tourism organisations can utilise also on the Internet.

5.2 Further decentralising and including other parts of the industry

The next steps for the project is to further decentralise the solution and extend the network to local tourist boards and tourist companies all over the country.

Incorporating main commercial actors within the industry, will also be one of the challenges in the next steps.

5.3 Business related information and ad hoc communication

In the next phase the solution will be extended to business related information and to using the Internet as a tool for communication, increased efficiency and development of the tourism business itself.

References:

Aanonsen, Kari "The travel industry as a part of the National Information Network in Norway".1996. Proceedings in *Information and Communication Technologies in Tourism*, p. 143-150 Springer-Verlag

Morrison, A.M "Hospitality and Travel Marketing". 1989. ISBN 0-8273-2938-5.

"National Information Network. Twelve Areas of Application". The Research Council of Norway. Division of Industry and Energy. 1995. ISBN 82-12-00463-5.

Olaisen, J "Information Management - A Scandinavian Approach", Scandinavian University Press, 1993

"Prototype Distributed Database". Deliverable 34, Report form the TIM project. Authoring companies: Telesystemes, NR, Saritel.

Sheldon, Pauline J. "Destination Information Systems", 1993. University of Hawaii, Annals of Tourism Research. Vol. 20, pp. 663-649

Schertler,W et. al. "Information and Communication Technologies in Tourism".. ENTER 95 Springer Verlag. ISBN 3-211-82669-6

"TIM Usage: Impact of System Design approaches to system introduction". Deliverable 42, 1994.

Tjoa, A.Min, Werthner, Hannes" Interfacing WWW with Distributed Database Applications in the field of Tourism" 1996. Proceedings in

Information and Communication Technologies in Tourism, p. 78 -85 Springer-Verlag

Tschanz, Nico, Klein, Stefan "Internet Strategy Development in Tourism - Concept and Case Study", 1996, SISnet Research Workshop 'Internet and Business', Lisbon, Portugal

In Norwegian:

"Nasjonalt informasjonsnettverk (NIN)" Norges forskningråd, Industri og energi. ISBD 82-12-00426-0

"Reiselivsproduktet - En introduksjon til den regionale besøksindustrien" Georg Kamfjord. Reiselivskompetanse A/S ISBN 82-992929-0-5

"Strategi for IT i norsk Reiseliv" Kari Aanonsen, Knut W. Hansson, Rapport 889. Norsk Regnesentral 1995.

"Destinasjonsselskap i Norge - bidrag til auka marknadsorientering eller keisaren sine nye klede?" Torstein Nesheim, Magnus Mjør. SNF-rapport 41/94

"Markedsføring av norsk reiseliv i internasjonale distribusjonskanaler". J.V. Haukeland, J.Ludvigsen, G.O. Ryntveit. TØI-rapport 253/1994

"Sats på reiseliv" Rapport fra intern arbeidsgruppe i Nærings- og Energidepartementet Oslo, mars 1993

"Prinsippmodell for finansiering av fellesgoder i reiselivet" Erik W. Jacobsen. SNF-rapport 42/94

The New Network in the Norwegian Tourism Industry - Marketing and Organizational Theories as Perspectives for a Study

Ingvar Tjostheim
NR, Norwegian Computing Center
Gaustadalleen 23, P.O Box 114, Blindern, 0314 Oslo, Norway
e-mail: ingvar.tjostheim@nr.no

1. Introduction

There are many ways of studing an ongoing project. From a broad perspective the new Norwegian Network project is based on three main pillars (shown in fig. 1.). The first, the technical element which for instance concerns the database-solutions, the technical infrastructure. The second, the organizational element having to do with the partners involved, how the organizational network is built and how the partners cooperate. The third, the financial element concerns the investments- and income model of the project - the expenses and earnings involved. To establish a network project like the one in question, solutions concerning all the three elements have to be made. In this paper the focus will be upon the organizational element. Using marketing and organizational theories as perspectives for studying this element, some questions arising from this view are investigated.

What is reported in this paper reflects the author's participation in this new project. This has some positive as well as negative consequences. Some types of information is for instance more easily available for an independent observer, and for some purposes quantitative data-collection methods are needed. The project is in an early stage, the question of a formal and comprehensive evaluation has a higher relevance in a later stage. The purpose of this paper is not to evaluate the project, but to relate experiences with the project to relevant theories and literature.

2. The project "Network for tourism in Norway"

The aim of this new project is to build a network within the Norwegian tourism industry that utilize information technology and the advantages of cooperation and joint solutions. The use of IT, and open networks like the Internet for information and marketing purposes, is based on the assumption that it can increase the profitability within the tourism industry. The grounds for the research funding to one of the partners, the Norwegian Computing Center (NR), are motivated by this argument. In reports by the Foundation for Research in Economics and Business Administration in Norway (Skalpe, 1995) it is documented that the Norwegian tourism industry has a high number of small and medium size companies - many of them with weak financial results.

Fig. 1. - **The Pillars of the New Network for Tourism in Norway**

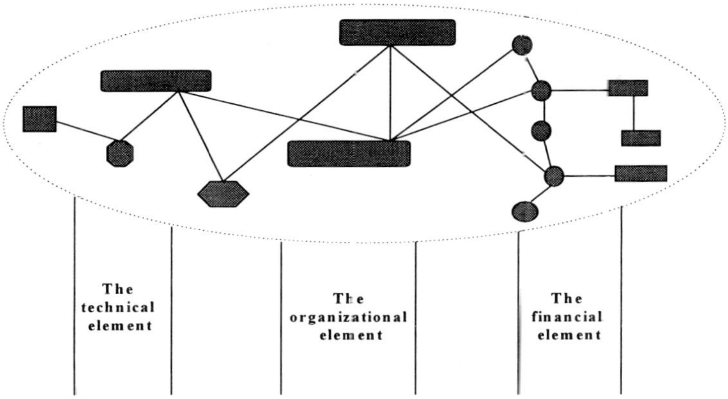

The main activity of the first phase of the project is to develcp 'Norway Tourism Guide' ('the guide') which is an Internet presentation of Norway as a tourism destination. The guide is going to have searchable information stored in databases linked together in a mirror database. The building of the databases, the core-design of the web-pages, the structure of the menus and information, are some of the major tasks in the first phase of the project.

2.1 The partners in the Network project

There are many ways of building a network. Normally a network or alliance is initiated by the companies themselves because they see some mutual benefits of the network. The pre-history for the new Norwegian network has relevance for this question, and some events are therefore mentioned briefly. The Norwegian Computing Center (NR) did at survey and report named 'A Strategy for IT in the Norwegian Tourism Industry' (Aanonsen & Hansson, 1995) and a project within the a national program run by a state department, the NIN (National Information Network) program. In the NIN-project, NR developed the demonstrator 'Norwegian Tourism Guide' (Aanonsen, 1996). Furthermore, a network model was developed and presented for the travel industry. Even though the ongoing Network project to a large extent was initiated by NR, it has had the necessary support by the tourism industry from the beginning. The NR initiative was based on the high priority given to IT (applied in different industries) by the Norwegian research council and one of the governmental/state departments. As a result, the Network project was established in May 1996 with the national and regional tourism boards as partners with their own financial schemes. Nortra, the national tourism board, took the overall responsibility for the project and for 'Norway Tourism Guide'. The tourism boards can be seen as parts of the public sector even though they also have other income sources than public funding. In relation to this, it is relevant to add that alliances between the private and public sectors seems to be particularly attractive in destination marketing (Palmer & Bejou, 1995). Against this background it is an interesting question to what extent the existing alliance or network is functioning as intended? In section four in this paper some issues of relevance to this question are raised and discussed.

Fig. 2. - the Partners of the Project

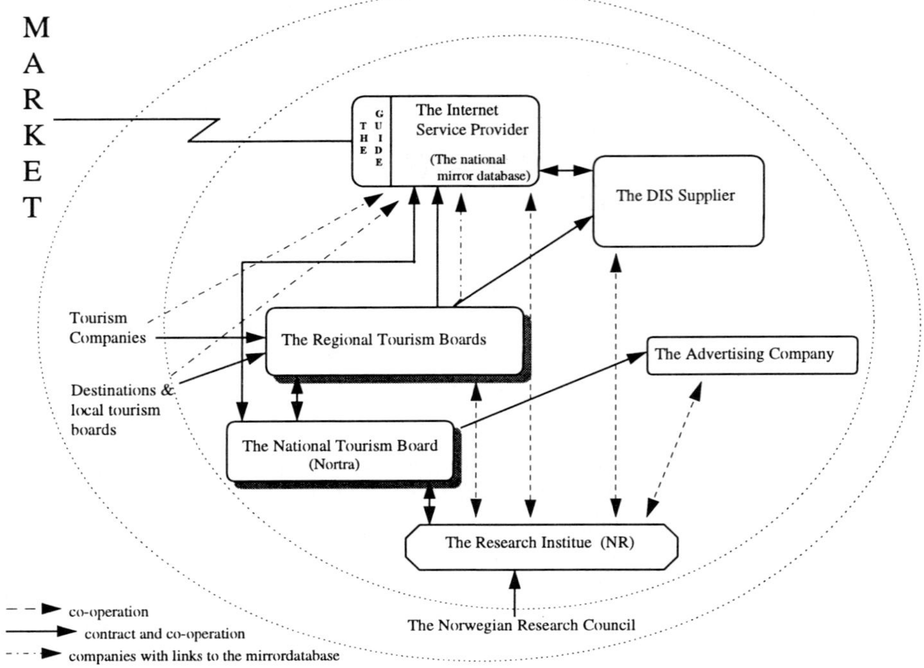

The inner circle shows the main partners or collaborators in the network. Even though the national tourism board of Norway (Nortra) is officially responsible for the project, the research institute NR is coordinating the project on behalf of Nortra. NR has research funding from the Norwegian Research Council.

One important purpose of the new project was to anchor it in the tourism industry. Thus, the five regional tourism boards in Norway were invited to join the network. Compared to Norta, they are closer to the destinations and the local tourism companies. As an ongoing activity they collect tourism information for their own brochures, and are therefore an obvious partner for this kind of network. They are responsible for regional brochures and marketing campaigns, and should to the greatest possible extent coordinate the marketing activities on this regional level. Internet marketing is a feature of the new project, and a high number of the tourism businesses in Norway do not have an Internet connection. One of the strengths of the project for the businesses in the industry is that through the new network they can get connection to a milieu with a relevant strenght and competence in this area.

Because of the high number of companies and tourism products in the industry, it is very challenging to maintain and update the information in a database on a national level. Compared to a central approach, it is easier to maintain to good quality of information using decentralized databases at a regional level. Thus, the five tourism boards play a very central role in the project. However, it takes time and effort to establish a position in a network (Mattsson, 1985) and it remains to be seen how it will develops.

The network also needed a Internet Service Provider (ISP) and a supplier of Destination Information Systems (DIS). The tourism boards together with NR decided to recommend the company Shibsted Net as the ISP. The Shibsted Net is one of the largest ISP's in Norway, and the company expressed a willingness to invest in this project as part of their positioning strategy on the Internet. The company Tellus was chosen as the DIS supplier. This firm is a DIS supplier with five years of experience in the tourism industry. Tellus has delivered and is operating a system (with a web application) for one of the tourism regions of Norway, the southern Norway. Moreover, they cooperate with a research institute in the Telemark region of Norway.

3. Information search and information sources in tourism

The importance of information for the tourists is emphasized by several scholars. This can be exemplified by the following citations; 'Information is the lifeblood of the tourism industry' (Sheldon, 1993, p.633) and 'For many vacationers, gathering, processing, and evaluating information is an integral part of the travel experience' (Snepenger & Snepenger, 1993, p.830) Very often, the tourists select among destinations on the basis of the total set of destination attributes. (Palmer & Bejou, 1995) A 'destination' consists of a number of commercial businesses in addition to public institutions and organizations. The marketing activities of the firms and organizations in the tourism industry varies in several ways. Some of them do not have the necessary resources for marketing - they are small businesses or have financial difficulties. Others do to not want to cooperate and coordinate their marketing activities - in relation to this more examples could be added From the travelers perspective, printed material, brochures, and different forms of advertisements from the tourism industry are very useful. However, to complement the picture other sources like books, newspapers, magazines, TV programs, the Internet, recommendation by friends, have to be added. The marketing activities co-works with other information sources, and as a results it is difficult to keep the effect of the sources separate from each others.

There have been carried out some relevant studies and reports on information search in tourism. Some of them are reported in an article by Kamfjord & Tjostheim (1996). However, the nature and effectiveness of various channels travelers use to obtain information have changed considerably over the past few years. (Fesenmaier & Bonifield, 1996) There is for instance much focus upon the Internet in the media, the total number of users is very high, the growth seems to continue etc. From a company's perspective, even though the Internet gives access to a global market, there are a lot of challenges to overcome (Quelch & Klein, 1996). There is considerable need for research on these matters.

To make an overview of relevant groups of information sources, and a model from a recent paper by Hoffman and Novak (1996) is used. In a tourism context, it is useful to divide between the two stages 'before leaving home' and 'when traveling' For instance, an important information source like welcome centers is of particular relevance in the second stage. (Fesenmaier et.al., 1993)

Fig. 3. - Groups of information sources.

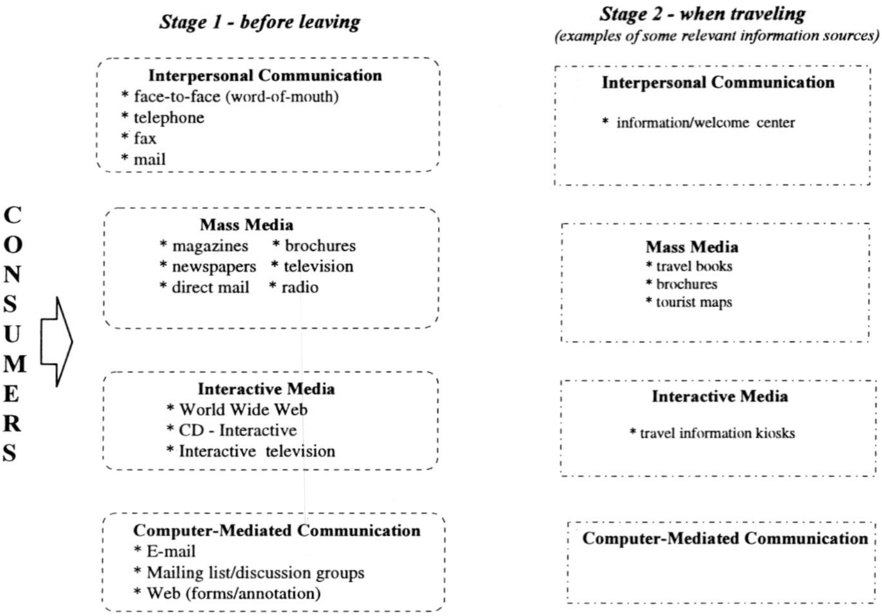

One of the features of the Network project is 'Norway Tourism Guide'. As a relative new marketing channel, the Internet competes with well established channels - some of them are shown in the model. It is likely that the Internet is a supplementary information source in most cases. However, the growth of the Internet is promising with regard to this matter.

In the Network project, the long term purpose is to store, maintain and make the information easily available for a number of tourism organization and businesses. This concerns the professionals point of view - backstage. For instance when (or if) the regional tourism boards in Norway start to use the databases in their travel counseling, at information centers, in production of printed material, for managerial purposes, interaction with other businesses and organizations and so on, the project has succeed in creating a tool of real inter-organizational importance. Even though high priority is given to the development of "Norway Tourism Guide" in the first months of the project, more resources will be used on supporting this kind of activities in the next stages.

4) Alliances and networks - insights from marketing and organizational theories.

In a research context it is important to report results and insights from the project. A fruitful approach in an early stage of a project might be to apply some different perspectives and theories as an analytical tool. Hence, the new network or alliance will be discussed on the basis of organizational theory (micro-economic theory) and marketing literature and theory. Relevant research and articles by the scholars; Webster, Williamson, Palmer and Bejou, Gemunden and Walter will be used in the following section.

4.1) Studies on marketing relationships and partnerships by Fredrick Webster

According to Webster (1992) new forms of business organization became common in the 1980s. The new organizations emphasized partnerships between firms, and could be characterized by flexibility, specialization, and emphasis on relationship management instead of market transactions. Webster refers to Thorelli (1986) who has studied the new networks. He stresses that they are dependent on administrative processes and negotiations, rather than market-based processes. Moreover, 'the purpose of these new organizational forms is to respond quickly and flexible to accelerating change in technology, competition, and customer preferences. (Webster, 1992, p.5)

Fig. 4. - *The Range of Marketing Relationships*

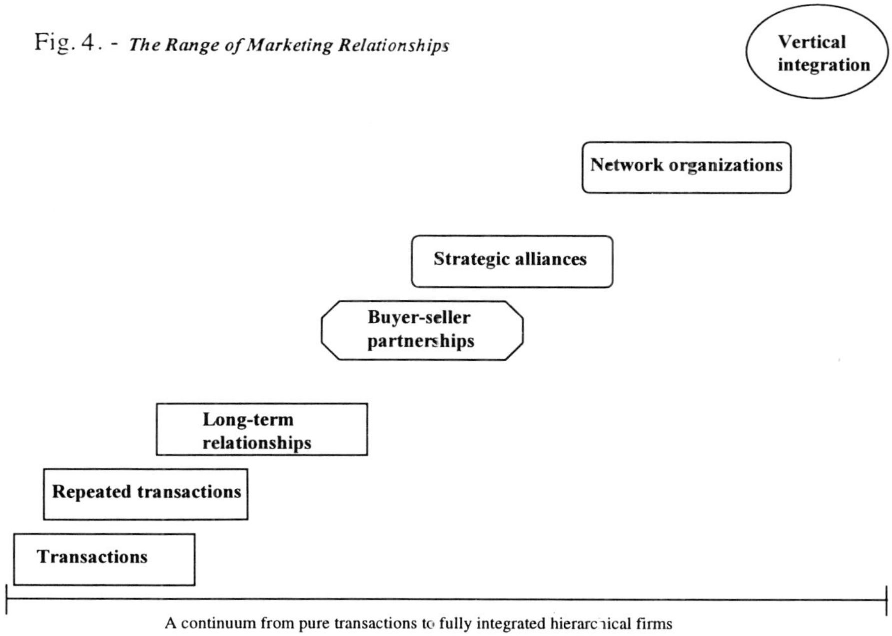

The model (fig. 4.) is based on Webster's article (1992) - the element 'distances' between the organizational forms are added in this version of the model.

A relevant question in this context is which label should be given to the new Norwegian network? Is it a strategic alliance or not? Strategic alliances are defined by the formation of new entity such as a product development team, a research project, or a manufacturing facility, to which the parties commit resources and which serves clear strategic purposes. Since network organizations are *corporate structures*, the Norwegian network cannot be given this label. The partnerships are one the other side of strategic alliances. In partnerships, according to Webster, each partner approaches total dependence on the other in a particular area of activity and mutual trust replaces the adversarial assumptions. It can be argued that the Norwegian network has some of the elements used to describe strategic alliances. However, to a large extent the partners act as they are independent of each others. The willingness to commit resources can be questioned - particularly for some of the partners. Some of them are focused upon specific tasks and not upon building an alliance as such. The partner NR, is a research institute funded by public money unlike the commercial partners. It is a reason to believe that the commitment to the project at NR is high. In addition, NR is coordinating or leading the project. However, if the process for instance slows down, a decision or request made by this kind of a partner is not given the same weight as a decision made by a (normal) commercial partner. To some extent, the effect is that the other partners are less concerned with the overall perspective or 'strategic purpose'. Thus, if Webster's model is used the new network in the Norwegian tourism industry should be described as a partnership and not as a strategic alliance.

4.2) Alliances in the theory of firms by Williamson

Oliver E. Williamson has developed an influential theory of firms named 'transaction costs economics' (Williamson, 1971 and 1985) It is relevant to compare his definition of alliances with the one used by Webster. Williamson (1990, p.7) writes;
'The *alliance* is a hybrid form of organization that is located between coalitions and mergers in its economic properties and differs from both in political respects. As compared with coalitions, alliances are more formal, have longer project duration, and are normally thought of as taking place between rather than within firms (although certain internal forms of cooperation might usefully be described as alliances). As compared with mergers, alliances preserve a greater degree of autonomy among the members' He cites Gerlach (1987, p.128) who remarks that one of the three integrating mechanisms on which these alliances rely is 'the creation of high-level executive councils which symbolically identify group members and the boundaries of the social unit, as well as provide a forum for interaction among group firms'. The project duration of the Norwegian network is three years, and it does not include the pre-project period and the period after the formalized project. If the database-structure functions as intended, and the project produces some other positive results for the collaborators, it is likely that the cooperation will continue. Thus, with regard to this criteria, the duration seems to be long enough. However, for the time being the project does not have a functioning 'high-level executive council'. Even though it seems that the term *alliance* as it is used by Williamson is broader than Webster's definition, it is questionable if the Norwegian network can be described as an alliance according to the definition by Williamson.

4.3) Marketing alliances in tourism - a recent study on by Palmer and Bejou.

Palmer and Bejou (1995) underline the fact that organizations involved in producing and promoting tourism destinations are interdependent. This provides a basis for the development of co-marketing alliances. One of purposes of the authors is to show that 'alliances between the private and public sectors are particularly attractive in destination marketing' Again, it is of interest to study how they define an 'alliance'. According to these authors 'a strategic alliance is a joining together (by means of dependency and collaboration) of two or more organizations over a given time period in order to gain a competitive advantage' (p.618.). This definition is somewhat pragmatic and broad. The focus is on marketing, not organizational theory as such. However, they also uses a taxonomy developed by Terpstra and Simonin (1993) to study tourism alliances. According to this taxonomy, an alliance has these four characteristics; coverage, form, mode and motive. Coverage refers to the extensiveness of an alliance in terms of functional competencies and geographical coverage. The form range from nonequity agreements to the opposite where alliance members buy equity stakes in other alliance members. The third characteristic, the mode, describes the intrinsic nature of the relationship among the members. In this connection, it is emphasized that commitment to the coalition from the senior management of collaborators has been seen as crucial to its success in the long run. Motives refer to the underlying reason for the creation of an alliance.

With references to this taxonomy the Norwegian network should be labeled an 'alliance'. However, by analyzing these four dimensions, a measure of the maturity can be established. The *coverage* of the alliance in question, is rather broad - it covers a number of important partners in the Norwegian tourism industry. The *form* is simple - the members of the alliance maintain their financial independency. The *mode*, the nature of the relationship can be described as somewhat unstable. As mentioned it does not have the commitment or involvement from the senior management level of all of the collaborators. With regards to *motives*, the partners are somewhere in the middle range if a scale from very low to very high motivation is used. To summarize, in this stage of the project the maturity of the new network is probably somewhere in the middle range.

4.4) The relationship promoter - a study on by Gemunden and Walter.

Within an organization the employees or members play different roles. Gemunden and Walter (1994, p.6.) define a relationship promoter as 'a person who actively and intensively advances inter-organizational exchange process by good personal relationships to key actors who dispose of critical resources.' The role of the relationship promoter is to bring the parties together and assist them in their dialogue. To a certain extent, the project manager at NR, has this role as co-ordinater and as a non-commercial partner. Normally project managers are granted possibilities (e.g. a time budget) in order to enter into and maintain contacts and relationship to external partners. In the article the authors divide barriers of inter-organizational cooperation in four groups; 1) the barrier of 'no knowledge of each other' 2) the barrier of 'no ability to co-operate' 3) the barrier of 'no will to co-operate' 4) the barrier of 'no allowance to co-operate' It remains to be seen to what extent some of these barriers will be of importance in the new network.

The network project is built around information technology and a number IT applications. The Norwegian Computing Center has relevant knowledge in this field and should be a useful partner in the network. Gemunden and Walter emphasize that 'numerous comprehensive studies on technologically oriented information and business relationships demonstrate that R&D co-operations with other enterprises, information exchange with clients, and technologically oriented exchange relationships with universities, colleges, and *research institutions* yield a strong influence on the technical and economic innovation success of enterprises.'

The relationship between organizations in a network takes time and effort to create, maintain, and develop. The Norwegian network is in its early stage and the contact or links will develop. According to Mattsson (1985,p.265) 'links between organizations can be seen as bonds of different types and strength, and he uses these six bonds; technical bonds, time-based bonds, knowledge based bonds, social bonds, economic bonds and legal bonds. Technical bonds concern the technical infrastructure and communication in the network or project. The knowledge based bonds can be described as the knowledge generated due to the exchange of information between the organizations over time. It can be hypothesized that the two bonds,technical and knowledge will be the most critical ones in this network project. However, it is a progess project and studies on this matter has to be carried out before any conclusions can be made.

5. Concluding remark

In this paper, literature and theories in marketing and organizational studies have been used to highlight some relevant issues with regard to the new Norwegian network. There are many important questions related to the building of networks, the Internet, the use of IT within the tourism industry, which are not addressed in this article. The new project is an attempt to build an interesting network. A complex set of factors will influence the results. Hopefully, some of relevant factors are mentioned in this paper. The need for further studies, from organizational- as well as other perspectives, is underlined.

The Future for Direct Retailing of Travel and Tourism Products: The Influence of Information Technology

Jacob Reinders and Michael Baker
correspondence to Michael.Baker@surrey.ac.uk

Introduction

Today, rising quality of information delivered to the remote terminal and the decline in the cost of the means have made direct purchase of travel and tourism products by electronic media a practical possibility. Middleton (1994) identifies the following methods of direct-response marketing: direct mail; telephone/tele-marketing; door-to-door distribution; travel related exhibitions; and interactive TV. The last was, at the time Middleton was writing, the only significant form of electronic direct selling (he had Minitel in mind) and it is this general field, in its present and expected future forms, which may revolutionise the way travel products are sold (Taylor, 1995), and which this paper is about. The phrase *direct* (ie principal to customer) *electronic retailing* (to distinguish it as sold to a consumer, rather than eg through EDI to an intermediary) has been coined to describe this. The research reported here explores the prospects for retailing by principals to end customers by electronic means, and the implications for the industry of this. It reports the results of a survey of senior people in the UK travel and tourism industry and its suppliers, carried out in 1995.

Technically, there are few if any obstacles to extending the electronic market in travel and tourism products and the development of electronic direct selling. There are numerous papers attesting to this, and many examples demonstrating it, including special networks such as those being developed in the SAM (Services and Applications for a World Wide Market in Tourism) trials, and for home shopping including tourism and travel products on the WWW (Byerley, 1996; Maartmann-Moe *et al*, 1994; Archdale *et al*, 1992; Sheldon, 1993). This paper therefore takes the availability of the necessary technical means as read.

Two factors particularly affect diffusion of the technology. First, few consumers will purchase expensive equipment just for home shopping. It must be bought for other purposes, such as leisure activities and entertainment (Economist Intelligence Unit, 1993; Schneider, 1995). However, as the necessary IT hardware has increasingly appeared in the home to meet consumers' leisure and entertainment needs, home shopping has expanded beyond the experimental stage. Second, diffusion will be inhibited until it is clear what the hardware platform of choice will be. The PC is likely to be the mainstay for some years to come (Sunday Times, 1995). However, there are alternatives, such as interactive television and the network computer (NC). Most consumers will wait to see which will be the market leader.

The Research Methodology

The research consisted of three phases: a literature survey, in which issues were identified; an interview programme, in which hypotheses about changes there would be and what might cause them were formulated; and a questionnaire survey, in which the views of a wider sample of industry experts were sought on these hypotheses. The methodology has much in common with mini-Delphi procedures (for an exposition of mini-Delphi and its use in tourism contexts, see Pan *et al*, 1996).

Initially, thirty-five experts were approached, fifteen of whom were interviewed in April, May and June 1995. These included representatives of the principals (airlines and car rental), the travel intermediaries (travel agencies and tour operators), system and network providers, tourist associations and consultants. Many interviewees requested that their views be treated as confidential, and for this reasons their contributions have not been referenced.

A novel feature of the questionnaire was to link main statements to subsidiary statements, reflecting views about cause and effect expressed by interviewees in the interview phase. So, for example, respondents were asked to agree or disagree about the extent to which single component travel products (eg point-to-point airline tickets) in the leisure market would be bought on-line; they were then asked, in the context of this, about matters which interviewees had indicated would inhibit or encourage this trend (cost advantages, the extent to which customers perceive these to be commodity products, the complex fare structures and so on).

Seventy-one questionnaires were sent to managers in or associated with the travel sector, and 31 valid responses were returned. The respondents were: six tour operators, seven systems suppliers, four consultants, eight travel principals, three from tourist associations and three travel agents. Nine of the respondents were either technical directors or heads of IT; eighteen were sales or marketing managers or marketing directors; the remaining four were in a variety of posts.

The Results: Literature Review and Interview Programme

The determinants of how much help a customer wants when buying a travel product can be considered along two dimensions: *Complexity* of the product; and *the requirements of the consumer,* with experienced travellers needing less information, and inexperienced travellers, or travellers to unfamiliar destinations or using unfamiliar or complicated travel arrangements, wanting more information and decision support to find out what they want and how to buy it. Point-to-point travel products like ferry or airlines travel are simple travel products, which leisure travellers may moreover buy quite frequently. The bulk (four-fifths) of the business travel products are of this sort, perceived as relatively straightforward and bought frequently. The other fifth can be classified as complex and bought less frequently. Off the shelf packaged tours are in general bought once or twice a year, but are perceived as being less complex.; but special interest tours and customised packages are seen as complex products infrequently bought. This analysis informed the division, for the purpose of the survey, of the market into: business travel; leisure mainstream inclusive tours; leisure special interest and customised tours; and leisure one component travel.

Commentators (Bruce, 1987: Schmid, 1994: Schertler, 1994: Baker & Baker, 1995) and personal interviewees suggest a number of factors which would encourage the growth of on-line retailing, and others which would inhibit it.

In the case of consumers, the incentives are those of *convenience*, *cost* and *choice*. If the consumers can accomplish all that a trip to the travel agent in the high street can achieve from home or office at any time of day or night, they may well prefer to book on-line, especially if they gain the impression that they, rather than the retailers' sales person, are in charge of the process. Cost advantages can be expected if on-line retailing provides greater market transparency, enhancing competition; or if the new distribution channel is less costly. Greater choice may be real or illusory. In a high street transaction the salesperson offers the customer a few choices at a time until one meets the client's expectations. To the extent that this filtering process eliminates (for whatever reason) options which the customer would have preferred to have considered, and which the on-line method would reveal, the increased choice is real. The disincentives are: *the absence of a well established home shopping platform; security; lack of product standards*; and *the habits of a lifetime*. Until one medium for home shopping emerges the clear favourite, few consumers are going to purchase the necessary hardware; and even then diffusion will take time. Security also remains a problem. The customer still perceives (probably rightly) payment by electronic means as insecure. Consumers may prefer to deal with a local travel agent rather than a computer if things go wrong. Product standards affect price comparisons as well as product descriptions. Travel products are "clouded" by multiple price structures and product descriptions which obscure comparison to the uninitiated. Interviewees thought that the lack of common price and product descriptions would continue to obstruct many customers. Finally, the pace of change is determined by the way the consumer adapts, and this is not likely to be the outcome of shopping for one holiday a year. This learning curve will be part of a wider revolution in retailing, as electronic shopping becomes part of everyday life.

In the industry, players in the distribution chain are increasingly looking towards IT for new ways of doing business. On-line retailing offers the travel principals large potential benefits. The most obvious is perhaps the prospect of *reducing distribution costs*. Total distribution costs per unit sold can amount to 30% of the total cost for hotels (Utell data 1992 in Beaver, 1992), 20-25% for airlines and 8-10% for tour operators. Bypassing agents would save 7-15% including overrides while bypassing videotext or airlines CRSs would in some cases reduce cost by 15% (Renshaw, 1992). Traditionally, tour operators have to allow for promotion to outlets, servicing them, brochure production and distribution as well as commission and possible overrides. These and the cost of a sales force serving traditional agencies can be trimmed if booking are made on-line. This looks bleak for the travel agent. But the distribution channel may offer travel agents the opportunity to develop an on-line presence (the *"virtual travel agent"*), obtaining a wider presence without opening additional high street retail outlets. It is a distribution channel which is likely to provide more, and more accurate *marketing information*. Operators and principals will gain information about their customers that otherwise would reach them only piecemeal or not at all. Currently, most customer data are gathered at the travel agents who have different data interests or do not have the skills, hardware or software to collect data systematically (Middleton, 1994). There are, however, vested interests in retaining the organisation of today's travel retail sector. The benefits of on-line retailing will be distributed unequally even between tour operators and travel principals, who in turn gain more than travel agents. There are heavy *sunk costs in the existing system*. The last few years have seen considerable concentration in the industry and vertical integration between tour operators and travel agents. Companies are unlikely to commit themselves to a strategy that reduces high street revenues. The dedicated nature of videotext links gives the larger operator competitive advantages over smaller competitors who can not afford the financial outlay. These tour operators will not wish to write off their investment in videotext as long as it still gives them a competitive edge in the market place. These interests might frustrate early growth of on-line retailing. Travel principals might be disinclined to adopt on-line retailing on a wide scale out of *fear of retaliation* by agents, through a sales boycott. The *cost to the principals* of distributing and updating elaborate information would be considerable. Some also fear that consumers will make reservations some time in advance which they will subsequently cancel when they can replace them with late availability bookings at bargain prices. Finally, the pace of the growth in electronic booking is being restricted by a *lack of standards* for terminology and communication processes.

The Results: Market-by-Market Analysis

Table 1 shows that in the *leisure market mainstream inclusive tours* market, a narrow

Table 1 Leisure market: mainstream inclusive tours

Statement	Mean[1] Score	%[2] Agreeing
More than a quarter of the mainstream package tours will be sold through home shopping systems in the foreseeable future (5 to 10 years from now)	2.3	57
Subsidiary Statements: Travel Industry		
The large tour operators will be reluctant to allow consumer access to the information on their databases	2.6	47
Travel agents would thereby be bypassed; the threat of retaliation by them will inhibit change.	2.65	47
The cost advantage of direct sell through on-line media will not be great. Travel agents will stay the most efficient way to sell standard package tours.	3.0	20
Subsidiary Statements: Customer		
People see most inclusive tours as commodity products that are not necessarily bought at a travel agent	2.75	36
The majority of the customers do not require sales support in terms of information and travel processing	2.9	27

1. Strongly agree =1; Agree =2: Disagree =3; Strongly disagree =4
2 . % strongly agreeing or agreeing (these definitions are used throughout)

majority of the respondents believe that more than a quarter of the mainstream package tours will be sold through home shopping systems in the foreseeable future. Some respondents suggested that 10-15% would be a more realistic estimate; a greater majority might have agreed to the lower figure. The potential reductions in cost are thought to be the main factor in favour of on-line purchase in the market, and the customers' continued need for sales support in terms of information provision and travel processing when buying a product they do not see as a commodity is the factor inhibiting the change. Only a minority think that most people see package tours as commodity products.

In the *special interest/customised tour* market, the results suggest a continued role for the travel agent. A majority (80%) of respondents thought that "customised inclusive travel and special interest holidays will continue to be sold through high street retailers" , although 80% also thought "on-line retailing will become a major marketing channel for specialised tour

Table 2: Leisure Market: one component travel products

Statement	**Mean Score**	**% Agreeing**
More than a third of single component travel ie airline tickets, hotel rooms, etc. will be purchased through home-shopping systems.	2.4	57
Subsidiary Statements: Travel Industry		
The cost associated with an increase in cancellations, whether caused by inexperienced customers or customers taking advantage of better prices and cancelling earlier bookings, will be prohibitive. Travel suppliers will therefore be reluctant to allow customers to make their own bookings.	2.8	19
Pressure on travel agents' revenue through commission capping etc. and the scope for price reduction by passing on savings on distribution costs will create a price gap between High Street and on-line prices.	2.4	62
Subsidiary Statements: Customer		
Most people see simple travel e.g. airline, car rental, as commodity products that do not need to be bought at a travel agent.	2.4	64
The complex fare structure and relative lack of product standards characteristic of most of today's travel market will deter the majority of customers from buying on-line.	2.6	48

operators". A larger majority (87%) thought that "home shopping systems for infrequent purchases and "complex" travel products will only satisfy some parts of the booking process with agents still needed for travel processing and information support". A majority (69%) believed that travel agents would themselves create holiday packages for their customer using the on-line facilities, but several commentators suggested that packaging by agents would be limited to the high end of the holiday market. It can be concluded that on-line retailing will make few inroads into this market. On-line travel services will be used by consumers to obtain information in the planning phase, and the increased functionality will mainly be exploited by the travel agents.

The third category, of point-to-point travel and other *single component products*, requires less preparation and usually lower expenditure, making these products *prima facie* suitable for on-line retailing. A small majority agreed. Several remarked that the absolute level of on-line retailing suggested ("more than a third") was too optimistic, so some who disagreed might have agreed to a lower figure. A majority thought that customers perceive these as commodity products. Forty-eight percent of the respondents thought that the complex fare structure and relative lack of product standards would be a deterrent, but some commented that the required expertise to make airline bookings, i.e. knowledge of the various bookings codes, prices structures and travel restrictions associated with a particular ticket status, would be resident in home buyers' intelligent software systems. Reviewing all the responses in this section, most respondents expect a shift towards direct selling of point-to-point travel, albeit not necessarily as great as suggested in the questionnaire.

A slight majority expected that half of the *business travel* will be sold on-line, but given that a considerable majority foresaw a quick take-up the results suggest that the disagreement

Table 3: Business market

Statement	**Mean Score**	**% Agreeing**
The majority of business travel(ie more than half) will be bought directly by corporate travel departments through the use of on-line travel services	2.25	55
Subsidiary Statements: Travel Industry		
Complex and infrequent travel itineraries will continue to be bought through travel gents	2.0	90
Subsidiary Statements: Customer		
The high frequency of travel combined with relatively well defined travel itineraries and needs in terms of late availability will make on-line bookings attractive and cause a quick take-up	2.1	75

recorded about the main statement is primarily about the level of change that can be expected ("more than half"). Almost all respondents agreed that the remaining 20%, comprising complex travel itineraries, will continue to be bought through travel agents.

The Impact of Electronic Direct Retailing on the Industry

Porter's (1980) framework of industry and competitive analysis was used to examine the effect on the industry, as in Schertler (1994). The interview programme covered all five of Porter's competitive forces and the questionnaire asked questions about four of them; all these are discussed below, including the fifth, *rivalry amongst existing firms*. However the table below shows only the responses to the main statements. Responses to subsidiary statements are touched on in the discussion..

On-line direct retailing offer opportunities for industry 'outsiders', traditionally strong in information technology and unhampered by old habits, high street infrastructure and systems. The new entrants would most likely be respected firms in the retail, media and entertainment worlds which could capitalise on their reputations (as Marks & Spencers and Virgin have done in other markets) and use their core competencies in distribution, customer databases or tele-selling know-how. There was strong agreement with this analysis; there was less agreement about where the invader might come from, with retail chains such as supermarkets being preferred to mail order companies.

Table 4: Competitive Forces

Statement	**Mean Score**	**% Agreeing**
The impact of recent IT developments, notably the reduced needs for specific travel infrastructure intermediaries and distribution networks, facilitate the entrance of industry outsiders in the travel market	1.9	89
The increased opportunities to deal directly with travel suppliers will increase the bargaining power of the consumer	2.3	54
The increased opportunities to deal direct with the customers will increase the bargaining power of the travel suppliers.	2.3	60
Alternative on-line travel services will grow in importance at the expense of traditional service providers.	1.9	95

The 'information highway' will provide consumers with a range of information direct from the supplier and with more choice as to who to place their business with and when to place it. A small majority of the respondents expected that the bargaining power of the consumer would increase.

A majority agreed that the increased opportunities to deal directly with the customers will increase the bargaining power of the travel suppliers. 91% of all respondents expected that direct access to the consumer market would lead to a reduction of commission levels and reduced margins for high street retailers.

Almost all respondents believed that the increasing IT functionality will stimulate growth of alternative services providers at the expense of traditional service providers. More than 90% expect that direct selling through on-line services by tour operators, hotels and other travel principals would lead to a 'shake-out' amongst travel agents; and all but a few (less than 5%) expect that the development of 'common carrier' information and communication networks will reduce the tendency for firms to be tied to one reservation or distribution system such as CRSs and videotext networks.

The question of continuing competition between the existing firms was covered in the interview phase, but not continued into the survey. Our interviewees suggested, however, that the rapidly reducing cost of IT was not thought to offer opportunities easy to grasp by the small businesses, as might have been expected, because it will be the companies that offer the most user-friendly and professional interfaces, and complete service at one site, which will do better than the rest. These are likely to be the major tour operators and principals as the combination of both skill and financial resources is likely to be decisive.

Conclusions

It should be noted that the conclusions are about the UK market, which has special features. The fastest growth will be in the business market with approaching half the purchases made directly on-line within the forecast period (up to the year 2005). Although growth of on-line booking will not be so fast in the leisure market, as much as a third of single component travel products (eg airline tickets and hotel rooms) might be booked directly on-line within the forecast period. In the rest of the leisure market, progress will be slower. Although as much as a quarter of straightforward mainstream packages might be booked on-line, it seems 10 -15% would be more realistic. Customised inclusive travel and special interest holidays will mainly continue to be sold through travel agents. The new technology will increase choice and competition, especially for the travel agents. This will come from: the principals' improved direct selling ability; entrants, for example the big supermarkets; and from greater transparency, of prices particularly, to the consumer. Given the extent of the expected shift to on-line direct booking across all markets, taking into account the current extent of direct, albeit non-electronic, bookings, it would be reasonable to predict a significant reduction in the number of high street retail outlets, by perhaps as much as 15% or 20%. Clearly the branches mainly serving the business market are the most vulnerable. In this context it is interesting to note that Poon (1993), it seems not anticipating the explosion of consumers' access to the electronic marketplace, believed that the position of travel agents would be strengthened *viz a viz* others in the supply chain rather than weakened, as this research suggests.

However, travel agents are likely to put pressure on other players to maintain the current status, and the recent vertical integration in the industry can be expected to act as a brake. Progress may be further slowed down by the absence of common standards for the required IT infrastructure. Similarly, customers have to overcome old habits, feelings of insecurity, fear for loss of privacy, and other constraints.

Although things look tough for the travel agents, those whose staff can use the electronic marketplace to find a better deal on a flight, a hotel nearer the beach for the same money, or a travel connection which saves more time, than the d-i-y netsurfer can, will continue to attract customers, especially if the staff, as well as being experts in the electronic tools, can talk authoritatively and accurately to their clients about destinations and transport modes.

References

Archdale, G., Stanton, R.M. and Jones, C.B. (1992), *Destination databases, issues and priorities, a paper prepared for the Pacific Asia Travel Association*, PATA Intelligence Centre, San Francisco

Baker, K and Baker, S. (1995), *Desktop direct marketing*, John Wiley & Sons, New York,

Beaver, A. (1992), Hotel CRS: an overview, *Tourism Management*, Vol. 13 (1), pp. 15-22

Bruce, M. (1987), New technology and the future of tourism, *Tourism Management*, Vol. 8 (2) pp. 115-120

Byerley, P (1996), *SAM: Past, present and future, Marketing travel over the information highway: the SAM opportunity*, Conference, Edinburgh 20-21 May.

Economist Intelligence Unit (1993), Home Shopping, *EIU Retail Business Quarterly Technology Review*, No 421, March, pp 4-9

Maartmann-Moe, E., Byerley, P and Guinano, R. (1994), *The TIM project: tourism information and marketing*, in Schertler, W, Schmid, B, Tjoa, A, Werthner, H, (eds), *Information and Communications Technologies in Tourism*, Conference Proceedings, ENTER 94, Innsbruck 12-14 January, Springer -Verlag, Wien, 35-42.

Middleton, V.T.C. (1994), *Marketing in travel and tourism, 2nd edition*, Butterworth-Heinemann Ltd, Oxford

Pan, S.Q., Vega, M., Vella, A.J., Archer, B.H, and Parlett, G.R. (1996), A Mini-Delphi approach: an improvement on single round techniques, *Progress in Tourism and Hospitality Research,* Vol 2, No 1.

Poon, A. (1993), *Tourism, technology and Competitive Strategies*, CAB International, Wallingford.

Porter, M. (1980), *Competitive Strategy - Techniques for Analysing Industries and Companies,* The Free Press, NY.

Renshaw, M.B. (1992), *The Travel Agent*, Centre for Travel and Tourism in association with Business Education

Sheldon, P. (1993), Issues in the development of Destination Information Systems, *PATA issues in tourism series,* No 1, PATA Intelligence Centre, San Francisco.

Schertler, W (1994), Impact of new Information technologies on tourism industry and businesses, Schertler, W, Schmid, B, Tjoa, A, Werthner, H, (eds), *Information and Communications Technologies in Tourism*, Conference Proceedings, ENTER 94, Innsbruck 12-14 January, Springer -Verlag, Wien,

Schmid, B. (1994), Electronic markets in tourism, *The Tourist Review*, No. 2, pp. 9-15

Schneider (1995), Shopping on the information highway, *Retail Insights*, Spring pp. ix-xi

Sunday Times (1995), *CableComms*, 3 September, pp 4-7

Taylor, P. (1995), ATM - Asynchronous Transfer Mode: technology of high promise, *Financial Times Review*, June, pp. VI

The Promise of Information Technology in the Travel Industry

Brenda L. Dietrich
IBM T. J. Watson Research Center, P. O. Box 218, Office 04-214
Yorktown Heights, NY 10598 USA

Jane L. Snowdon
IBM T. J. Watson Research Center, P. O. Box 218, Office 06-132
Yorktown Heights, NY 10598 USA

JoAnn B. Washam
IBM Corporation, 1133 Westchester Avenue, Mail Drop MD362
White Plains, NY 10604 USA

Abstract

Two words best characterize the future for the travel industry: growth and change. Many global forces have driven companies in this industry to adapt quickly to survive and remain competitive. Information technology plays a vital role in the way the travel industry responds as the world is "getting connected" at almost every level. The purpose of this paper is three-fold. First, the effect of network computing, the combination of electronic ticketing and smart cards, corporate travel management systems, and other trends that are transforming the travel industry will be examined. Second, the public policy issues involved such as the liberalization of government regulations, affordability and ease-of-use, and data security and privacy will be examined. Finally, the future direction of underlying technologies such as displays, storage, and microprocessors and advancements in the global networking infrastructure, mobility, and speech recognition will be presented.

Keywords: Corporate travel management, data mining, displays, electronic ticketing, information technology, kiosks, microprocessors, mobility, network computing, smart cards, speech recognition, storage, travel.

1 Introduction

Two words best characterize the future for the travel industry ... growth and change. The World Travel and Tourism Council have estimated the gross output of travel and tourism at $3.4 trillion in 1995. By 2005, this figure is expected to more than double to $7.2 trillion. Consumer expenditures, the largest piece of the industry, were $1.9 trillion in 1995 growing to $3.9 trillion in 2005. In this same period, capital investment is expected to grow by nearly 70% and business travel will grow 54%.

Many global forces have driven companies in this industry to adapt and change quickly to survive and remain competitive. For example,

- There is a massive shift from business to leisure travel as frequent, short breaks become the norm and as middle income wealth is created in emerging economies;
- Capacity is being rapidly added, particularly in Asia, where, over the next decade, it will climb from 34% of all international travel to over 50%;
- Competitive differentiation is a key issue to avoid commodity-like status. Hence, continuous improvement to raise product and service quality is important;
- Insightful and creative approaches to market segmentation and loyalty management are being used to improve load factors and occupancy rates;
- Continuing price and margin pressures are driving programs to reduce cost while improving productivity and asset utilization;
- And, more fundamentally, the competitive battlefield is changing with industry consolidation, global alliances, de-regulation, private ownership, and rapidly emerging electronic distribution channels.

Information technology will play a vital role in the way the travel industry responds to these global forces. Total spending on information technology will reach $1.3 trillion by the year 2000 according to the Gartner Group -- a 16% compound annual growth rate in information technology spending. The travel industry is spending approximately $20 billion per year on information technology today.

The massive changes in the travel industry raise numerous questions for both the customer and the industry. For example, if customers can use the Internet to make and pay for reservations directly, what happens to the traditional role of the travel agent as the intermediary? If kiosks, voice recognition units and smart cards become the norm for airport, rental car, and hotel check-in processes, how do service providers quantify and plan for proposed changes to achieve lower costs, better utilization and gains in customer loyalty?

The purpose of this paper is three-fold. First, the effect of these and other trends that are transforming the travel industry will be examined. Second, the public policy issues involved will be examined. Finally, future directions of the underlying technologies will be presented.

2 Information Technology Transforms the Travel Industry

The travel industry has been exploiting information technology products for more than 30 years. While the travel industry is already an intensive user of information technology for reservation, accounting and inventory management functions, information technology is becoming even more important in helping companies survive and grow as costs and service expectations rise, cycle times compress, and market share becomes harder to maintain. Information technology plays an essential role in lowering the cost of business processes, such as inventory management, scheduling, pricing, and customer service. Global scale and scope, seen by many as the key to survival, can only be managed with real-time information and constant communications. Information technology is central to any new distribution approach, especially those going directly to leisure customers and those providing travel management processes to corporations. Information technology can also differentiate between products, for example self-service check-in, trackable freight, or inflight seat-back systems. The point is that the use of information technology is moving from the "back office" to the face of the consumer and that novel uses of information technology become a competitive advantage.

Increased demands for information technology will come in the following areas:

- Reengineering of legacy systems which are obstacles to change;
- Best-of-breed, open industry applications for both operational efficiency and revenue generation;
- Network computing applications to supplement existing distribution channels and enable global alliances;
- Outsourcing services as companies return to core competencies and seek to gain operational savings.

The technology is available today to affect fundamental change in the travel industry. Three applications that will revolutionize the travel industry are described here: network computing, the combination of electronic ticketing and smart cards, and corporate travel management systems.

First and foremost, the move to network computing, where applications and data reside in the network, represents a powerful computing evolution. Network computing allows: companies to connect to their customers, suppliers, and business partners; individuals at work, home, or school to share information over interconnected networks; and businesses, governments, and educational institutions to reach new markets, offer new services, and lower costs. There are now 40 million people connected to the Internet. The number is doubling every nine months and is expected to reach between 400 and 800 million in the year 2000 when there will be over 1 million networks and 100 million connected computers.

A WebCrawler search on "travel" now produces over 60,000 hits, up from only 5,000 in August 1995. One industry analyst, Julius Maldutis of Salomon Brothers, has predicted that the airlines' growing use of the Internet will produce the third revolutionary change in the industry, comparable to the impact of jet aircraft in the late 1950s and airline deregulation in 1978.

Greg Conley, IBM's General Manager for Travel and Transportation, states that "travel distribution is the industry that invented electronic commerce and network computing over two decades ago." He believes that travel distributors need to be at the cutting edge and forefront to drive network computing to be a more broad-based model. For example, hotel and car rental companies should aggressively embrace the Internet to reach new customers, to give existing customers what they demand, and to lower costs. Furthermore, providers of customer reservation systems (CRS) and global distribution systems (GDS), being in the transaction processing business, should aggressively embrace the Internet as an additional source of income. Lastly, travel agents should also aggressively embrace the Internet to reduce costs and increase productivity. Examples of increased productivity include moving the booking process to the consumer, advertising specials worldwide to gain new customers, developing a database of information on customers to understand their buying patterns and preferences, eliminating the need to stock and distribute brochures which can now be accessed on-line, on CD-ROM, or via self-service kiosks, and finally, establishing links to other web sites in which your customers are interested. This last example could be extended to web agents that create "on the fly" links to other web sites by observing a user's patterns of access to aid the Internet search process.

The second technology that will revolutionize the travel industry is the combination of electronic ticketing and smart cards. This combination will eliminate a large percentage of the cost from travel distribution through on-line settlement, revenue accounting, and the elimination of physical ticket handling or ticket delivery. Airlines, hotel chains, and rail companies across the world are evaluating ways in which smart cards can help them to provide a more competitive service to their key customers -- the frequent flyers, repeat hotel guests, and regular rail passengers who travel mainly for business reasons. With 20% of air passengers responsible for 60% of the journeys, airlines are keen to build on their existing frequent flyer relationships.

Worldwide, some 760 million smart cards were issued in 1994, and the number is expected to grow to 2.5 billion in the year 2000. France has pioneered smart cards as telephone cards, and Lufthansa has used smart cards for frequent fliers on German domestic routes. The potential for the travel industry is significant. For instance, a single smart card carried by a customer could serve as:

- Reliable identification for fast, ticketless travel and for instant, kiosk-operated hotel registration and check-out;
- Reliable identification for immigration and customs;
- An updatable database of the traveler's preferences for seat location, accommodation, catering and amenities;
- A medium for implementing incentive and reward schemes;
- An electronic purse for pre-boarding, on-board, or in-hotel purchases;
- A co-branded link to a payment firm (e.g., credit card, debit card, or travel and entertainment charge card);

- A link to a travel business partner (e.g., a car rental company);
- An automatic, electronic record of the traveler's itinerary and expenses;
- A key to making secure payments for Internet bookings.

In October 1996, American Express and IBM announced plans to pilot a multi-purpose corporate smart card. Initially, these smart cards will be tested with American Airlines' enhanced gate readers, which are now installed in 21 U.S. airports, for ticketless travel.

At Southwest Airlines, which became the first major carrier to use ticketless travel systemwide in January, 1995, officials have estimated that the innovation saved the company $25 million last year. Beyond that, in the first two years of the program, the airline has gone from selling 40 percent of its tickets directly to consumers to 60 percent, with large savings in commissions to travel agents. In the Los Angeles Times' October 27, 1996 issue, the following North American carriers were described as deploying electronic ticketing in some form or another: Alaska Airlines, American Airlines, Continental Airlines, Delta Airlines, Northwest Airlines, TWA, United Airlines, and USAir. Clearly, this is a technology that airlines are embracing.

As soon as there is an accepted and published resolution from IATA/ATA on standards for electronic ticketing and smart cards, airlines will commit to the wide deployment of smart card systems. Installation of check-in kiosks and the integration of card readers with existing systems will be a key implementation step.

Self-service kiosks can dramatically reshape the relationship between the service provider and the customer by providing "direct customer access." Available kiosk solutions include: airline reservations and ticketing, airport check-in, seat assignment and changes, issue of boarding passes, receipt printing in a ticketless environment, arrival and departure information, loyalty program administration, passport documentation check, hotel room reservations and selection, concierge services, rental car reservations, casino coupon and complimentary redemptions, casino player card issuance, electronic shopping, and gathering customer feedback. Specific examples include Promus Hotels piloting a versatile kiosk system that provides all the information relevant to a guest's stay and Harrah's Casino Hotel in Atlantic City introducing Quick Comp multimedia kiosks as part of their strategy for improving customer loyalty.

The third area where information technology will have a profound impact on the travel industry is in corporate travel management. Total corporate travel management will expand from booking reservations to expense reporting and disbursement. Corporations will look to travel managers for ways to control escalating costs and expenses. Travel agents will become more differentiated, specializing in specific travel offerings, for example African safaris, and will extend their capabilities to other value-added services such as arranging to stop the mail, pick up the dry cleaning, walk the dog, and so on.

American Express, Apollo Travel Services, Carlson Wagonlit Travel, Nationwide Insurance, Rosenbluth International, SABRE and United Airlines are a few of the long list of suppliers and travel agencies racing to develop a comprehensive, all-in-one product that automates the book, pre- and post-approval, and expense reporting processes. Most of the products currently available are off-line resulting in a delay from when the traveler makes a request for travel information and when it is received.

SABRE's Business Travel Solutions (BTS), which is currently in beta test, is distinct in that it offers real-time access to SABRE. A booking module with real-time access to SABRE offers airline, car rental, and hotel reservation options based on the corporation's travel policy and preferred supplier program. An expense reporting module can automatically match a traveler's completed expense form to the corporate policy before it is sent through for approval to identify any errors, expenses that may require receipts or those that exceed what is allowed in the policy.

Internet Travel Network, a leading provider of on-line travel reservation services via the world wide web, announced in August, 1996 the debut of Internet Travel Manager, a full-service Internet travel booking and management product. The Internet Travel Manager is the first service deploying commerical Internet purchasing technology into a corporate intranet environment to support travel booking by employees.

3 Public Policy Issues

One of the drivers of growth and development in the travel industry has been the liberalization of government policies which encourage open competition. "Open skies," and policies which are resistant to regulation, government monopoly, subsidy, and other barriers to trade have reshaped the industry. For example, eleven European govements have signed "open skies" agreements with the United States as a prelude to new alignments. As a result, United entered into an alliance with Lufthansa, Northwest with KLM, and Delta with Sabena, Swissair and Austrian Airlines. Pending in 1996 is a marketing alliance between British Airways and American Airlines.

Affordability and ease-of-use are increasingly important. To date, despite much progress, estimates are that only 1% of the world's population uses on-line services or the Internet to arrange business travel. According to USA Today, this number is expected to increase to 20-30% in Europe within several years and number 30 million users worldwide by the year 2000. Keys to the adoption of these services are electronic ticketing, smart cards, improved Internet search facilities, and other user-friendly, front-end reservation applications.

Data security and privacy are pressing issues for travel, as for most other industries. As corporations move more of their business to the Internet and more private networks, called intranets, are linked to the Internet, the need for secure communications, backup, and disaster recovery will grow. These issues are not unique to the travel industry, but they do have special implications where the individual privacy of the traveler is involved, where distribution channels might be disrupted, or where mission-critical deliveries are involved.

Public key cryptography is a technology that is becoming very widely used and extremely important in a wide variety of areas. It is both different from the older secret key, or symmetric cryptography, and is complementary to it. In secret key, both the sender and recipient share the same secret key and use the key to encrypt or decrypt messages. This leads to the problem of how to share the secret key securely.

Public key cryptography solves this problem and also adds some very important features secret key cryptography doesn't handle including:

- *Authentication.* Using public key cryptography, a message recipient can authenticate that the message was created by a specific person and was not forged or created by an impostor.
- *Integrity.* Using public key cryptography, a message recipient can determine that the message was not changed or corrupted in any way since it was created by the originator.
- *Nonrepudiation.* Using public key cryptography, a message recipient can prove that a message was definitely created by the originator even if the originator "repudiates" the message, claiming that someone else created the message using their name.
- *Privacy.* Privacy is best provided by the well-established symmetric cryptography systems and algorithms, however public key cryptography makes the key transmission, sharing, and management much simpler and more secure, and thus a combined public/shared cryptographic system is now widely used to provide privacy.

4 Future Directions in Technology

This section describes the future direction of underlying technologies such as displays, storage, and microprocessors and advancements in the global network infrastructure, mobility, and speech recognition. Implications are given for the impact to travel.

4.1 Displays

Display technology is moving very quickly and will drive several new classes of applications over the next decade. First, flat panel displays will surpass the cathode ray tube (CRT) in image quality and approximate or even exceed the image quality of paper. This advancement will open up new uses such as reading complex documents, viewing high quality images, studying detailed diagrams and graphics, and exchanging lengthy communications.

The second major class of applications involves the new ability to miniaturize displays and make novel and ultra-portable devices. Display chips result from the marriage of silicon integrated circuit technology and some form of optical transducer (e.g., liquid crystal cells, movable mirrors, or light emitting materials). When display chips are combined with a suitable optical system, they can project images directly on the retina and give the visual appearance of a conventional-sized display. The display will no longer be the limiting factor in size reduction of personal clients.

The third set of new or expanded applications involves large screen displays for group viewing, or individual viewing, of very large amounts of information. The technologies will be based on display chips with projection optics and high brightness illumination sources, possibly including lasers, or on flat panel technologies like plasma which economically scale to large areas. These new display systems will be able to present large, detailed images and diagrams or multiple screens of information and video.

4.2 Storage

Storage devices such as hard disks, tape libraries, and optical disk libraries are on an evolutionary track that is well-established. For example, hard disks have improved by six orders of magnitude in areal density in the last forty years. In the next ten years, about 30X density improvement for hard disks is attainable. Magnetic storage will continue both its rapid advance in storage densities and its domination as a storage medium.

Optical storage in CD-ROM and digital video disk (DVD) technologies will have stable capacities, as consumers will not be willing to exchange their players and entertainment libraries on a rapid basis. In the next ten years, about 10X density improvement for optical storage and tape is attainable. Publishing using hypertext markup language (HTML) and DVD technology will become the new medium for disseminating large amounts of content.

4.3 Microprocessor Performance

Parallel and multiprocessor systems will become very important in the various server regimes. Low power microprocessors for personal digital assistants, controllers, and consumer devices will become both widespread and inexpensive.

4.4 Global Networking Infrastructure

The computer application environment is primarily TCP/IP-based, and direct ATM to the desktop is expected to only gradually encroach on this. As ATM in wide area networks is the better protocol for delivery of real-time and multimedia traffic, a heterogeneous network environment is ensured. Multimedia traffic will greatly increase, and cause degradation of network service, which will lead to novel caching solutions, and private toll roads for guaranteed levels of network service. The delivery of significant bandwidth to the home has several paths, including telephone systems (e.g., ISDN, ADSL), cable, and satellite. It is unclear which mode, if any, will become widespread and dominant, as the deployment depends on significant financial commitments in this turbulent arena.

4.5 Mobility

Mobile systems will continue to shrink in size and grow in power. A personal communicator is one example of a mobile system that combines a personal information manager with universal access and interaction using speech, pen, and graphics. An electronic book is another example. Early applications include maintenance repair manuals and parts manuals.

1995 was an excellent year for paging, which is evolving from one-way to two-way, and with increasing ties to computing and the Internet. The future will see smart cellular phones that offer some computer functions, but are primarily phones.

There are new devices for paging networks. A few carriers and resellers have introduced web sites that can be used to enter messages for transmission over paging networks. Several companies are developing new strategies to connect paging with electronic mail.

4.6 Speech

Continuous, large vocabulary, speaker independent speech recognition will be available by the year 2001. There will be a wide range of resource versus accuracy trade-offs, allowing deployment in a variety of systems and usage contexts. Speech synthesis will emerge as a strong, complementary technology. Together, both technologies will enable applications such as reliable automatic dictation, voice-mail acquisition, digital libraries (information retrieval), domain-specific control, home appliances management, and others.

Some of these advancements will be enabled by better recognition algorithms, larger vocabularies, and better training methods, but a significant part will be driven by advances in natural language processing (understanding and generation). Thus, "language aware" algorithms that support syntax and semantics, intonation, and articulation will be necessary to complement raw speech recognition and generation capabilities. Speech may also be used to authenticate travelers using 60 seconds of sampled speech. Other biometric advances include fingerprints and hand geometry.

4.7 Implications of Future Technology on Travel

The future directions for the role of information technology on travel include:

- Enhanced voice recognition to further simplify the reservation process;
- More powerful and easier-to-use Internet search engines;
- Internet tools that "mine" customer data to reverse the equation around, i.e., push offers to consumers instead of waiting for the consumer to ask "what is available at this time and at this price?"
- High quality videoconferencing capability at the individual work station;
- WebTV and network programming tools such as Java to expand the Internet customer base.

5 Summary

To summarize our view about tomorrow's landscape, the future of travel as enabled by information technology, we will make six predictions:

1. Travel Distribution Systems will evolve from GDS-dominated to truly interconnected networks.
2. Customer information systems will support highly personalized service as a basis for competitive advantage.
3. There will be a revolution on the revenue side of the business, with finer segmentation, better yield management, and expanded market reach.
4. Travel sales channels will undergo fundamental structural change, and disintermediation will create winners and losers.
5. Legacy systems and business processes will be radically redesigned to gain cost advantages.
6. Individual enterprises as well as entire industries will upgrade and connect their information technology instrastructures to support collaborative computing and electronic commerce.

Each organization must take great care to ensure that information technology plans are perfectly aligned with business strategy, information technology projects are carefully prioritized within the limits of available skills and resources, and performance measurements are in place to ensure adequate return on information technology investment.

To understand what all this means to the traveler, let's look at a future scenario.

The year is 1998 and identical twins Bill and Ted are about to celebrate their 40th birthday with a golf vacation in Bermuda. The logistics are complicated because their careers have led them in different directions. Bill lives in Vancouver, Canada, where he is vice president of a pharmaceutical company, and Ted lives in Melbourne, Australia, where he is a physician.

No problem. With collaborative computing, they can view airline schedules and travel brochures together, even though they are 8,000 miles apart, hunt for bargain fares and competitive hotel rates, and book their flights and rooms on-line. Smart cards pay their fares by debiting their bank accounts electronically through readers built into their personal computers. Their smart cards also contain a travel profile which tells the airlines their seat and meal preferences, and the customs and immigration officials their passport and other data -- no standing in line.

Problem: They're about to set off when a hurricane builds up in the Atlantic and their flights are canceled. Once again collaborative computing comes to the rescue. Using self-service kiosks at their respective airports, they agree to change their destination to Scotland and book new flights, rooms and rental cars.

In the air, Bill uses his computer to plan his next business trip through a corporate travel system which not only makes reservations but also keeps track of his business expenses. The system files his expense report and transfers the reimbursement to his bank account. It also ensures he is complying with his corporate travel policy and taking advantage of the company's preferred supplier agreements.

Ted checks his home security system, since his wife and son are also away on vacation, and finds the electronically controlled garage door is open. A simple keystroke closes the garage door.

One problem remains. Bill planned to surprise Ted with a custom-made set of golf clubs from a manufacturer in Arizona -- ordered and paid for through the Internet -- and now the clubs are on their way to Bermuda instead of Scotland. Bill plugs his PC into the communication port on his airline seatback, connects via the Internet to the supplier's freight management company, and tracks the clubs to a truck heading for Phoenix airport. The truck driver locates the clubs by means of the radio frequency identification tag affixed to the package and updates the tag to divert the clubs to Scotland.

Ted decides to relax on the flight from Australia to Scotland. The flight attendant gives him personalized customer service by bringing his favorite magazine and a bottle of fine French wine. His special meal is also on-board the plane. All this information was gathered from his smart card and from records the airline accessed in their database about frequent flyers. As Ted starts to unwind and relax, he decides to play video games and gamble on the in-flight system in the seat back.

Bill and Ted's smart cards contain all their passport information, which speeds them through customs and immigration in Scotland. They must do a hand scan for identification verification. Some airports have implemented biometrics where the hand scan identifies versus verifies the passenger; however, the Edinburgh airport has not yet migrated to this technology.

Bill's active smart card notifies the rental car company that he has passed through customs. The car is brought to the terminal. The rental car agency has run a decision support system to find the optimal route for Bill and Ted to take to the hotel. A bad accident has forced the authorities to close the major highway so Bill and Ted take the back country roads. The map, optimized route and directions are depicted on the on-board computer in their vehicle.

At the hotel in Scotland, they go straight to the check-in kiosk in the lobby and swipe their smart cards through the reader to get their room assignments and keys.

This is not just a fantasy of the future. Information technology is already making such ease-of-use and efficiency possible at lower cost, and end-users like Bill and Ted are the beneficiaries.

Federating Databases and Files: An Approach to a Uniform and Comfortable Interface to Heterogeneous Tourist Information Systems*

Michael Höding
Institut für Technische Informationssysteme
Otto-von-Guericke-Universität Magdeburg
Postfach 4120, D–39016 Magdeburg
E-mail: hoeding@iti.cs.uni-magdeburg.de
Tel.: ++49-391-67-12220 Fax: ++49-391-67-12020

Abstract

In this paper we sketch an approach to the integration of heterogeneous tourist information systems into database federations. Since many tourist information systems offer WWW-interfaces based on HTML-files and furthermore, many local information systems are based on specific file formats, we have to consider such file associations. For the integration of these file associations, the main problem is the lack of meta-information. Therefore, we have to derive an appropriate local conceptual schema. The derived conceptual schemata and the schemata of other tourist database systems will be integrated into a federated schema. Moreover, we briefly discuss the benefits of a federated tourist information system.

Keywords: federated database design, file integration, legal systems, reengineering, tourist information systems, WWW.

1 Introduction

Considering the situation in many organizations and enterprises we discover a large number of different information systems in use. Most of these information systems store their important data in databases. Database management systems are often used for this purpose. Beside this, many information systems are based on files. Here, we use the words *file association* to refer to a set of files which are used by one or more application systems to store data persistently. In conclusion, there are now different kinds of heterogeneous databases with different kinds of interfaces and many independently developped application programs.

The integration of the independent databases is a promising way which can help the organization to be competitive while the market is changing faster

*This research was partially supported by the German State Sachsen-Anhalt under FKZ: 1987A/0025 (Federating Heterogeneous Database Systems and Local Data Management Components for Global Integrity Maintenance: **SIGMA**$_{FDB}$)

and faster. The possibility of a uniform access to data from different databases allows a comfortable and uniform representation of the information. Thus, more efficient methods for information mining are possible. Especially in case of the large amount of information presented on the Internet, such a uniform interface is very useful [TK96].

In this paper we will try to illustrate the problem and a possible solution for the integration of different tourist information systems [Buh96]. These systems are good examples for heterogeneous information systems. We propose the use of a federated database system (FDBS) with an external HTML-interface for coupling the heterogeneous local data sources. FDBSs [HM85, SL90] provide a uniform and transparent access to heterogeneous database systems by means of a global interface. Thus, it is possible to establish new applications that can deal with the integrated data, e.g. a tourist information application providing an integrated view to the distributed tourist information systems of a hotel chain, a railway company and different airlines.

In the following we will discuss the application of federation techniques for tourist information systems. First, the problem of heterogeneity in tourist information systems is shown. Then, we illustrate the basics of FDBS by the help of some examples. After this we focus on the problem of file integration. Therefore, our approach is demonstrated by the transformation and remodeling for a WWW-based information system. We conclude with a brief discussion of the possible benefits of a federated tourist information system.

2 Heterogeneity in Tourist Information Systems

Computer based tourist information systems can be designed in a quite different fashion. First, we find systems that have been established for a long time and are based on data base management systems, e.g. the booking system of an airline or a hotel chain. Next, there are specially designed systems which store data by using the file system. Last, we find more and more information in HTML-files accessible with any Internet browser. Therefore, similar or related information is available, e.g. booking information of a hotel chain or a local tourist board. Moreover, the planning of a holiday activity is a complex and integrated task. The planner needs simultaneous access to information about different transport facilities, accommodation and other more or less structured information, e.g. about cultural events.

Access methods using WWW-pages and search machines solve the problem only partially. On the one hand, a comfortable and intuitive access to a large pool of information is available in this way. On the other hand, the quite different presentation of similar or related information by several WWW-servers leads to problems in efficient searching for relevant data in the huge amount of information. Since this task has still to be done by the human user, one is often lost in cyberspace. Moreover, a great part of the available WWW-pages are created by hand or by batch jobs that export data from other information systems. As a result, sometimes redundancies lead to incorrect information. Other systems provide an online access to a real database with a WWW-interface. In these cases, problems occur with write transactions to critical databases due to missing authentication mechanisms in the WWW. As a result today many write operations via Internet interfaces, e.g. the reservation of accommodation via the Internet, are still carried out via e-mail.

Nevertheless, a global interface, which allows a comfortable, intuitive and uniform read and partial write access to up-to-date information, would be very useful. Because of the independency of the enterprises and organizations, which provide the information systems, an integration into one homogeneous database is not possible. The existing systems must be preserved. In other words, the demand of local autonomy is important.

3 Basics of FDBS

Federated database systems are an appropriate solution for the integration of heterogeneous information systems. In this section, we will present the basic issues of FDBS by means of an example of a federated tourist information system.

For the integration of heterogeneous local information systems, a special software layer, called federated database management system, is added to the heterogeneous database systems. In figure 1, a federated tourist information system is sketched. The system integrates three local information systems: a "Railway Company Information System"(RCIS) based on a "real" database system, an "Airline Information System" (ALIS) based on virtual WWW-pages resulting from queries to a hidden database system, and an "Accommodation Information System" (ACIS) storing data in files of its own.

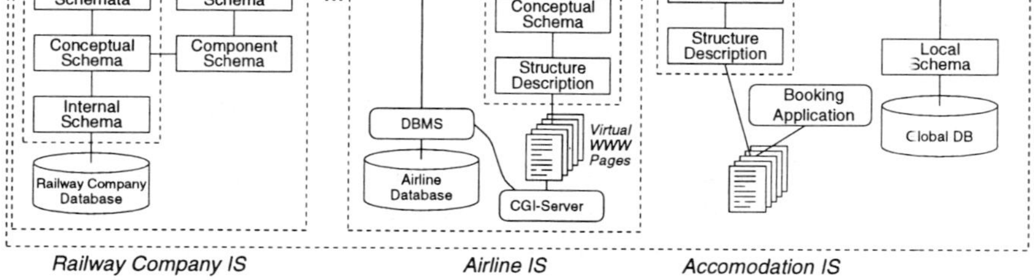

Figure 1: Schema architecture for FDBS

For FDBS, the five-level schema architecture of Sheth and Larson [SL90] is generally accepted (cf. figure 1). First, we take a look at the "Railway Company Informationsystem"(RCIS). We see a database, which is managed by a database management system that in turn contains the tree schema levels *internal, conceptual* and *external* schema according to the three-level-architecture of ANSI/SPARC. Such systems can be integrated into an FDBS in the common way, presented in [SL90]. Therefore, heterogeneity on data model and schema levels has to be overcome in the FDBS design process [BLN86]. Thus, we can distinguish the *schema transformation*, which transforms the local conceptual schema into an equivalent *component schema* in the common data model of the FDBS, and the *schema integration*, which integrates the transformed schemata into one *global (federated) schema*. The possible derivation of *external schemata* provides user specific views to the integrated data [SS96].

Whereas -in the case of the RCIS- we have to deal with an information system that is based on a "real" database system, the other example systems provide only an access to a file based interface. These systems do not have an available schema expressed in a well-known data model. The ALIS provides an interface that presents (virtual) WWW-pages on the Internet. Therefore, the user can fill in a form, which transmits a query to the WWW-server of the ALIS. The result is a human readable WWW-page containing information from the flight schedule. The question is: How can we integrate the query language and the resulting WWW-pages in an FDBS ? It is necessary to derive an appropriate schema, which is a valid view to the conceptual schema hidden in the database system or the CGI-application.

The problem is similar in case of the ACIS. This system uses files of its own in a specific file format. The schema, this means the information about the possible structure of the files is hidden in the application program. For the integration of the data of the ACIS, we need a formal description of the structure of all possible files.

We propose the derivation of two different kinds of schemata for local file associations (cf. figure 1). The first schema level is a *structure description* of the file association. The second schema level provides an improved *conceptual schema*. The schemata of this schema level should be based on the common data model of the FDBS to prevent an additional transformation step. Since the individual derivation of schemata for file associations is quite costly, we now sketch an approach to an efficient derivation of local schemata for file associations.

4 Derivation of Local Conceptual Schemata

Obviously, the transformation of data from a system that uses files with specific file formats is a difficult problem [ACM93, Höd96]. Now we take a closer look at the HTML-files provided by the ALIS (cf. figure 2). This example file contains the airline's flight schedule from Berlin to London. Due to the space restrictions and the very extensive HTML-format, we present only an extract, which we will use for an illustration of problems and possibilities for the schema derivation.

```
---EXAMPLE-FILE : schedule_ query_result.html--------------------------------------------
<HTML>
<HEAD> <TITLE>British Airways Online Timetable</TITLE> </HEAD>
....
<H2>Berlin - Tegel To London Heathrow</H2>
....
<TABLE>
 <TR>
  <TD>Depart</td> <TD>Arrive</td> <TD>Day</td> <TD>Flight</td> <TD>Price</td> <TD>Phone</td>
 </TR> <TR>
  <TD>07:35</TD><TD>08:35</TD><TD>1234567</TD><TD>BA981</TD><TD>124</A></TD><TD>030 4242</TD>
 </TR> <TR>
  <TD>10:15</TD><TD>11:15</TD><TD>123.5.7</TD><TD>BA983</TD><TD>124</A></TD><TD>030 4242</TD>
 </TR><TR>
  <TD>16:45</TD><TD>17:45</TD><TD>1...5..</TD><TD>BA985</TD><TD>124</A></TD><TD>030 4242</TD>
 </TR>
</TABLE>
....
The carriage of passengers, baggage, and cargo by air will be subject to the
Conditions of Carriage and Regulations of the carriers concerned which
may be seen on request at any British Airways Travel Shop.
....
</BODY>  </HTML>
-----------------------------------------------------------------------------------------
```

Figure 2: Example for a (virtual) HTML-file of the ALIS

As mentioned above, a local schema in a well-known data model is not available for both of the file associations in our example scenario. Commonly, the schema of such a database is hidden in the source code of application programs. The meaning of data or structure is only available in form of user manuals or program documentations or it is knowledge of the user. For the integration of such files, it is necessary to generate an equivalent schema and a mapping to physical file structures. Therefore we have to analyze *the structure of example files*, *information about access operations to the file association*, *the background information from documentations or interviews with users* and (if available) *the source code of application programs.* The analysis and derivation of the appropriate schema is a quite costly task, which has to be done for any different system separately. To make this process less expensive and to derive schemata with a higher quality, we need a schema derivation methodology for file associations. Furthermore, a set of tools, which are based on this methodology, should support the designer of the derived schema with condensed information.

Our approach for a systematical and efficient schema derivation is based on the idea of classifying the large number of file formats in use into classes of similar file formats. Therefore, we can use the dimensions *kind of structure*, *degree of intensional export*, *degree of extensional export* and *degree of writing rights.* Beside this, we have proposed a more empirical classification in [Höd96], which was mainly motivated by the examination of PC-based systems for factory planning. The classes are:

- record-oriented file associations,
- xBase file associations,
- completely structured text with meta-information,

- partially structured text with meta-information, and
- hypertext.

These classes, in our opinion, cover a wide range of file associations in use and are discussed in more detail in [Höd96].

Now, we will briefly present the steps of our approach:

1. *Pre-Analysis*
   The goal of this step is the condensation of basic information.
   (a) Background Knowledge Formalization
   Obviously a method catching the common and specific background knowledge is necessary. Our idea is to use a dictionary that contains common information, and which could be enhanced during the analysis of background information.
   (b) Class Assignment
   Based on the enhanced dictionary and the other information sources, e.g. example files and access information, the file association is assigned to one or more classes of a predefined set of classes, such as presented above.
2. *Specific Analysis/Structural Design*
   The goal of this step is the derivation of a structure description for a local file association. Due to the fact that neither the example files nor the other information sources contain enough information of the semantics of the local file association, this process cannot be done automatically. Mostly, this process has to succeed manually. We suggest the use of SGML[1] for the structure description of the local file association.
3. *Conceptual Design*
   The conceptual design is the last step of the derivation process. We assume that a correct structure description of the local file association is available. The goal of this step is to design an improved and more comprehensible schema expressed in a well-known data-model. The use of the common data-model is a good choice because in this way we can avoid one transformation step.

Now, we will illustrate some steps of our approach for the two file associations of the local tourist information systems ACIS and ALIS. First, the analysis of background knowledge leads to the enhanced specific dictionary, sketched in figure 3. The essential idea is that data or meta-data in example files refer to meta-data that we are looking for. The dictionary contains a common part with data patterns, e.g. `date_pattern_1`, common example data, e.g. `jan, feb, ...` and common meta-data, representing synonym links, e.g. `flight_number` → `flight`. In the enhanced part of the dictionary specific data of the application scenario (e.g. examples for flight numbers: `BA981, BA983...`) and specific meta-date (e.g. `arrival_time`) can be found.

[1] Standard Generalized Markup Language [ISO86]

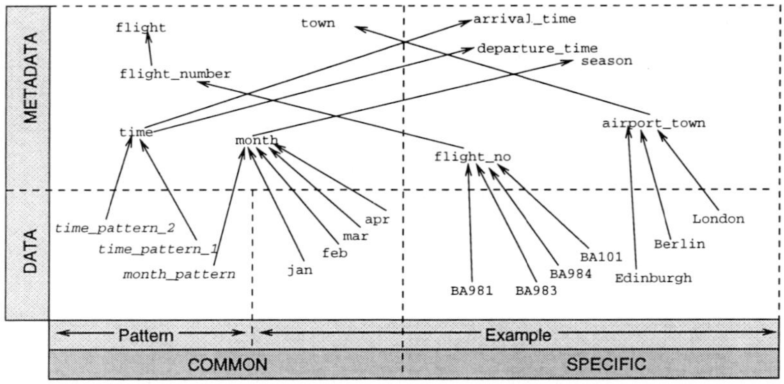

Figure 3: Common and specific meta-data and data in the enhanced dictionary

The ALIS is based on the well-known HTML-format. Therefore, we can assign the file association of the ALIS to the class "hypertext". Due to the partially structured components, which can be discovered by processing the example files with an assigning-tool using the enhanced dictionary, it can be assigned as well as "partially structured text with meta-information". However, HTML only defines the meaning of some expressions for the interpretation by an HTML-browser. These expressions are only of partial interest to the integration and should not be added to the integrated schema. On the other hand, they can help to find the more important information about structure and meaning of included text data. For instance, HTML-links are often used to express structure, e.g. possible destinations for a selected airport. Another example is the repeated occurrence of similar HTML-tables which give hints for existing relations.

The ACIS of our example scenario is a system that stores records containing information about bookable rooms of a local tourist board. Formalizing background knowledge in an enhanced specific dictionary and applying the mentioned assigning-tool with the dictionary to the example files, we can classify the ACIS database as a "record-oriented file association".

As mentioned above, some constructs in the HTML-files of the ALIS can give hints for the derivation of the conceptual schema. In our example (cf. figure 2), the table construct is used to show a (selection from a) relation from the hidden database. We find candidates for attribute names (meta-data) in the first line (the head of the HTML-table). The following lines contain tupels with their attributes corresponding to the attribute names. The application of a specific tool using the enhanced specific dictionary provides the designer with condensed information about the possible structure of the file association. Therefore, the designer can formulate a structure description as sketched in figure 4. This simplified description says that the ALIS file association consists of one or more files. Each file consists of informal text, followed by a schedule table and by more informal text. The interesting information is contained in

```
---> SGML-DTD:airline-information-system <----------------------------------------------
<!ELEMENT alis_file_asso - - (alis_www_file)+                                          >
<!ELEMENT alis_www_file  - - (any_text,"<TABLE>",schedule_table,"</TABLE>",any_text)   >
<!ELEMENT any_text       - - #PCDATA                                                   >
<!ELEMENT schedule_table - - ("<TR>",schedule_head,"</TR>",(flight_record)+)           >
<!ELEMENT schedule_head  - - ("<TD>",(attribut_name)+,"</TD>")                         >
<!ELEMENT attribut_name  - - #PCDATA                                                   >
<!ELEMENT flight_record  - - ("<TR>",dep_time,arr_time,day,flight_no,price,phone_no"</TR>") >
<!ELEMENT dep_time       - - ("<TD>",#PCDATA,"</TD>")                                  >
<!ELEMENT arr_time       - - ("<TD>",#PCDATA,"</TD>")                                  >
<!ELEMENT day            - - ("<TD>",#PCDATA,"</TD>")                                  >
<!ELEMENT flight_no      - - ("<TD>",#PCDATA,"</TD>")                                  >
<!ELEMENT price          - - ("<TD>",#PCDATA,"</TD>")                                  >
<!ELEMENT phone_no       - - ("<TD>",#PCDATA,"</TD>")                                  >
----------------------------------------------------------------------------------------
```

Figure 4: Structural description of the ALIS file association

the table which consists of a head with the attribute names in the first line and records in the following lines. To find the table and parts of the table, the tokens <TABLE>(table begin), <TR>(row begin), <TD>(field begin), and their corresponding end-tokens are used. However, we have to add information about data types to the structure description, e.g. in table form (cf. [Höd96]), due to the missing data type concept in SGML.

An appropriate structure description has also to be derived for the ACIS. In [Höd96] we have shown the derivation of a SGML-DTD for "record-oriented file associations". The method sketched in this paper can be applied to the ALIS file association, too.

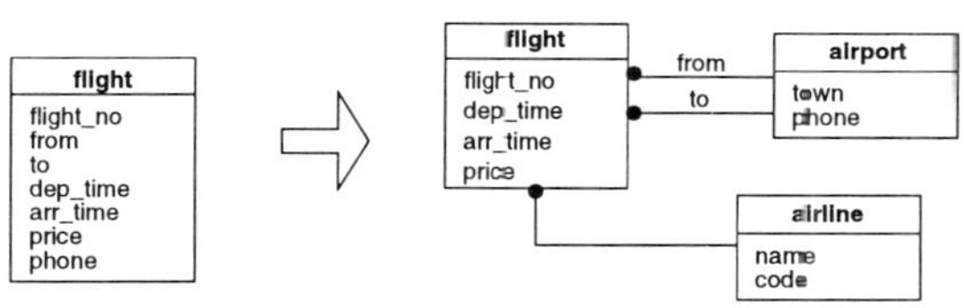

Figure 5: Improvement of the ALIS schema in the conceptual design step

The last step of the schema derivation is the improvement of the structure descriptions for the local file association as a conceptual schema. This design step is related to the well-known conceptual design step of the database design process. However, in contrast to this methodology, we could and should consider the information sources mentioned above, especially the example files that represent possible extensions for the schema. Moreover, the analysis of the example files can help us to find violations of the first normal form in the local file associations. The problems of modelling mistakes and incomprehensibility in the local structure description have to be overcome in the conceptual design step. The improvement of a conceptual schema is shown for the ALIS' OMT model in figure 5. This process can be denoted as semantical enrichment.

5 Schema Integration

When all local schemata are transformed in a common data model, the transformed schemata have to be integrated into one schema. For this purpose, we propose the integration methodology of [SS96] using the Generic Integration Model GIM. The detection and solution of all possible conflicts and the derivation of an integrated schema commonly cannot be done automatically. However, it can be assisted by an integration tool. We have sketched the integration process for our example in figure 6. Moreover, the methodology of [SS96] allows the derivation of different external schemata based on different data models. In that way, it is possible to provide a comfortable query interface to the integrated data as well as a HTML-interface for enhanced WWW-applications.

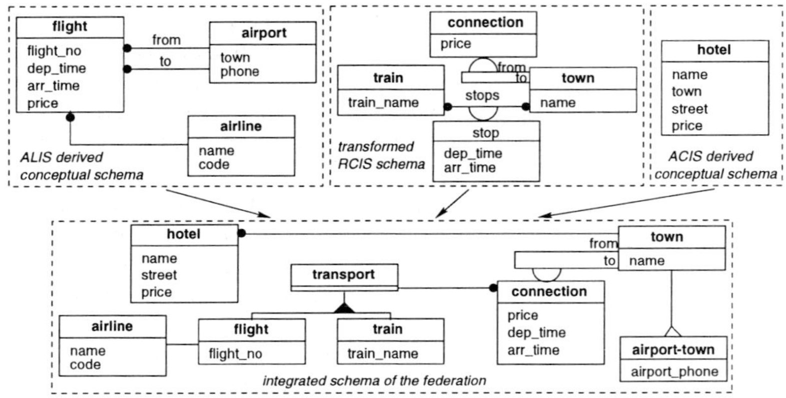

Figure 6: The integration of the component schemata

6 Conclusion and Outlook

In this paper, we have discussed the application of federation techniques for heterogeneous and distributed tourist information systems. Since in this application area many file- or WWW-based information systems are in use, we have focussed on the problem of file association integration. Due to the main problem of missing schema information, we sketched an approach to the derivation of conceptual schemata for local file associations. Thus, we can establish a federated tourist information system that provides a comfortable SQL-like query interface to the integrated data as well as an HTML-interface.

Indeed, it is possible to use the query interface of the FDBS for complex queries. For example, a tourist can search for possible rail connections from an airport town to a destination town with an available hotel accommodation. In this case, the query is decomposed into subqueries to the RCIS for the train schedule, to the WWW-interface of the ALIS for the flight schedule as well as to the ACIS of the local tourist board for information about available accommodation.

Moreover, FDBSs provide concepts for write access to the local databases [SL90, ACM93]. Therefore, the authentication mechanism and the user management of the federated database management system can be used to support more safety in the web. The FDBS itself is a special authorized user which connects the local information systems using Internet services like FTP. In this case, a complex transaction is possible, for instance booking of flight *and* accommodation.

However, the federated database design process is difficult and costly. Nevertheless, we think the use of federation techniques is an interesting approach to support a uniform and, as a result, comfortable access to certain information on the Internet.

Acknowledgements: We would like to thank Andreas Christiansen, Stefan Conrad, Andrea Diekmann, Klement Fellner, Andreas Köller, Mark-Andre Krogel, Gunter Saake, Ingo Schmitt, and Can Türker for useful remarks and hints.

References

[ACM93] S. Abiteboul, S. Cluet, and T. Milo. Querying and Updating the File. In *Proc. of the 19th VLDB Conference*, pages 73–84, Dublin, Ireland, August 1993.

[BLN86] C. Batini, M. Lenzerini, and S. B. Navathe. A Comparative Analysis of Methodologies for Database Schema Integration. *ACM Computing Surveys*, 18(4):323–364, December 1986.

[Buh96] Dimitros Buhalis. Enhancing the Competitivness of Small and Medium-sized Tourism Enterprises. *Electronic Markets - International Journal of Electronic Commerce*, 6(1):1–6, 1996.

[HM85] D. Heimbigner and D. McLeod. A Federated Architecture for Information Management. *ACM Transactions on Office Information Systems*, 3(3):253–278, July 1985.

[Höd96] M. Höding. An Approach to the Integration of File Based Systems into Database Federations. In *Proc. of the Int. Workshop on Heterogeneous Information Management*, Prague, Czech Republic, November 1996. *To appear.*

[ISO86] ISO (International Organization for Standardization). *The Standard Generalized Markup Language.* (ISO 8879), 1986.

[SL90] A. P. Sheth and J. A. Larson. Federated Database Systems for Managing Distributed, Heterogeneous, and Autonomous Databases. *ACM Computing Surveys*, 22(3):183–236, September 1990.

[SS96] I. Schmitt and G. Saake. Integration of Inheritance Trees as Part of View Generation for Database Federations. In B. Thalheim, editor, *Conceptual Modelling — ER'96, Proc. of the 15th Int. Conf., Cottbus, Germany, October 1996*, pages 195–210. LNCS 1157, Springer-Verlag, Berlin, 1996.

[TK96] N. Tschanz and S. Klein. Web-enabled Co-operative in Tourism - A Case Study from the Lake Constance Region. In *Proc. of EMOT Workshop - Inter-Firm Networks: Outcomes and Policy Implications*, Modena, September 1996.

Clustering Distributed Complex Objects - Management and Performance

Vlad Wietrzyk, Mehmet A. Orgun, John M. Hughes *

Macquarie University, Department of Computing
Sydney NSW 2109 Australia
*University of Technology Sydney
A/Vice-Chancellor
vlad@krakatoa.mpce.mq.edu.au

Abstract. Modern enterprises have to be effective, efficient and innovative in order to sustain in a competitive environment. This complex situation calls for advanced electronic infrastructures for the flexible exchange of all sorts of information.

Rapid advances in computer network technology have made it possible for disparate computers and database management systems to transparently share data and applications at more than one sites on the network.

Due to the increased complexity and multidimensionality of distributed database management systems the issues, theories, and implementations are quite diverse and vast. Distribution is a natural outcome of the business practices, therefore the need for distributed databases is inevitable. Corporate locations are distributed and so is corporate data.

Object-oriented technology is now a topic of intense study as the major candidate to successfully meet the requirements of advanced applications that require data management services of the tourism and hospitality industry. However, a close study of these applications reveals that they are distributed in nature and require data management support in a distributed environment. Thus, these systems require the development of distributed object management systems (DOMs). The purpose of this research paper is to provide an overview of some of the issues that need to be addressed in the development of this technology.

Rapid access to information is one of the key ways in which telecommunications can contribute to the effectiveness of a business operation.

1 Introduction

Issues in distributed multimedia systems environment the overall process of spatial and temporal data integration over a network poses serious engineering challenges. The major challenge due to the distributed environment is the real-time coordination among information storage sites to control the overall process of integration. Currently very little is known about the nature of the problems encountered in such process.

As the need for diverse services and the number of users increase of the tourism and hospitality industry, the next generation of database technology will essentially be distributed heterogeneous multimedia database systems.
We address several important issues that arise in the development of Massive Digital Database Systems (MDDS) in which data is being added continuously and on which users pose queries on the fly. News-on-demand and document retrieval systems are examples of systems that have these characteristics. Given the size of data, metadata such as index structures become even more important in these systems - data is accessed only after processing the metadata, both of which will reside on tertiary storage. The focus of this paper is on query and transaction processing in such systems, with emphasis on metadata management.
Digital information services are library services that address a group's information retrieval needs. These services differ in several ways from those required of digital libraries or information systems that meet, for example, education or entertainment related needs.
First, travelling groups or individual tourist want to retrieve information, rather than documents per se. Because the answer to a query may be in more than one document, or even in textual form, tourists require information systems that can perform powerful, complex retrieval and analysis of heterogeneous objects.
Realizing tourism digital information systems requires a broad technical agenda that includes:

- document image analysis, natural language analysis, and computer vision analysis for effective information extraction;
- new user interface paradigms and authoring tools for better accessing of multimedia information;
- improved protocols for client program interaction with repositories;

These systems additionally allow the user to jointly navigate in local or distributed multimedia databases and to jointly create or edit multimedia documents. At its most ambitious, a group of users can jointly create Multimedia Documents while having simultaneous multimedia communication.
This increased functionality in comparison to stand-alone computers or connected PCs is not only a sum of its parts. It gains a quality which not only enables the user to do more things in different ways, but it presents the user with new challenges in handling and using such a system. The specification of the user interface plays the paramount importance and should be done in respect of a user who wants to obtain multimedia information about holiday locations in Victoria or Queensland, states, so that they can decide where to go. The user should be able to specify their interests (sports, restaurant, culture, etc) as well as their criteria (beauty, price, etc) thus forming a database query.

We need to better understand the database issues involved in managing these distributed collections so that tourism digital information services can be used by tens of thousands of multiterabyte servers.

2 Media access and retrieval

2.1 Files and indexes

In the MDDS environment database storage structures, indexing schemes, and index files will have to take on totally new appearances. For one thing, we will now have incredibly large collections of binary data floating around be it aural, visual, tactile, or other various and sundry binary data encodings. The issue will be how are we going to manage, at the logical and physical level, very large collections of very diverse types of data files very widely dispersed over the MDDS nodal network? How does one decide on the data organization a MDDS setting? It can be argued that the success of DBMS technologies depends largely on sophisticated data placement strategies and data organization techniques. But in a MDDS setting with many diverse users, many and varied data types, widely varying usages, and either a lack of schema or multiple schemata, it is not clear what data placement and data organization strategies would lead to efficient query processing and information delivery. Data placement and data organization questions when answered correctly, as in traditional DBMSs, can yield significant performance increases and when answered incorrectly can yield significant performance obstacles. Databases practitioners have struggled with these issues for years and solved them for very specific environments with well-defined schema and narrow classes of users. In the MDDS context, however, under widely varying loads and conceptual linkages, and extremely diverse data formats we have yet to find efficient methods which will yield acceptable performance.
Indexing poses similar challenges. In a multi-media setting we will need to move to feature-based indexing systems that look not at information headers and labels but inside the data, scanning its content. Open questions abound. Will the traditional indexing schemes such as primary, secondary, hashing, and hierarchical scale up to the new challenges of the MDDS or will drastically new ones need to be created and adopted? Will we need to redefine the definition of an index key now that these values might be images or sound-bytes or tactile encodings? The only help that the MDDS will have, again, is the frequent low-volatility of many of the data sets. The traditional database problems of adjusting sorted index files as the data files are updated will largely go away in these predominantly read-only data arenas.

2.2 Clustering

One of the major problems with even a reasonably sized hypermedia system is that the screen gets cluttered with too much information for many given queries. One way to overcome the problem is to reduce the information that needs to be shown on the screen through suitable abstractions. These abstractions will show the user the overall information space without cluttering the screen. Hence we may form an abstraction which clusters all files in the same directory into a single node.

Clustering is possible for the abstracted views also. For example, if the number of directories are large, even the abstraction which represents each directory as a single node may be too confusing for the user. Hence there may be further clustering based on, for example, the topic of the pages. These abstracted views would be able to show the overall information space on a single screen without much cluttering. Hence the user would have a better feel of the global information structure from these abstracted views. Depending on the intentions and goals of the users, different kinds of abstractions would be useful to them. For example, one user may want to see an abstraction of the information space based on the topic for example: historical monuments in the particular city. Some other user may want to see the abstraction based on the author of the pages. Hence, for the abstracted navigational views to be really useful, we should allow the interactive specification of the abstractions.

2.2.1 Types of clustering

Two types of clustering are useful. They are as follows:

- *Structure-based clustering*: In structure-based clustering, the hypermedia structure is searched for subnetworks that match a given pattern. These subnetworks form a cluster. The simplest form of this type of clustering is *link-based clustering* in which nodes that are joined by a particular class of links are clustered together. For example, in an overview diagram we may want to form clusters of nodes that are linked by annotation links.
- *Content-based clustering*: In this type of clustering the nodes of the hypermedia are considered individually and their attributes are examined to determine the clusters. The simplest form of content-based clustering is when all nodes whose attributes satisfy certain properties form a cluster. Sometimes the user would like to cluster all nodes that have the same value for a certain attribute. If the user has to generate a clustering command for each value of the attribute it would be quite tedious. Therefore, *attribute-similarity-based clustering* is allowed in which for each value of the specified attribute, clusters of nodes having that value is formed. For example, all monuments built by the same architect can be clustered together.

To allow the clustering to be done interactively, it is essential that the clustering algorithms are efficient. We discuss our clustering algorithms in [2]. Note that clustering can be combined with binding so that information about the clusters can be shown visually. For example different shapes can represent different architectural objects.

2.3 Global query evaluation

The main processing effort in a MDDS is comprised of query statements from one query representation to the one supported by the location where the requested information resides.

For being able to perform such transformations, a number of facilities have to be supported globally, allowing for the utilization of the information pertinent to the particular tourist request and for the analysis of globally arrived queries in terms of the global definition of the locally supported query languages.
The globally supported mechanisms for the distribution of selections and projections over the local data, must guarantee the reduction of processing costs for the evaluation of global queries, and the optimum exploitation of local services offered by the underlying MDDS. The global system will be responsible for the generation of query execution plans and execution of inter-site joins which guide the further execution of sub-queries and lead to the integration of the intermediate results.

2.3.1 Basic notions of the query model and object join

Let N be a countable set of symbols which can be used as names. We consider a database schema Σ for complex objects as a collection of schema rules of the form $R = (A_1,..., A_n)$ ($R(A_1,..., A_n)$ for short), where $R \in N$ is called a complex scheme type and each $A_i \in N$ ($i =1,..., n$; $n \geq 1$). We call A_i ($i =1,..., n$; $n \geq 1$) a *higher-order name function*, if it is an atomic set-valued function (which returns a set of atomic values for each input), or if it appears on the left hand side of some schema rule (which returns a structured value, or a set of structured values, or a set of atomic valued identifiers for each input). Otherwise, A_i is called *zero-order name function* of R (which returns a single atomic value for each input).
The list $(A_1, A_2,..., A_n)$ includes (zero-order and higher-order) names of all functions locally defined in type R or in supertype of R. We call them *inherent functions* of type R. If A_i is a higher-order name function whose output type is of the form $R_i (A_{i1}, A_{i2},..., A_{im})$, then R_i is a constituent type of R. To avoid ambiguity, names on the right hand side of the same rule are by no means unique and no two rules can have the same name on the left hand side in a given database schema Σ. In the sequel, for a scheme type R, we use $Lfn(R)$ to denote the list of (zero-order and higher-order) function names of R. The order of them is not important. If R is one of the base types or a reference type, we have $Lfn(R) = \varnothing$. Otherwise, for any $R(A_1, A_2,..., A_n) \in \Sigma$, let $sup(R)$ be the set of supertypes of R. We may define $Lfn(R)$ inductively as follows:

(i) $\text{oid}(R) \in Lfn(R)$;

(ii) $\{ A_1, A_2,..., A_n \} \subseteq Lfn(R)$;

(iii) $\cup_{T \in sup(R)} Lfn(T) \subseteq Lfn(R)$.

We use $Zfn(R)$ to refer to the set of all zero-order function names in the list $Lfn(R)$ and $Hfn(R)$ to denote the set of all higher-order function names in it. We have $Lfn(R) = Zfn(R) + Hfn(R)$. We may refer to a scheme type $R(A_1, A_2,..., A_n)$ simply by its name R when no ambiguity occurs. Besides, let r be a class of type R.

We refer to the pair (R, r) as a complex object scheme and denote it by $\mathfrak{R} = (R, r)$ for brevity. Let t be an object variable ranging over a class of type R, for any function (say A) applicable to R, we have $t[A] = A(t)$. Suppose $X = (A_1, A_2,..., A_m)$ and $1 \leq m \leq n$, then $t[X] = (t[A_1], t[A_2],... t[A_m])$.
A *function space* of a given type R, denoted by $Fspace(R)$. is defined inductively as follows:
- if R is one of the base types (e.g. integer, real, string, date), then $Fspace(R) = \emptyset$.
- if R is a set type, say $\{T\}$, then when T is one of the base types, $Fspace(R) = \emptyset$. Otherwise, $Fspace(R) = Fspace(T)$.
- if R is a reference type, say *ref T* or $\{ref\ T\}$, then $Fspace(R) = \emptyset$.

Let A be a member of the set $Fspace(R)$ and $Path(A[R])$ denote the path expression of function A in type R. Let $\oplus$ be the path connection operator such that $R.R_1 \oplus R_1.A = R.R_1.A$. Then we define the path expression of function A with respect to type R as follows: if $A \in Lfn(R)$, then $Path(A[R]) = R.A$.

2.3.1.2 Object join

Unlike the conventional data models, in object-oriented models many relationships between entity objects can be represented within the objects themselves, such as IsA relationships (subtyping), 'is-part-of' relationships (aggregation references) and arbitrary associations between entities. Hence the join operations are used less frequently in the object algebra than in the relational algebra. However, the existing objects in a database may not explicitly reflect all the relationships required by the queries. An explicit join is still needed to handle the cases when a relationship being queried upon is not defined within the object types. Such a relationship is called *value-based* relationship in contrast to those specified explicitly in the object model through object reference types.

For example, a query '*find all the architects who had build castles in their home city*?' requires a value-based join between classes the ARCHITECT'S GUILD and ARCHITECTS through guild's *location* and architect's *home-address*, because this relationship is not explicitly modelled in the interface defined on either type GUILD or type ARCHITECT. The explicit join is also very important in building large knowledge bases where knowledge is often obtained by building new relationships among existing facts. We below provide an *O-join* operator to realize the queries on those objects which have *value-based* relationships.
Note that the definition of O-*join* operator is formally extended in comparison with the (nested) relational join. It allows combining objects of two scheme types at all levels of nesting, and creates a new type as an immediate subtype of the most general type OBJECT, holding the result of join. The definition is also intended to preserve the associativity of the operator, which we expect to be used in the query optimization in more or less the same way as it is used in the relational query optimizer.

Before presenting the formal definition of object join, we need to discuss under which conditions two functions are comparable.

Two functions A_i and A_j are comparable, if and only if the domain of the output type of A_i is identical to the domain of the output type of A_j, or there is a subclassing relationship between domains of the output types of A_i and A_j, or domains of the output types of A_i and A_j have a common user-defined superclass.

Let $\Re_1 = (R_1, r_1)$ and $\Re_2 = (R_2, r_2)$ be two object schemes respectively. Assume the join condition F is of the form $t.A\delta t'.B$, where A and B are comparable functions, $t \in r_1$ and $t' \in r_2$. If A and B are both zero-order functions, δ is in { =, ==, <=, >=, $\neq$ }; otherwise δ is in { $\in$, ==, =, $\subset$, $\subseteq$, $\supset$, $\supseteq$, $\neq$ }. The O *-join* of $\Re_1$ and $\Re_2$ at the instance level, denoted by $\otimes_F(r_1, r_2)$, is a collection of new objects satisfying the join condition F:

If the join condition $F = \emptyset$, then $\otimes_F(r_1, r_2)$, is a *Cartesian Product* of r_1 and r_2.

$$\otimes(r_1, r_2) =_{\text{def}} \{t \mid \exists t_1 \in r_1, \exists t_2 \in r_2 : t[Lfn(R_1)] = t_1[Lfn(R_1)] \wedge t[Lfn(R_2)] = t_2[Lfn(R_2)]\}.$$

2.4 Timed-output queries

In this section, we indicate how the algebraic language described above can be expanded to handle certain kinds of queries called *timed-output queries* (or TO-queries, for short).

A *timed query* is a query where all state variables occurring in the query have an associated *time annotation.* In any query *Q to a MDDS's database,* certain variables range over *states* - in particular, any variable, *S*, that occurs in an atom of the form *frametype*(*S*, -) is a state variable, ranging over states. When answering the query *Q*, the state variables typically get "instantiated" to some state-constants. The idea is that the physical objects (*e.g., picture of the hotel, sound-frame of the regional music, text file describing historical development of the region or city, etc*) represented by these named states are to be "brought up" or "output" on the appropriate output device. Thus, in the query:

$$(\exists S_1, S_2)\ \text{frametype}(S_1, \text{video}) \wedge \text{frametype}\ (S_2, \text{document}) \wedge \text{(castle-historical building-tourist attraction)} \in \text{flist}(S_1) \wedge \text{(castle-historical building-tourist attraction)} \in \text{flist}(S_2)$$

a request is being made to find a video frame of the '*tourist attraction*' and documentation concerning the '*tourist attraction*'. The substitution $S_1 = v1$, $S_2 = d1$ is an answer to this query.

However, this answer may not always be satisfactory - for example:

Temporal Relationships, the user may wish to watch the video frame *v*1 first, followed by the document *d*1, etc.

We present a somewhat more general picture of how state variables may be used to define constraints over different types of domains - we will start by showing how state variables may be used to define a broader class of timed-output queries.

To express *temporal relationships*, we may define a very simple temporal constraint language over the integers or the reals (as deemed appropriate). We do this as follows:

- For each state variable S, we have two constraint-variables $start(S)$ and $end(S)$.
- The language contains the nonnegative real numbers as constants.
- The language contains the operator symbol +.
- The relations in the language are = and $\leq$.

Constraints can now be expressed easily in the form $t\langle OP\rangle t'$ where t, t' are either constraint-variables or real numbers (or summations, thereof). Thus, for instance, the user may enhance the query:

$$
\begin{aligned}
(\exists S_1, S_2)\ &\text{frametype}(S_1, \text{video}) \wedge \text{frametype}(S_2, \text{document}) \\
&\wedge (\text{castle-historical building-tourist attraction}) \in \text{flist}(S_1) \\
&\wedge (\text{castle-historical building-tourist attraction}) \in \text{flist}(S_2)
\end{aligned}
$$

above by applying to it, the constraint Ξ given as:

$$\text{start}(S_2) \geq \text{end}(S_1) + 0.001.$$

This would specify that tourist wishes to see the document only after seeing the video. Thus, in general, a conjunction of one or more constraints may be appended to queries to specify the temporal relationships between the outputs (on media or output devices) generated by a query. *This option is of paramount importance in high performance oriented MDDS technologies in tourism.*

3 Query optimization strategies

The main task of query optimization is to create an equivalent expression for a query, which can be processed more efficiently than the original expression. The well known strategies (heuristics) for determining equivalent but more efficient algebraic expressions for relational queries include performing selection and projection before joins, combining several selections into one selection, finding common subexpressions, and so on. Rules about the commutativity and the associativity of the operators are very helpful in finding equivalent expressions. In fact, most of the heuristics and rules used in optimizing the algebraic expression of a relational query can be extended and applied to the operators of our object algebra.

This is obviously one of the main advantages for developing the object algebra in the spirit of (nested) relational algebra, since we can readily adapt or extend the results previously obtained on the algebraic equivalence and the optimization heuristics in the development of object algebra optimization. *This has far reaching practical and beneficial consequences in application to MDDS technologies in tourism.*

The big problem with query optimisation for OODB application is that much of the time it is not "query-only". Instead there is a mix of updates data access. Now, any kind of update changes the state of the database and thus destroys the referential transparency of set expressions referring to the data. Evaluating the same expression before and after the update can produce different results. Thus it is possible to adopt our proposed strategy, where pieces of relational algebra representing sub-queries are combined together to allow overall optimisation of a large relational algebra expression through the use of rewrite rules.

This can give very significant performance gains through eliminating repeated evaluation of common sub-expressions.

4 Conclusions: issues in distributed multimedia management

In this paper our discussion has been centered around query processing and optimization. The more query processing activities that local query sites can do, the better for the MDDS. The biggest costs in MDDS query processing, however, will probably not be the standard ones which exist today of query decomposition, compilation, and re-assembly. We expect the biggest cost in MDDS query processing to be in the resource location and synchronization processes. Also, access to some object may require the intervention of the payment mechanism.

Efficient clustering of multimedia objects will significantly increase useability of distributed interactive multimedia information systems in tourism industry.

Even database constraints will need to evolve to meet the MDDS technologies in tourism challenge. *For example, requiring that the meaning of a picture (i.e. the architectural object - monument captured in the image) be related to the MDDS database representations of all of those pictured objects and to no others will be critical.*

It will be the new MDDS technologies in tourism challenges of optimizing queries in spite of the overwhelming scale (in terms of the number of nodes, the number of objects, the huge sizes of objects, the number of distinct data types, the number of conceptual mappings, and the number of mnemonic mappings), billing verification, and multi-media delivery synchronization that will have to be focused on.

5 References

1. Lin, S. H. Son: Concurrency control in real-time databases by dynamic adjustment of serialization order. Proc. 11th Real-Time Syst., 1990.

2. Wietrzyk, V. I., (1996), "Performance Evaluation of New Clustering Algorithm in Object-Oriented Database Systems". In Proceedings of The 7th International Conference on Database and Expert Systems Applications DEXA '96 Zurich, Switzerland, September 1996.

3. Wietrzyk, V. I., (1994), "QUELLE: A Knowledge - Based Approach to Systems Development". In Proc. of The Fourth International Conference on Information Systems Development - ISD'94; Methods & Tools - Theory & Practice. (1994).

4. J. Rothine, T. Lozano: Attribute based file organization in a paged memory environment, CACM v 17 n 2, 1974.

5. J. W. Stamos: Static grouping of small objects to enhance performance of a paged virtual memory, ACM trans. On computer systems, v2, N 2, 1984.

6. I. Williams, M. Wolczko, T. Hopkins: Dynamic grouping in an object oriented virtual memory hierarchy, Proc. of European conference on object oriented programming, 1987.

7. M. M. Tsangaris and J.F. Naughton: A stochastic Approach for Clustering in Object Stores, In SIGMOD, 1991.

8. J. G. Hughes: Object-Oriented Databases, Prentice Hall, 1991.

9. M. O. Albertson, J. P. Hutchinson: Discrete Mathematics with Algorithms, John Wiley & Sons, 1988.

10. S. Kundu, J. Misra: A linear tree partitioning algorithm, SIM J. Comput., Vol. 6, 1977.

11. H. F. Korth, A. Silberschatz, Database System Concepts, McGraw Hill, 1986.

12. C. H. Papdimitriou: The Theory of Database Concurrency Control. Computer Science Press, 1986.

13. J. R. Haritsa, M. J. Carey, M. Livny: Dynamic Real-Time Optimistic Concurrency Control, IEEE Real-Time Systems Symposium, Orlando, Florida, 1990.

14. J. Huang, J. A. Stankovic, K. Ramamritham, D. Towsley: Experimental Evaluation of Real-Time Optimistic Concurrency Control Schemes, VLDB Conference, Barcelona, Spain, 1991.

15. L. Sha, R. Rajkumar: A real-time locking protocol. IEEE Trans. Comput. 40, 1991.

Mugla Tourism Information System

Sinan Neftçi, Mugla University
Cevdet Dengi, M.E.T.U.
Cercis Ikiel, Mugla University
Shahpour Thaerzadeh, Mugla University

November 15, 1996

1 Introduction

Mugla is a province of Turkey located in the south west corner of the Anatolian peninsula.

It is estimated that, the geographical area we are concerned with (which is described with greater details in a section below) generates close to half of the tourism revenues of the whole country.

The work that is described here is an ongoing effort mostly within the Mugla University carried in close cooperation with local tourism organizations and state tourism authorities.

The subject of our work, MUTIS (Mugla Tourism Information Systems) has started in response to a request from the local State Tourism Authority to create a tourism inventory of our region. Although the initial objectives are still valid what is aimed now is well beyond a monolithic database for local resources (and potential resources) of the region. The current purpose of the MUTIS (Mugla Tourism Information Systems) is multi-faceted but the the primary objective is still to come up with a system to assist the researchers, decision makers and strategic planners in the Tourism domain. But the system that is being planned is also expected to provide all the underpinnings of a dynamic network environment to glue all the major players of the Tourism business.

This paper defines the software and the network architecture of a system that will be the basis for;

- Reliable Tourism Data Collection which is automated to the fullest possible.

- Data base to serve as a repository for all local tourism data.
- Distributed
- Forecasting of tourism demand.
- Marketing research
- Long term strategic planning
- Investment planning
- Guidance for State Regulations and Control
- Provide a technology base to the large and small Tourism enterprise
- Advertisement and Internet access
- Web content provisioning
- Media tools and advising

All of which aimed at serving the tourism sector.

2 Region

Since we are taking a markedly regional approach, the establishment of a sound Geographic information Structure is of very high importance.

As noted earlier Mugla region is a highly valued tourist region of Turkey located in the south west Anatolia.

The south of Mugla is surrounded by Mediterranean Sea while the Aegean sea takes its place at the West. With respect to its natural environment, Mugla has a very peculiar combination of geographic characteristics. Mesozoics, Tertialy and Quarternary fields existing all at the same place is a good examples of such characteristics. The development of lands forms has been appeared at the neotectonic period. The shore line is characterized with an intricate systems of bays and coves creating a huge potential for yacht tourism. Behind the regional coast there is main plateau named Mentese, which extends through the region and regional coast with average hight of 700-800m. This whole region is sparkled with a multitude of tiny villages creating a rich folklore base. At the surface of Mentese plateau many steep faults can be found. Generally, the hight of Mentese mountains, which expands from north-west to south - east, is higher than 1000 m. The highest point being the Oyuklu mountain which is approximately 2000 m. Among the mountains there are many depressions, particularly on the surface of plateau. These depressions were generally formed by tectonic movements in neotectonic period. All these characteristics of the terrain creates an excellent ground for a largely un-exploited sports tourism. Due to the natural barriers, the region is well preserved both from environmental and cultural point of views, up until the very recent boom in tourism. The internal regions are still highly under-developed largely preserving its authentic character. Al these variability in the geographical character is overlaid by an even richer clusters of archaeological sites. It is estimated that there is a total of more than 200 such locations (some still unexplored) in the region. Considering the importance of all three of these dimensions (i.e. natural, historical. and folklore) the necessity of overlaying an Geographic information Structure and the Tourism Information Systems becomes obvious. Two main result can be driven from such a database; First, the expansion (due to tourism) can be balanced with a reasonable yet far sighted protection plan. Second, some form of optimality can be achieved in allocating and developing the resources with tourist potentials.

Well known site location selection models like LOCAT [2] or AHP (Analytical Hierarchy Process) model can be used in a modified manner by considering the current attraction points and the points between them [1] .

One other benefit of coexisting geographic data with tourism data can be observed in obtaining the correlation between such data and use the results in market predictions. In Mugla region, Mediterranean climate can be seen and thus characterized as mild and rainy winters with hot and dry summers. The mean annual temperatures varies between 14.8C - 18.7C. The mean January temperatures changes between 5.3C - 11.1C. The July temperature is over 25C, the highest value attains 40C-45.5C. The mean annual precipitation ranges from 640 to 1190 mm. The half of the precipitation falls during the winter period. But, there is high variability of precipitation from year to year. For instance, the high relative humidity means as above 50In simple correlation analysis we have determined that the climatic variations is a prime determinant of the tourism volume and that the tourism facilities reach their peak during the time when mean daily temperature is above 20C (between May- October)while at the same time the sea temperature is above 20C. So far we have accumulated a large climatic data and we are pursuing further correlation analysis of similar nature on these data. Aside from geo-physical considerations, the administrative center of the region is Mugla city which is in the hinterland and much smaller compared to the major tourism centers located on the Mediterranean coast. The three major tourism centers of the region are Bodrum, Marmaris and Fethiye such centers are followed by smaller centers of interest like Datca, Gokova, Dalyan, Koycegiz etc. When considered from the networking advantages Mugla can be seen almost as the logistic center of gravity of all these points. This topology has created the backbone of our networking with the major hub placed in Mugla (which also hosts all the servers and support services) and connected to Marmaris, Bodrum and Fethiye with direct links where smaller server configurations are planned in the future.

3 The Hardware and Software Infrastructure

The University has both academic and commercial Internet service accesses which is provided as a service to the whole region, and primarily to Marmaris, Bodrum and Fethiye via leased-lines and later hopes to establish X.25 connections to the smaller cities.

The server structures are a combination of SUN SPARC Stations and Pentium based Linux systems. The central hub located in Mugla is served with a router (cisco 2500) connected via 64Kbs lines to the outside world.

Sharing the common destiny with many similar software projects MUTIS has an early prototype phase that is put together hurriedly to satisfy some anxious users. This prototype is implemented with ACI's 4D data base server systems hosted by pentium PC's. The ease of use and PC/Mac availability of 4D database systems has made them an attractive test bed for the initial phase. The 4D servers are accessed from 4D clients (also hosted by PC's) connected via the (Ethernet) LAN's or using dial-up connections to a LAN connected terminal server which has a rather circuitous way of connecting to the servers through some network entities.

However the target environment that the who system is currently migrating to, is a true multi-user (Oracle) based UNIX server (sun SPARC-station) with TCP/IP connectivity. The main servers are again backed-up by Pentium PC based Linux systems as secondary servers. Standard Internet (Netscape) browsers are used to implement the complete user interface with still some occasional 4D client accessing the Oracle over DAL (Data Access Language. of ACI) sessions. The uniform HTML based browser interface is of extreme importance for the standardization of the data entry and user access of the Tourism Information System. The current CGI based access to the database is expected to change somewhat with JAVA based implementations gaining more ground.

As part of our charter to be a technology lead for the local tourism organizations we have also set up a multimedia laboratory equipped with the following hardware; 3 PowerMac's (8200,5200,6200) a QuickTake-150 digital camera and Umax VITA s-12 Color scanner with transparency Unit. All these equipments are dedicated to the content providers WEB page generation efforts. One of the preferred authoring environment chosen is the Acrobat Writer/Distiller while the animated graphics are usually generated on a Macromedia Schockwave.

4 Data Collection

The reliable data collection is probably the most difficult part experienced in building the Mugla Tourism Information System. Most of the organizational data is collected by close cooperation with local state tourism agency which has some legal control over the regulatory aspects of the tourism operations. Also the state enforced data collection at the custom entrances which is again supplied to us via the State Tourism Agency is of great value in obtaining the base volume estimations. However there is little or no data available about;

- Tourist consumption habits
- In country mobility
- Satisfaction an quality measurements
- Market segments
- Historic, Recreational and Cultural activity demand
- National and in-land tourism

So far the only reliable data collection on the issues seems to be directed sampling studies with questionnaires and trained pollsters. Although coupons and randomized (prized beacons) experiments are also being discussed.

5 Forecasting

In MUTIS only an auto-regressive time-series prediction is applied to the monthly tourism data so far. Although there are some useful information that can be learned from such predictions, this simplistic approach must be augmented with a better multi-segment, trend based, seasonally adjusted forecasting model. One such approach can be the Box-Jenkins univariate forecasting method [3]. But some intuition must be developed about the problem setting before we sail in this direction. In the mean time there is a strong drive in providing more environmental and external correlation based studies like the climatic example showing the clear-days to occupancy rate correlation.

6 Interplay with other systems

The competitive advantage of most tourism organizations today will arise from their ability to adopt to the fast and chaotic evolution of the Information Technologies of our age. The evolution of CRS (Computer Reservation Systems) will most probably continue well into the next decade but it is inconceivable to think that the CRS will be unaffected by the explosive growth of the Internet. One of the purposes of the MUTIS is to provide the an inter-operable common ground for such systems. The triangle between the tour operators, Travel Agents and Accommodation/Transportation service providers should be a better, and more tightly knit, yet open information structure for the better and more competitive offerings by all parties.

7 Some Suggested New Technologies For Tourism Industries

Contemporary enterprises in the tourism sector are forced to develop new services and reduce the costs under the pressure of global competition. Enterprises face these challenges by re-engineering and automating their business processes. Automation necessitates new generation of information systems and architectures for distributed computing.

7.1 Interoperability

Interoperability among loosely coupled heterogeneous, autonomous components is an important problem in distributed computer architectures. Distributed Object Management (DOM) [4] platforms are merger of object oriented programming and distributed computing technologies which have emerged in the last decade and influenced contemporary software to a great extent. Common Object Request Broker Architecture (CORBA) is a well established DOM technology which has been used successfully for the implementation of complex systems including multi-databases [5, 6]. CORBA is the communication mechanism of the Object Management Architecture (OMA) [9] proposed by Object Management Group (OMG). It provides a virtually local and homogeneous environment over the actual distributed, heterogeneous environment. Changes in object implementations, or object relocations are transparent to the client. Object interfaces are defined using Interface Definition Language (IDL). An IDL interface declares a set of client accessible operations, exceptions, and typed attributes. To use or implement an interface, the interface must be translated, or mapped, into corresponding elements of a particular programming language. OMG defines a set of standard *Object Services* which provide the main functions for implementing basic object functionality using ORB. Life-cycle services, Object Transaction Services (OTS) and Naming services are examples of the object services defined by OMG.

DOM technology promise plug-and-play Interoperability among distributed software components. apparently it will be the base technology for encapsulation of legacy systems and for new distributed applications. Although DOM implementations are very new, their usage in industrial strength applications seems to be possible in near future.

7.2 Inter-networking

Nowadays, computer networks has become a primary way of finding information and doing business. Enterprises try to give access to their services for both their clients and their business partners through international computer networks in order to improve the quality of their service. Web technology allows hypertext integration of huge amounts of information distributed over millions of computers. More and more media types such as sound, video clips, animation are supported by web browsers. Search engines and subject lists allow users to find the information easily. Emerging Java technology allows business transactions become a part of Web navigation. Even mobile agents (such as a person with a pocket telephone) will be able to run Java applications and do business through Web. Since advertisement and fast service is very important, integration of tourism information systems with the Web technology is inevitable.

7.3 Workflow Support

Tourism information systems frequently need to cooperate with other information systems. For example, an ordinary holiday reservation needs cooperation among information systems of hotels, car rental services, airlines and credit card services. Autonomy and heterogeneity of these component information systems does not allow tightly coupled architectures such as multi-databases. Furthermore, classical transaction systems fail to support long duration or co-operative activities which are found in typical tourism information systems. For that reason, tourism enterprises are one of the best application areas for Workflow Management System (WFMS) technology. Workflow Management Systems completely define, manage and execute Workflow processes through the software whose execution order is driven by a computer representation of the Workflow process. There is a standardization effort for WFMSs. Workflow Management Coalition (WfMC) provides a Workflow Management Reference Model specification [7].

7.4 The Scenario

A DOM based Workflow management system architecture integrated with the Web [8] can be a good framework for large scale tourism information systems.

In such a system, the client searches the alternatives for her holiday from the web. After seeing a couple of multimedia advertisements she connects the CORBA based Workflow management system of a tourism agent using the Java application down-loaded from the web server of the enterprise. She fills a web based form and gives the necessary information such as name, preferences, credit card number in a secure way. WFMS of the tourism agent starts a Workflow process instance for her that tries to make airline, hotel and car reservations interacting with other Workflow management systems. Finally, the approves the schedule and the WFMS starts couple of activities which prints the tickets and have them mailed to the client. She can change the schedule interacting with the WFMS afterwards. After she has returned back to home her Workflow process instance is committed.

8 Conclusion

The experience we are having with the building, maintaining and user coordinations of a Tourism Information System, is mandating us to use some of the most advanced technologies in the network and database research domains. There are some clear economical benefits to be gained from applying such techniques to the marketing of touristic resources.

References

[1] Saaty, Thomas L. *Hierarchy Process* MacGraw-Hill (1980)

[2] Moutinho, L. Robert, P. Site Selection Analysis in Tourism: The LOCAT model. *The Service Industries Journal*, vol. 11, no. 1, pp. 1-10. (1991)

[3] Box, G.E.P., Jenkins, G.M. *Time Series Analysis:Forecasting and control*,Holden-Day. (1970)

[4] Özsu, M.T., Dayal, U., Valduriez, P. *Distributed Object Management*, Morgan Kaufmann, 1994.

[5] Dogac, A., Dengi, C., Özsu, M.T. Building Interoperable Databases on Distributed Object Management Platforms, to appear in *Communications of the ACM.*

[6] Dogac.,A., Dengi, C., Kilic, E., Ozhan, G., Ozcan, F., Nural, S., Evrendilek, C., Halici, U., Arpinar, B., Koksal, P., Kesim, N., Mancuhan, S. A Multidatabase System Implementation on CORBA, in *Proc. of RIDE'96.*

[7] Hollingsworth, D. *The Workflow Reference Model.* Workflow Management Coalition Specification, TC00-1003 (Draft 1.0), 1994.

[8] Miller, J., Sheth, A., Kochut, K., Wang, X. CORBA-Based Run-Time Architectures for Workflow Management Systems. in *Journal of Database Management*, 7(1), 1996.

[9] Soley R.M. (ed.), Stone C.M. *Object Management Architecture Guide*, Third Edition, John Wiley & Sons, 1995.

An Intelligent System for Tourism Information

A. S. Monat, PhD - Lecturer - monat@iprj.uerj.br
G. Kern - Msc Student - kern@iprj.uerj.br
P. Werly - Msc Student -phwerly@iprj.uerj.br
S. Montenegro - System Programmer - sylvio@iprj.uerj.br
State University of Rio de Janeiro - IPRJ/UERJ

Abstract

Nova Friburgo is a Brazilian tourist city which attracts different types of visitors. This article describes an intelligent system to be installed at a tourist information centre for assisting those who need directions to this city's major attractions. Currently, there is an ordinary multimedia application for tourism support but such system is being expanded to include a knowledge base of profiles of tourists. On the base of the tracks or paths a user has already performed during his/her consultation, the system selects the profile which most closely reflects the options taken. According to this selected profile, the system is expected to suggest various routes and destinations based on the tourist's possible interests, in addition to provide information on hotels and restaurants he/she will likely be interested in. The knowledge base which provides this type of inference is being developed in a hybrid system involving Prolog and C++.

Keywords: Intelligent Hypertext System, Technology for Tourism, Expert Systems.

1 - Introduction

The city of Nova Friburgo receives thousands of tourists every year. These visitors come interested in knowing the several tourist attractions available in the city, mainly the waterfalls, rivers, mountains etc. The tourists also look for the remains of the european immigration in Brazil. Nova Friburgo, in portuguese, means New Freibourg, and this has a historical reason. The city was founded by German and Swiss immigrants, invited by the Portuguese king and later by the Brazilian emperors of last century.

Such unique history and natural landscape attract the curiosity of many, but some come for different reasons. Nova Friburgo is also a busy commercial city, which offers since typical goods, as delicatessen products (mainly preserves) and handcrafts, to highly industrialized textiles products. This latter type of commerce is mainly based on the coming of low income businessmen (normally from Rio de Janeiro city) who stay less than one day, acquire as many textiles products as possible, and returns spending the least possible.

In brief, Nova Friburgo receives many different types of visitors. In this work, we introduce a system (called HyperFriburgo) which intends to provide general information for those visitors. HyperFriburgo should also figure out what type of visitor is navigating through the hypertext and provide specific information for the profile of this user.

This is achieved by a knowledge base of profiles of tourists who visit the city. On the base of the tracks or paths a user has already performed during his or her consultation of the hypertext, the system selects the profile which most closely reflects the options taken. Henceforth, we are going to call these selected tracks or paths as the "*user's choices*".

The system is expected to suggest various routes and destinations based on the tourist's possible interests, in addition to provide information on hotels and restaurants he or she will likely be interested in. This approach is consequence of an observed necessity for more intelligent orientation for complex hypertext systems [5]. Basically, complex hypertext systems may offer such a vast amount of information to their users that more individualized orientation has to be included into the system.

2 - Contents of the Hypertext and its Intelligent Capabilities

We are going to use a formal methodoly in order to describe HyperFriburgo contents. This methodology is called RMM - Relationship Management Methodology [4]. According to RMM, the first step in designing a hypertext is to ellaborate its Entity-Relationship (E-R) diagram. This diagram, for HyperFriburgo, is shown in Figure 1. Those interested in reviewing E-R technique are suggested to read [3].

An important entity-type of the system is "*Tourist Attractions*" which means all relevant places for visiting. This entity-type has four sub-classes: "*Caves*", "*Rivers and Waterfalls*", "*Monuments and Museums*" and "*Parks and Squares*". "*Tourist Attractions*" has a relationship "*is_at*" with "*Geographical Areas*", which is an entity-type about the districts of Nova Friburgo and its main areas. Nova Friburgo is a hilly city, with very independent and separated districts. Therefore, tourist oriented tours may be quite distinct according to the area choosen. Other three entities-type, "*Restaurants*", "*Hotels*" and "*Shops of Textiles and Shops of delicatessen*", also have "*is_at*" relationship with "*Geographical Areas*".

We are not going to describe step 2 of RMM methodology (called entity design phase) since it is not relevant for our analysis. The third step of this methodology is the navigation design. In this step, we ellaborate the paths a user may perform while using the hypertext. A review of RMM is beyond the scope of this article. Nevertheless, in Figure 2 we have the system for Nova Friburgo described according to it

The main feature of HyperFriburgo is its ability to anticipate what is the main interest of the user. This is achieved according to the *user's choices*. The time spend on these options, and the number of times he or she 'clicked' the same item is fundamental for deciding the issues the user is interested in, and therefore his or her profile.

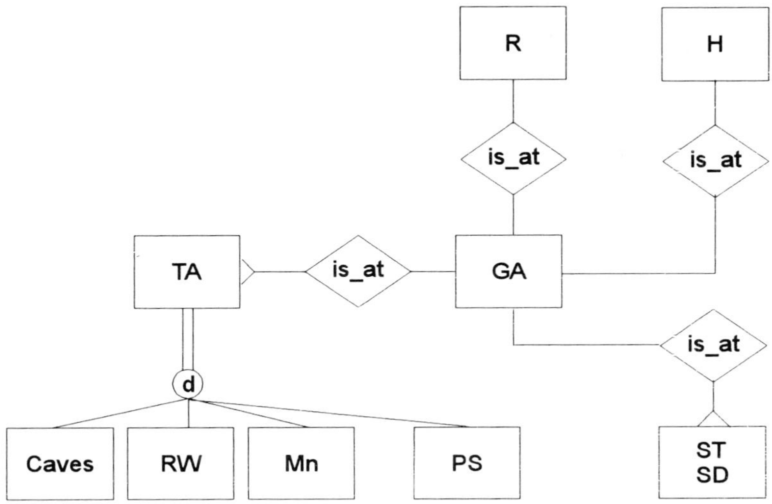

Notation: TA = Tourist Attractions
RW = Rivers and Waterfalls
Mn = Monuments and Museums
PA = Parks and Squares
Mt = Mountains
GA = Geographical Areas
R = Restaurants
H = Hotels
SD = Shops of Delicatessen
ST = Shops of Textiles
SM = Sub Menu

Figure 1 E-R Diagram for HyperFriburgo

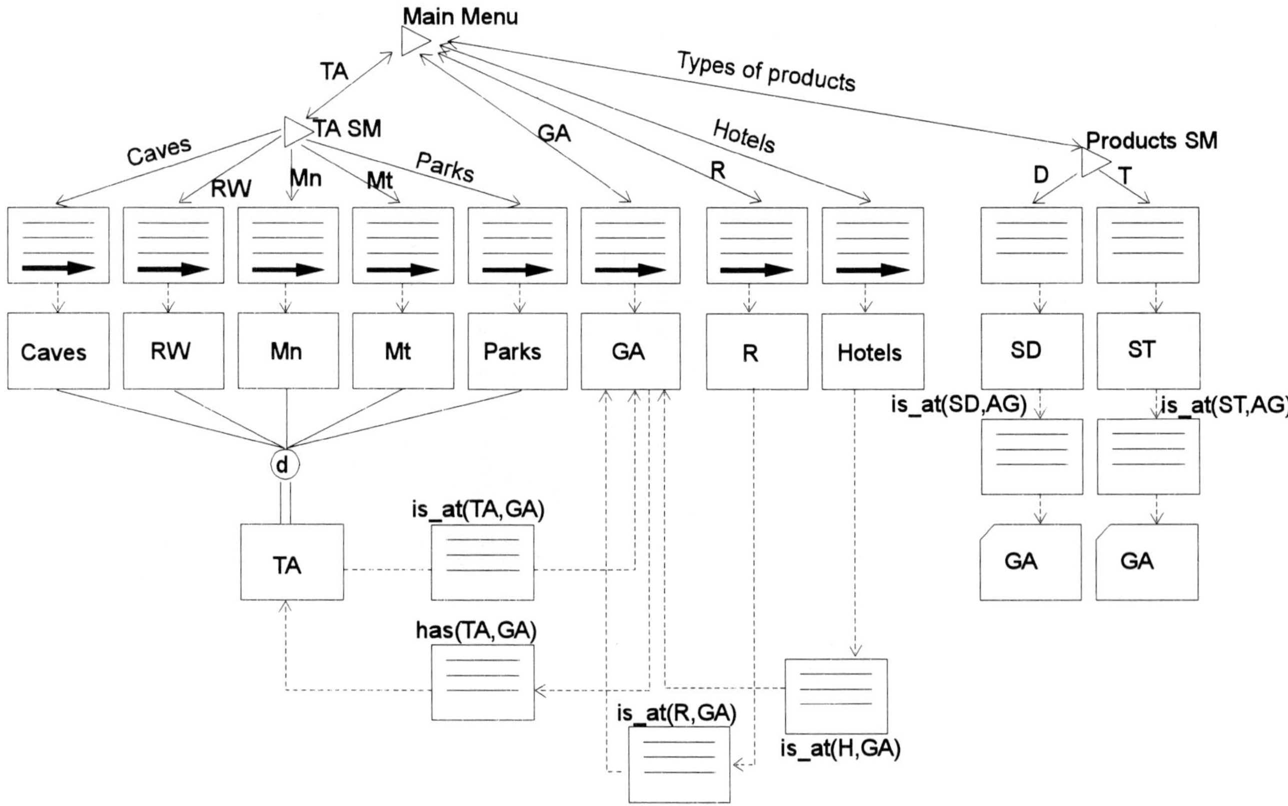

Figure 2 RMM Diagram for HyperFriburgo

Basically, HyperFriburgo selects among a set of possible profiles of tourists, which one fits most the *user's choices* and present him (her), a selected menu of options for his (her) specific navigation. The following are the profiles of tourists that are already encoded within the system:

1 - The Textile Buyer (TB): This type of visitor comes to Nova Friburgo in order to buy a large amount of textiles. Normally a very economic man, he seldom stays overnight, and arrives very early in the morning. He also leaves Nova Friburgo to Rio de Janeiro city before dinner time. He uses to look for inexpensive restaurants, and stays either in N. Friburgo downtown or in the district of Mury, where the main textiles shops and factories are found.

2 - The Ecology Lover (EL): This one is attracted by the rivers, waterfalls, mountains, flora and fauna of the city. This type of tourist spends a lot of time reading about environment. He or she uses to look for natural food restaurants, and little alpine style hotels in Lumiar and São Pedro da Serra, which are the best districts concerning natural resorts in Friburgo.

3 - The Sportsman (SP): Similar to EL, but more interested in campings than hotels. Not attracted by natural food but for fast (and cheap) food. Normally, this type of visitor is an adolescent.

4 - The Family Boss (FB): This visitor normally travels with his or her familly, and has a more typical list of places to visit. This user tends to exhaust all possibilities in downtown. This type of visitor normally eats in a steak house or a tipical german food restaurant. This tourist is likely to be interested in the history of the city as well. His or her favorite hotels are also in downtown.

HyperFriburgo calculates, mainly using the time spend on each available item within the hypertext, an index value, between 1 and 10, measuring how much the current user fits each profile. Then, it takes the highest value (if it exceeds a minimum of 5) as the most probable profile. This is calculated only after the user has taken at least 6 different choices. If the system 'decides' the user has a tipical profile then it asks the user if he or she wants a specific orientation. For instance, if the profile is EL, the system prompts the user about his or her interest in looking at a oriented tour through the rivers and waterfalls of the city.

The system also decides about a geographical profile and a economical profile of the user. If the user searches mainly for items in a specific geographical area (i.e. a district), HyperFriburgo decides that any suggestion will concern this area. Therefore, if the profile is TB, and there is a clear interest in the district of Mury, HyperFriburgo will suggest a oriented tour for visiting textiles shops and cheap restaurants in this district rather than through the whole city. If no personal profile is found (neither TB,EL,SP nor FB), a generic tour through Mury is suggested.

HyperFriburgo establishes an economical profile for the user if most of time he or she searches for one category of items (cheap, afordable or expensive). These items are mainly hotels and restaurants. Therefore, suppose someone only looked for the expensive hotels available within the system. When navigating through the section about restaurants, this user will be asked by HyperFriburgo if he or she wants a list of sophisticated restaurants. Basically, interest in expensive hotels suggested interest in expensive restaurants as well.

In HyperFriburgo, there is a database about the items available in the city. We understand as items all relevant places for a tourist as restaurants, hotels, tourist resorts etc. Besides the items, the system also includes the notion of groups and divisions. A division is one of the options of the main menu. For instance, *"Tourist Resorts"* is one division. A submenu of this main menu is called a group. For instance, when we select the option restaurants in the main menu, another menu divided by types of cuisine is presented. Therefore, we have that German food , Italian food, etc, are examples of groups of the division *"Restaurants"*.

This is an important classification since a user may spend some time looking at the options of a group (or division) without selecting a specific item. Nevertheless, this time may be very important for the decision concerning the visitor profile. As illustration, suppose someone 'click' *"Hotels"* (therefore a division), but spend some time before choosing a specific group. This means to spend some time before selecting among options as Alpine Style Hotels, Large Hotels, Camping etc.

Obviously, this time should be computed as time spend looking for hotels since it shows an interest in staying overnight. Nevertheless, time spend looking at *"Hotels"* diminishes the index rate of a profile as TB. Therefore, the time spend by the user looking at divisions and groups, may be as important as the time spend looking at specific items.

When the user selects an option, the lists of already selected items, groups and divisions must be updated. Then, based on these lists (that represent *user's choices*) the knowledge base of profiles selects one profile (or none) and decide which oriented tour should be suggested. This system was designed in PROLOG and VISUAL C++. In the next section we describe both versions.

3 - Implementing HyperFriburgo

Therefore, the knowledge base in prolog, takes as input a list of items, groups and divisions already selected by the user, during the current consultation. This list also has the amount of time spend looking at each selected item, group or division. As illustration, suppose a consultation that have produced so far, the following list of choices:

[el(sh,5),el(ts,5),el(sh1,20),el(sh2,30),el(sh1,15),el(rt,5),el(ff,10),el(ff1,12)]

This is a prolog list (see [1] for explanations about prolog), whose first element is el(sh,5). This element, according to our notation, means that the user spend five seconds in the division about "*Shops of Textiles and Shops of Delicatessen*" (whose symbol is sh).

The second element (el(ts,5)) means 5 seconds spend looking at the group of textile shops (ts) without taking any decision. The third element, el(sh1,20), means that the same user spend 20 seconds looking at the specific shop of textile represented as sh1.

The fourth element shows 30 seconds spend looking at a shop of same type (shop sh2) and the fifth shows that the user decided to return to sh1 and spend there further 15 seconds. Later, the same user spend 5 seconds looking at the division restaurants (rt), and kept looking at the group fast-food (ff) for 10 seconds. Finally, in the last choice so far in his or her consultation, the user decided to look at an specific fast-food restaurant (ff1) for 12 seconds.

The first step of the knowledge base is to accumulate the total amount of time spend on each item. If an item is visited more than once, the total time spend on looking at an item is multiplied by the number of times this item was selected. We believe that a tourist returns to a previously selected option only if he or she has a very special interest in that item. This is why we establish this rule of multiplying *time_spend* by *number_of_selections*. Therefore, the list above becomes:

[el(sh,5),el(ts,5),el(sh1,70),el(sh2,30),el(rt,5),el(ff,10),el(ff1,12)]

The main core of the knowledge base is the decision about the profile of the user. This is performed by a predicate which takes the maximum index HyperFriburgo calculates to each profile. For instance, for the textile buyer, the index (variable IndexTextileBuyer) is calculated by a specific predicate (tbuyer) as shown bellow::

```
tbuyer( TTI_MR,TT_poor, TT_afordable, TT_rich, TTTextile, IndexTextileBuyer):-
IndexTextileBuyer is
(

(

2*(TTTextile - TTHotel)/TTI_MR +

 TT_poor/ (TT_poor+TT_afordable+TT_rich)

 ) / 3

)*10, IndexTextileBuyer > 0, !
;
IndexTextileBuyer is 0.
```

In this predicate TTI_MR is the total time spend on items that are not restaurants, and TT_poor the time spend in cheap items (not expensive restaurants and hotels). TT_afordable and TT_rich are the same for more expensive items. TTTextile is the time spend looking at items (shops) which sell this type of product.

According to what was exposed, IndexTextileBuyer for the previous consultation is:

IndexTextileBuyer=(2*(((5+5+70+30)-0)/110+22/22)/3)*10 ∴
IndexTextileBuyer =10

which shows a clear decision (actually the maximum value) for this profile. Therefore, at this moment of the consultation, HyperFriburgo would present a dialog box asking:

"Do you want a oriented tour showing Textiles Shops?"

If the answer is positive, this tour would highlight the textile shops, cheap restaurants, bus time table to Rio de janeiro and all relevant information for a textile buyer.

Hyperfriburgo is also being implemented in C++, following the object-oriented and event oriented-programming aproaches.

In Figure 3 we present how HyperFriburgo comunicate with users and manipulates his or her options. The system presents 'dialog boxes' with explanations about Friburgo and options available for further navigation.

There is a main window for the system, in which the dialog boxes are shown. Part of the system (function WindowFunc) receives mouse messages concerning the *user's choices*, and through a loop of messages dispatches these messages. Then, a specifically designed function for each message may be activated (for further explanations about programming in *Windows 95* see [6]).

Therefore, whenever an option is taken, the system activates the function which deals with that message coming from the dialog box being presented. Then, the following steps are performed:

1 - Updating the arrays of items, divisions and groups that have already been selected during the current consultation. This updating concerns time spend on each option and number of times the option was selected.

2 - According to what was exposed, the system activates the function ExpertAnalysis to check if there is a profile concerning geographical area, economical status or type of visitor. This function performs as the knowledge base.

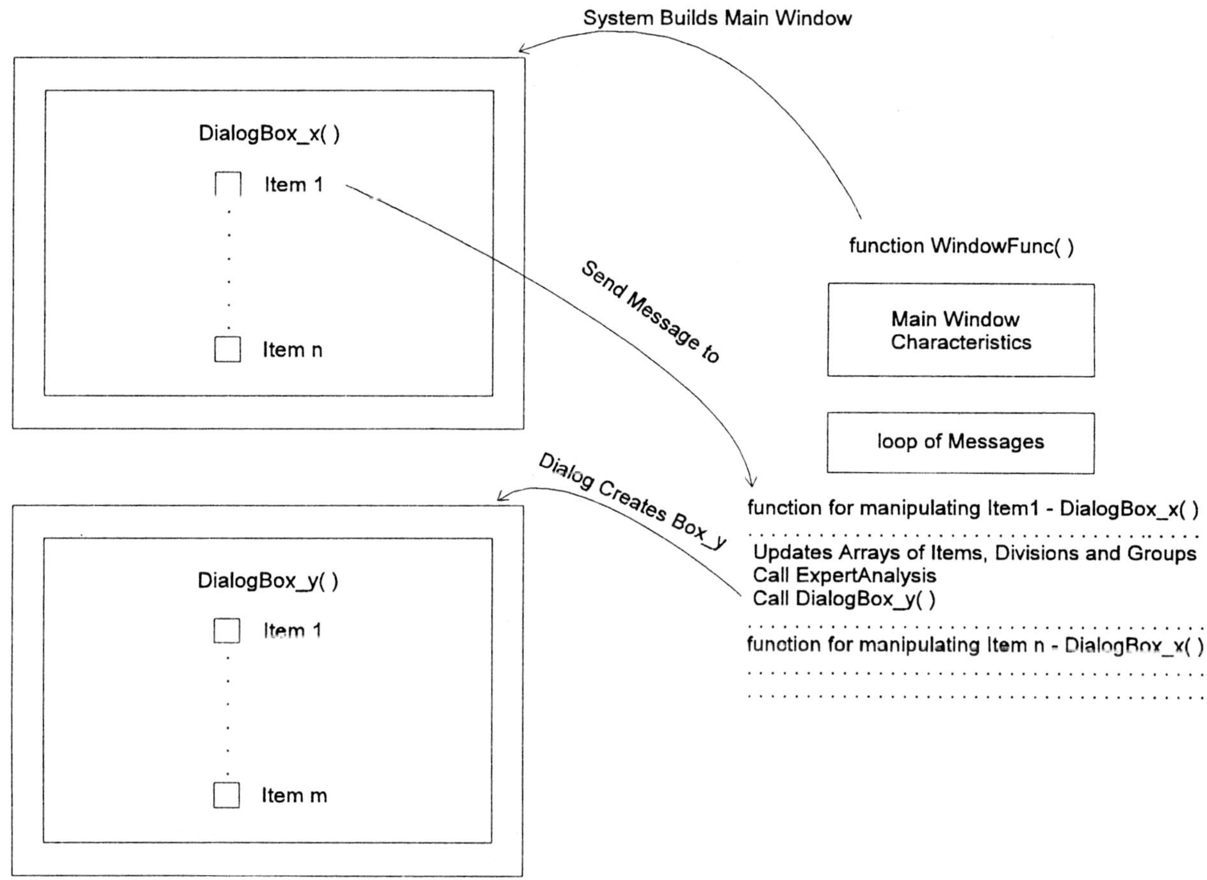

Figure 3 -Designing HyperFriburgo in C++ for Windows

3 - The system presents a new dialog box concerning the last option of the user. It also presents a suggestion for a specific oriented tour if a profile was detected.

Therefore, C++ version of HyperFriburgo is conceptually identical to the Prolog version. The differences are only in the programming techniques that we applied.

4 - Conclusions and Further Research

The main goal of this article was to present an approach to include some intelligent inference capabilities in a tourist information system.

The architecture adopted for such system is relatively simple and straightforward and seems to be able to reproduce the work of the most experient tourist guiders of Nova Friburgo.

The authors are now interested in performing a statistical analysis of the system efficiency, and to improve the knowledge base involved. This improvement follows two main directions: First, to encode more different profiles of visitors into the system, and second to ellaborate more accurately the procedures involved in choosing one of them.

So far, HyperFriburgo only suggest oriented tours according to a possible profile of the current user. Nevertheless, if such suggestions are not accepted, HyperFriburgo should review its conclusions about the profile selected. In a next future a truth maintenance system [2] is likely to be added into the system in order to deal with this inneficiency.

Acknowledgements: The authors are deeply in debt with Professor Valéria Iório and her collaborators from ESDI (Industrial Design College/UERJ) for their insights about HyperFriburgo and valuable suggestions.

6 - References

[1] Bratko, I. "Prolog - Programming for Artificial Intelligence". Addison Wesley. Second Edition .1990.

[2] Doyle, J. "A Truth Maintenance System". Artificial Intelligence. Vol 12, pp 231-272. 1979.

[3] Elmasri, R. and Navathe, S. B. "Fundamentals of Database Systems". *The Benjamin/Cummings Publishing Company, Inc.* 1989.

[4] Isakowitz, Tomás; Stohr, Edward A. and Balasubramanian, P. "RMM: A Methodology for Structured Hypermedia Design". In *Communications of the ACM*, August 1995. Volume 38, Number 8.

[5] Thüring, Manfred; Hannemann, Jörg and Haake, Jörg M. "Hypermedia and Cognition: Designing for Comprehension". In *Communications of the ACM*, August 1995. Volume 38, Number 8.

[6] Schildt, Herbert. "Schildt's Windows 95 Programming in C and C++". *McGraw-Hill*. 1995.

TIS@WEB - Database Supported Tourist Information on the Web

F. Burger, P. Kroiß, B. Pröll, R. Richtsfeld, H. Sighart, H. Starck

FAW - Research Institute for Applied Knowledge Processing, Hagenberg, Austria
Softwarepark Hagenberg, 4232 Hagenberg, Austria, e-mail: birgit@faw.uni-linz.ac.at

Abstract

The currently running tourism-WWW-server TIS@WEB makes touristic information about Austria available world-wide to the public under the WWW-address **http://www.tis.co.at**. The extraordinary feature about TIS@WEB is that you get latest tourist information because it is generated out of the touristic database TIS, which is updated frequently. Due to this approach the effort to actualise the data (a well known problem of WWW-Servers) is minimised. In the present paper we will give you a closer look on the architecture, the realisation, and the functionality of TIS@WEB. Our approach of "batch-generation" and "on-the-fly" generation of Web-pages is described. Afterwards, requirements for the next generation of tourism WWW-servers are discussed.

1 Introduction

In the 70s and 80s the TCP/IP based Internet was mainly used for the exchange of information. This service was primarily used by academics and research people. With the development of the World Wide Web (WWW) by Berners-Lee´s team [1] the growth of the internet began and is still holding on in an unbelievable rapidness. Today more that 100 million people use the Internet. Following research institutes, first industrial companies and recently also private households discovered the usefulness of the Internet. Especially the latter are the customers for the tourism industry. Besides pure Internet "surfers", people who run a Web-site and provide information have changed from researchers to people from all industries. Also the number of these so-called "Web-sites" is growing rapidly. Especially for the marketing industry the Internet created new ways to reach its customers. Besides self-presentation of companies with the main purpose of providing information (e.g. telefon numbers), advertising and selling of products are the topics of the electronic marketing field. The tourism area is one of those industries which is able to gain synergy effects from the use of electronic networks.

The Internet is a perfect platform for the tourism industry to bring information about their products to the customers all over the world in a direct, cost minimising and time effective way. Moreover, WWW adds features which are perfectly designed for the tourism industry, namely Hypertext, to navigate through the information (an very adequate means for geographical navigation) and multimedia, which is of most importance to attract the tourism customer to visit destinations and to buy touristic products like rooms or tickets for events. "The Internet, Hypertext, and Multimedia - Three important technologies of the Nineties. Bring them all together, and you have the World Wide Web." [2]

Although nobody can neglect the success of the Internet and the WWW for the tourism industry, there are still a lot of problems to be solved. One main deficiency is the poor performance of most WWW-Servers. Tourism information providers have to care, that their customers get the required information in a reasonable time, otherwise they will change to another faster tourism-site. Much can be done on the WWW-Server site. Static pages can be accessed faster than pages that are generated out of a database at runtime. Even if databases can be tuned and database servers can be scaled up, this carries weight especially if the database stores a huge amount of data and lots of people try to access the same data at the same time. Approaches have to be defined how to make up-to-date Web-pages accessible faster. Network performance is the second factor which is a drawback. Much research is done in this field. E.g. the EU project MNET evaluates the usefulness of ISDN and ATM for the purpose of reducing the time to distribute huge amount of multimedia tourism data to tourism service providers all over the world. [3].

Another characteristic of tourism-WWW-servers is that they are never "finished" because new features, like real-time-reservation, are strongly demanded by the tourism providers on the one hand and by the tourism customers on the other hand. To be on top of the development, today´s tourism servers are subject to a permanent improvment and extension of the software. Currently, tourist Web-sites spread over the Internet and providers have to invest in ideas to offer better information (which means up-to-date and well organised data) and high standard features to beat their competitors. Meanwhile there is no company who holds a monopol on providing tourism information on the Web, but many companies are trying to get this position. TIS GmbH. (the developer of TIS@WEB) is one of the companies holding a leading position on the European tourism market. Even if it is not much time that they started their work with the Internet, TIS@WEB already got an ENTER 1995 award.

2 Combining WWW and databases: Existing Approaches for Tourism WWW-servers

Most of today´s tourism WWW-servers are "simple WWW-Serves".

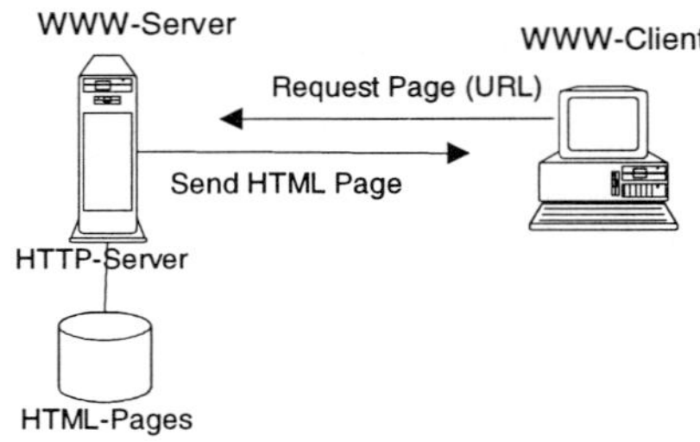

Fig. 1: "Simple WWW-Server"

They provide touristic data on so called "static" pages. The term "static" indicates a net of linked ready HTML-pages which are send to the cutomer who requests them. The HTML-pages of these often "passive servers" [4] include text and pictures and are perfectly designed to the imaginations of the information provider. Figure 1 shows the architecture of a "Simple WWW-Server". The WWW-client sends a request in form of an URL (uniform resource locator) to the HTTP-Server. The HTTP-Server searches the requested HTML-page in his directory and sends it to the WWW-client.

The advantages of this approach are:

- possibility of perfect design and layout
- individual presentation regarding content and layout
- acceptable performance as the pages are "ready to go" to the customer´s client

In spite of these advantages the touristic WWW-site cannot fulfil the advanced requirements of the touristic community and the touristic customer. This is mainly due to the following drawbacks:

- the information is never up-to-date
- maintaining the information is a hard job because HTML-coding is required
- even if HTML-editors exist, using them means restrictions in layouting the pages

A lot of Web-Servers meet these shortcomings by connecting their WWW-Server to a database and generating Web-pages out of the database. This approach combined with a suitable data acquisition tool makes it much easier to hold the information up-to-date. But it means restrictions to individual content and layout of the pages. Furthermore, interactive servers are demanded: WWW provides a concept called "Forms". With this, Graphical User Interface (GUI) elements such as text entriy, buttons, check boxes, radio buttons, and drop-down-lists can be added to WWW-pages. When the user pushes a submit button the server reads the user-entries and decides on further actions, e.g. a complex search. [4]. Currently, database system companies like Oracle on the one hand and web server/browser companies like Netscape on the other hand try to bridge the gap between HTML, the standard language of the web, and SQL, the standard query language used to access relational DBMS. First solutions are already available and are mostly based on the cgi-paradigm. Figure 2 shows the components of an "interactive WWW-Server".

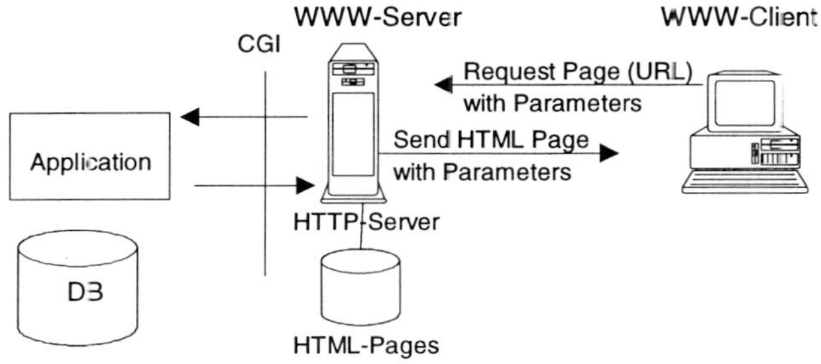

Fig. 2: "Interactive WWW-Server"

The client's request is extended with parameters needed in the application that e.g. selects data out of a database. A detailed description on how a Web application can be connected to a relational database (e.g. DB2) is given in [5]. Contributions and problems of Internet-related database research can be found in [6].

The next chapter is dedicated to the approach of combining the existing tourism system TIS 2.x supported by the relational TIS-database and the Internet.

3 Architecture and Functionality of TIS@WEB

3.1 The TIS 2.x System

The company TIS GmbH (tourism information systems, Innsbruck, Austria) is engaged in the tourism area since 1991 and besides other activities develops, runs and maintains the Austrian tourism information system TIS - version 2.x. TIS is based on the concept of distributed maintainance of tourism data. Data acquisition and update is done via a user interface which is installed at local and regional tourism organisations. The co-operating tourism organisations are permanently sending present-day data (accomodation data, snow and lake data etc.) via modem to the central TIS-database, where this data is checked for correctness and consistency and distributed to national and international partners, who request this data: ÖAMTC - an Autrian automobil club, cable TV stations, etc. TIS decided for the relational database system PROGRESS. The user interfaces is realized by using the PROGRESS programming environment, too. The system runs on a PC basis.

3.2 TIS@WEB

In 1995 TIS decided to make the huge amount of touristic data stored in their database available worldwide through the Internet technology. The challenge was not to produce "static" HTML-pages but to generate HTML-pages out of the already existing tourist database. To be open to the Internet, the existing TIS 2.x System had to be extended by a WWW-server with connection to the TIS-database and by a second database called "WWW-database" which stores additional data used for the WWW-part of the system. The overall TIS@WEB architecture is shown in Figure 3.

For the most part the touristic data is maintained at the local tourist office and transferred to the TIS 2.x database. Additional data is maintained by the TIS@WEB administrator and stored in the HTML-database. Frequently, a generator produces WWW-pages and data-files used for interactive searching by selecting data out of both databases. Whenever a client requests information over the internet, he accesses the TIS@WEB-WWW-server. Depending on the request, the WWW-server sends an HTML-page back to the client or starts the searching CGI-procedures. Fax/email CGI-procedures are responsible for the fax/email-handling. These issues are described in more detail in the following subsections.

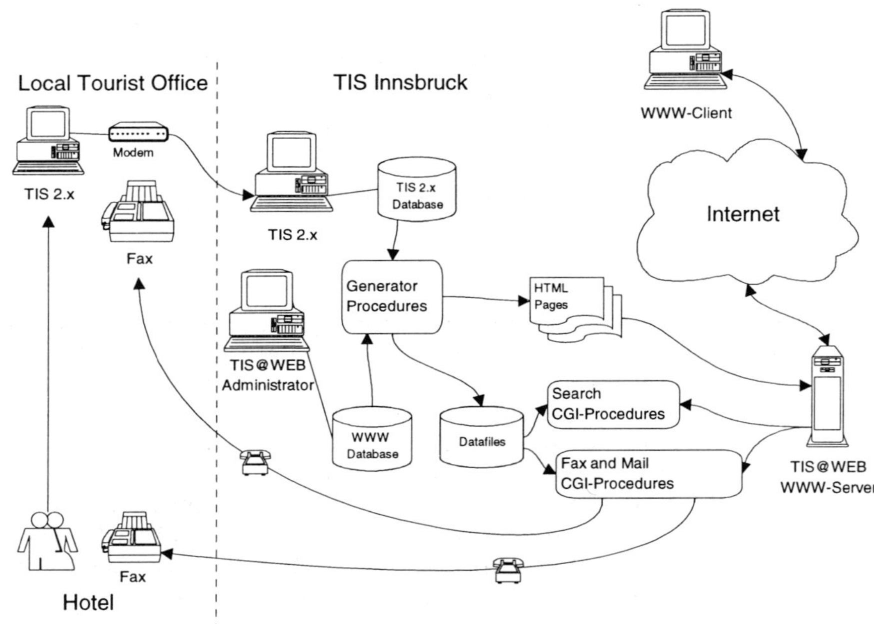

Fig. 3: Architecture of TIS@WEB

The WWW-database

The WWW-database, maintained by the TIS@WEB-administrator, stores data that was not available in the TIS database but is relevant for the WWW-server:

- Multimedia data

The TIS 2.0 system is not construed to acquire and store multimedia data. But, pictures are a must for an attractive tourism WWW-server. The solution was to hold scanned .gif and .jpg files in an intelligent directory structure with the purpose to reference them easily.

- Multilingual data

A contribute you have to pay to the internationality of the Internet is, that you have to provide your information in at least two languages (as far as english is not the local language). The translation of data is done at the TIS@WEB administrator´s site. The translation tool supports the translators by marking touristic objects (e.g. a village) whenever some data has been changed by a local tourism office and has to be translated.

Gateway between the PROGRESS-databases and the WWW-server

As PROGRESS did not support a connection to WWW under Windows at this time, a solution for the exchange of data between the PROGRESS-databases and the WWW-applications had to be realized. The best way (that is the most performant one) to do this, revealed to be on the basis of ASCII-files. A PROGRESS-application has to select the demanded data out of the database and puts the result of the query on a file. A Perl [7, 8] procedure parses the file and builds up the HTML-pages which are send to the WWW-Client directly or are stored in a specially designed directory-structure.

Generating HTML-pages - "batch generation" versus "generation on-the-fly"

TISWEB provides an intelligent solution for the compromize between acceptable performance and actuality. Due to performance optimization, HTML-pages are not generated "on-the-fly", that means at request time out of the database. Whenever a client requests information e.g. about a village, he gets a "static" HTML-page. To keep information up-to-date all the "static" HTML -pages are generated out of the database once a day and stored in a semantic directory hierarchy. This holds for HTML-pages which present touristic objects like towns, villages, accomodation facilities, events, etc. in an attractive way to seduce the Internet surfer to visit their locality and to buy their products. These HTML-pages are designed attentively by integrating extensive amounts of information and multimedia data (see Figure 4). Generating these pages at runtime spends too much time.

Besides surfing through "static" presentation pages, TIS@WEB supports a "dynamic search" for accomodation and holiday packages. This search does not access predefined pages but the required data is rather collected by invoking a query against a predefined ASCII-datafile. The resulting HTML -page is generated on-the-fly". The query must not be put against the database itself, even if the gateway would allow it, because this could lead to a mismatch between search-result and "static" HTML-pages. The datafile is generated whenever the "static" HTML-pages get updated.

To give an example, the surfer may start a search for all hotels in the village Kitzbühel and gets a list of names and basic data on the hotels as result. Following the link to the detailed description of a hotel, the surfer arrives at a "static"- HTML-page again. He may also search for all holiday packages suitable for a family between Feb 3rd and Feb 16th 1996. Figure 5 is a hardcopy of the page to define the search.

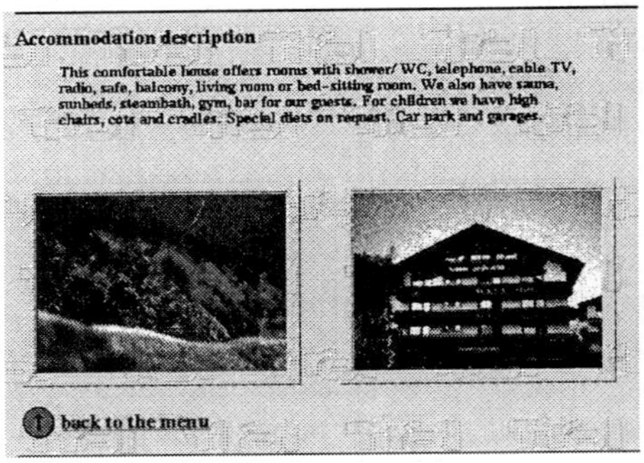

Fig. 4: Presentation of a Hotel

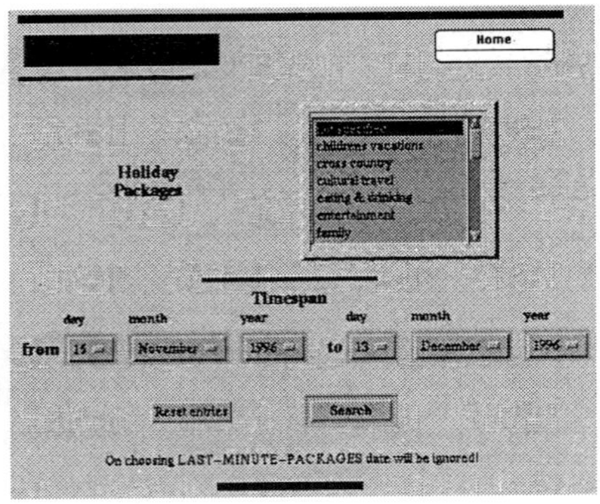

Fig. 5: Searching for a Holiday Package

Realizing Reservation through an integarted Fax/Email Solution

A major problem of today's tourism WWW-sites is the realization of a reservation system. TIS@WEB provides an integrated fax/email solution as a first step to fulfil customer's requirements. The customer may put an "reservation-request" for rooms, holiday packages, and camping to the system. The "reservation-request" is passed on to the responsible tourist office automatically. Figure 6 gives a view on this process.

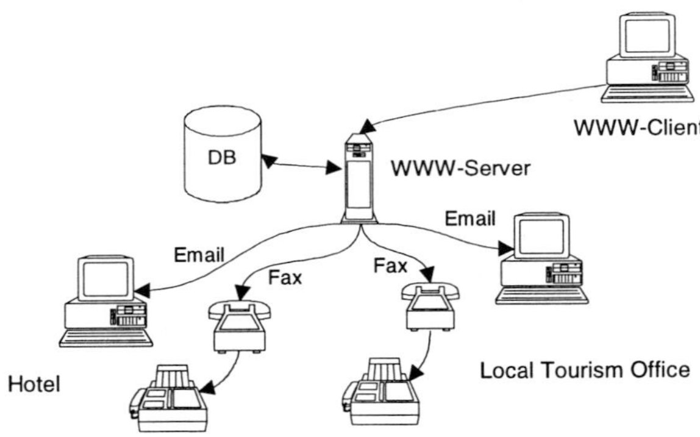

Fig. 6: Integrated Fax/Email Reservation

A WWW-client is able to fill out a fax/email-form on his WWW-browser and send this fax/email to a specified hotel or local tourist office. On submission of the form the WWW-server checks whether to send an email or a fax to the specified destination, depending on the availability of an email address of the accomodation provider. In case of an reservation via email an email-file containing the customers request is created and sent to the specified email-address by using the email-programm postmail. If a fax has to be sent, a file is created out of the form-data and archived. The fax-solution of TIS@WEB is based on a faw-program which behaves like a daemon. It checks if a new file is created and automatically sends the fax to the hotel or the local tourist office. If the transmission was not successful after repeated attempts of sending the fax, the WWW-site provider gets an email.

Time-Critical Reports

WWW-pages on time-critical information like snow and water reports, weather information and traffic news have to be generated more than one time a day because some hours later this information might not longer be relevant. As it is involving great expense to get this data on time, there are few WWW-servers which offer this information in a serious way. Especially for Austria the accurate snow report revealed to be very attractive and during winter times gets updated several times a day.

4 Summary and next generation requirements

The described architecture of TISWEB fulfils the requirements on a tourism information system: of the customer´s side broad, complete, and up-to-date information at reasonable performance and at the Web-site provider´s side: less effort to update the data as most of it is automatically maintained by the local tourist offices. TIS@WEB started in October 1995. During the last year new requirements have been analysed with the main challenge to extend TIS@WEB to all European countries.

Extended Use of Multimedia

In TIS@WEB pictures are added to the pure structured data of the TIS database. In addition, movies, animated gifs and VRML graphics should be supported.

Complex searching facilities

TIS@WEB supports search for accomodation and holiday packages by defining the location and the type of accomodation or the relevant time and activity type for a holiday package. Searching facilities should be extended to other types of touristic objects like available rooms, events, sporting facilities, etc. Advanced filters (e.g. price limits) are demanded which allow to specify the required result in more detail.

Individuality

Due to the database approach, HTML-pages for one type of touristic object (e.g. the homepage of accommodations) look similar because the underlying layout is the same. But, tourism information provider want to differ from each other and present their data in an individual way (layout). New generation tourism-WWW-servers shall integrate approaches to comply with this requirement.

Internationality and multilinguality

TIS@WEB supports German and English. As the challenge is to provide information of different countries and present information to different nationalities worldwide, great importance has to be attached to the aspect of internationality. What is required is so called “open multilinguality“, meaning the support of any kind and any number of language. Moreover, different currencies etc. must be supported.

Real-time-reservation and on-line-payment
Real-time-reservation of rooms, appartments etc. is a very sophisticated module. The number of available room has to be administered and calculating prices gets difficult, as soon as all aspects of reality are consider, like different prices for different seasons or price reductions for children. The aspect of security of "on-line-payment" by credit card is still up for discussion.

Data Maintenance - a homogeneous solution
An obvious disadvantage of TIS@WEB is the data acquisition and maintainance via the PROGRESS user interfaces, which must be installed at the local sites. Data maintenance could be done in a better way by using the WWW-technology again. WWW is dedicated to serve as homogeneous user interface for geographically distributed information providers, who can access and update their data as soon as they have an Internet connection and the required authorization. System administration could be done via the same user interface, too. This approach demands a complex authorization system.

Distribution of WWW-Servers and Database Servers
Instead of a European wide centralized WWW-site (one WWW-server, one database-server) multiple WWW-sites should be supported. One the one hand this will have technical advantages like better reliability of the system and a reduced amount of data to be stored and maintained at each site. On the other hand distributed systems will deal better with the individual requirements of European countries. Approaches how to integrate the WWW-servers and how to solve queries on distributed databases are necessary. An approach on using distributed databases for tourism can be found in [9].

References

[1] T. Berners-Lee, R. Cailliau, A. Luotonen, H. F. Nielsen, A. Secret. The World Wide Web. in: Communications of the ACM 8/1994, pp. 76-82

[2] J. December, N. Randall. The World Wide Web Unleashed. Sams Publishing, Indianapolis, 1995.

[3] Y. Lindsjörn, N. Regnesentral, P.F. Byerley. Broadband Network in the Electronic Market place in Tourism. in: Proceedings of ENTER 1996. Information and Communication Technologies in Tourism. ed: W. Schertler, B. Schmid, A M. Tjoa, H. Werthner. Innsbruck, Austria

[4] R. Maier, W. Röckelein. Electronic Networks: New Challenges got Tourist Organizations? in: Proceedings of ENTER 1995. Information and Communication Technologies in Tourism. ed: W. Schertler, B. Schmid, A M. Tjoa, H. Werthner. Innsbruck, Austria. pp. 112-123

[5] T. Nguyen, V. Srinivasan. Accessing Relational Databases from the World Wide Web. in: Proceedings of the 1996 ACM SIGMOD - International Conference on Management of Data. Montreal, Quebec, Canada. June 1996. acm press. pp. 529-540

[6] P. Apers. Identifying Internet-related Database Research. in: Second International East-West Database Workshop. Extending Information Systems Technology. ed: J. Eder, L. Kalinichenko. Sep. 1994. Klagenfurt, Austria.

[7]. E. Quigley. Perl by Example. Prentice-Hall Inc., 1995

[8] R.L. Schwartz. Einführung in Perl, 1995

[9] A M. Tjoa, H. Werthner. Interfacing WWW with Distributed Database Applications in the Field of Tourism. in: Proceedings of ENTER 1996. Information and Communication Technologies in Tourism. ed: W. Schertler, B. Schmid, A M. Tjoa, H. Werthner. Innsbruck, Austria

Electronic distribution of holiday and business hotels

Dr. Carl H. Marcussen

Research Centre of Bornholm, Stenbrudsvej 55, 3730 Nexoe, Denmark.
Fax: +44 56 49 46 24. Tel.: +44 56 44 11 44. e-mail: marcussen@rcb.dk

Abstract. There are a number of typical differences between holiday and business hotels with respect to price level, seasonally, size etc., having implications for which electronic information and sales channels are most economically feasible to utilise. For a number of reasons the basis of revenue for holiday hotels is limited, and therefore also costs - including spending on telematics - must be kept at a moderate level. Hotels may be distributed as single products or as part of packages through a number of links in a chain, which may be conventional *or* electronic. Five strategic issues for hoteliers pertaining to electronic distribution of hotels are discussed. The Internet is a relatively cheap medium with great reach, which may be regarded as coming to the rescue of many small holiday hotels as far as electronic distribution is concerned. It is discussed to what extent this is really the case.

1 Introduction

The last couple of years the main emphasis in the academic literature on inter-organisational tourism technology has shifted from CRS' (GDS') [2,5,6,8] and Destination Information/Management Systems (DIS') [1,4,12,13,14] to the Internet, specifically the World Wide Web (WWW). This has gone hand in hand with the advent of the latter. From 1996 a number of articles on tourism marketing via Internet/WWW have started appearing in tourism journals [7,9,10,11,15]. GDS', DIS' and proprietary on-line systems are forced to make moves in the direction of Internet. This paper focuses on hotels only, taking the perspective of the hoteliers. There will be a discussion of a number of strategic decisions facing these hotel managers and owners. A distinction is made between holiday and business hotels, hotels as individual products and hotels as part of packages, and finally between sales through intermediaries (via CRS') and sales directly to the consumer (via Internet/WWW). It is discussed if the Internet has eliminated differences in visibility and availability between hotels in electronic distribution channels for hotels.

2 Characteristics of holiday vs. business hotels

The mere fact that hotels may be oriented towards the holiday rather than the business travel segment hardly implies any limitations as to their utilisation of telecommunications based information and sales channels. There are, however, a number of typical differences between holiday and business hotels having implications for which electronic information and sales channels are most economical to utilise.

Table 1. Typical characteristics of holiday and business hotels.

Parameter/factor:	*Business hotels*	*Holiday hotels*
Location	Cities and towns, 'industrial' areas, and junctions.	Cities, cosy small towns, near attractions, and at resorts.
Price level	High - medium.	Medium - low.
Open	All year.	Only in season.
Number of rooms per hotel	Relatively many rooms per hotel	Relatively fewer rooms per hotel.
Individual products or packages	Hotel nights primarily as individual products.	Both individual products and as part of package tours.
Number of nights per guest per visit	One or two nights: as individual products.	About one week: Relatively long stay.

Holiday hotels may be located, for example, in beautiful and interesting cities like capitals or at sea side resorts. The purpose of the stay is to see sights, visit attractions, relax, and enjoy nature. Business hotels, on the other hand, are located near or in centres and industrious areas, including areas with service industries and public organisations/administrations, or at motorway junctions and airports.
The price level at holiday hotels are typically a little lower than at business hotels, since price sensitivity is lower at the latter because the bill is passed on to the firm. Holiday hotels are to a greater extent than business hotels part of package tours, which also contributes to reducing the price per night. Furthermore, the former have peaks at one or more seasons, and their average size is smaller. For these reasons the basis of revenue for holiday hotels is limited, and therefore also costs - such as spending on telematics - must be kept at a moderate level.

The price of computer equipment for internal use and the price of participation in telecommunications based information media and booking systems are independent of the size of the hotel. The typical holiday hotel finds it more difficult than the typical business hotel to financially justify investments in computer equipment and participation in telecommunication based information media and booking systems.

3 Conventional and electronic distribution channels for hotels.

Marketing and sales (distribution) of hotel nights - individually or as part of packages - takes place through a chain, which consists of the following links:

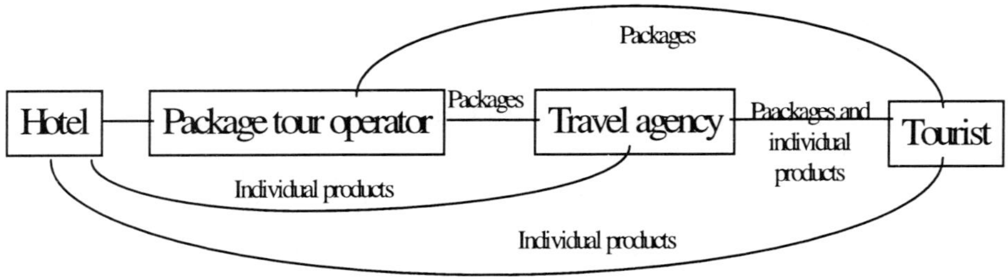

Figure 1. Distribution of hotel nights.

The distribution chain may be electronic. In that case the distribution chain is in principle the same, except that an extra link is required for the mediation of the electronic communication. A distinction between hotel nights as single products and as part of packages also including transport and other services must be made, since there are different types of computerised reservation systems (CRSs), which are being employed for the two purposes.

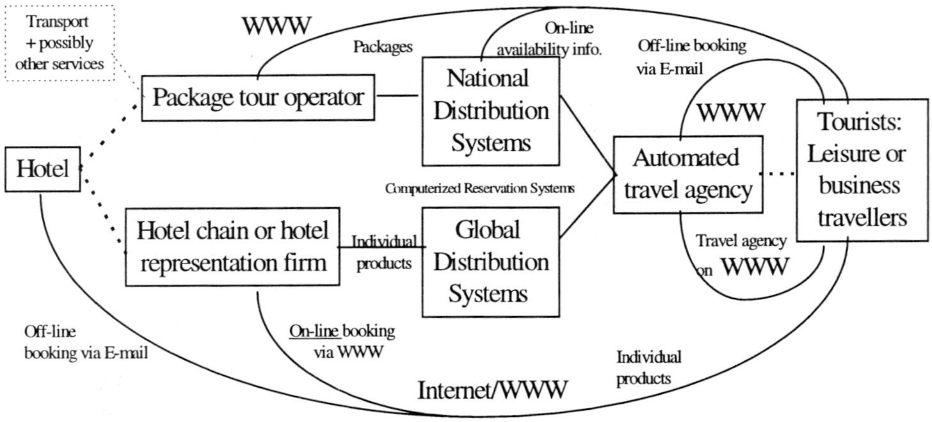

Figure 2. Electronic distribution of hotel nights.

Until a couple of years ago the telematics based systems only comprised those reservation terminals (now PCs), which are used by travel agencies for booking of primarily air tickets, secondarily *hotel nights*, and thirdly and car rentals. The very fast development of the Internet, which has occurred recently, has meant that there is now a rather common electronic channel directly to the individual tourists, that is leisure or business travellers.

4 Strategic issues for hoteliers pertaining to electronic distribution.

There are a number of questions, which hotel managers must consider:

1. To what extent will we target holiday and/or the business segment?
2. Will we make it possible for travel agencies to book individual room-nights at hotels electronically?
3. Do we want to be a completely independent hotel or do we want to join a hotel chain?
4. To what extent do we wish to sell the hotel capacity as part of holiday packages and/or as individual products?
5. Do we want to convey information about our hotel via WWW directly to the travellers, and if so do we want to make it easy for the WWW user to make a booking and guarantee payment of stays at the hotel?

If you are a hotelier, which is targeting business travellers, then this is an argument for making your hotel 'bookable' by the systems, which are used by the automated travel agencies for making reservations of hotel-nights. Also for holiday hotels it will become increasingly more relevant to make themselves electronically bookable for travel agencies. Travel agencies work on thin profit margins. So, reservations for all types of products must be fast and easy to undertake.

There are two different ways whereby hotels can make individual room-nights 'bookable' at the CRS terminals, now typically PCs, at travel agencies. A hotel may decide to become a member of a hotel chain, which sees that all their members get listed in the travel agency computerised reservation systems for individual room-nights at hotels. This is the case for Best Western hotel chain. With about 3,700 hotels world-wide it is the world's largest hotel brand. If the hotel owner/manager wishes that their hotel should be independent of any brand or chain, then there are so-called hotel representation firms, which sees that a given hotel gets listed in all the world's four Global Distribution Systems, i.e. Amadeus/System One, Galileo, Sabre and Worldspan, in addition to a number of regional reservation systems.

As a general rule it is only holiday hotels, which are part of package tours, with a couple of exceptions. Firstly, 'incentive travels' may be sold as 'packages', which are assembled in terms of individual wishes. Secondly, business hotels often try to utilise excess capacity, for example, in weekends or during holidays by packages of tour operators. Holiday hotels will almost always want to sell part of their capacity as an integrated part of one or more tour operators' package programmes. When hoteliers give hotel rooms in allotment to package tour operators, it is in the hotelier's and the tour operator's common interest that the packages are as easily accessible for ordering as possible with national and international travel agents. The hotel owner must take an interest in the tour operator's electronic distribution of the package tour programme via national booking systems for package tours in the hotel's main markets. Unfortunately, unlike the hotel reservation systems for individual room-nights it is necessary for the tour operator to sign *one* contract for each main market. For example, one contract for Germany, the system Start, and another for Scandinavia, the system Smart, since each of the travel agency oriented CRS' for package tours tend to cover one or a few countries only. There is an economic barrier for tour operators, and thereby also for hotels, against participation in some of the national or regional booking systems for package tours, since it is expensive, and therefore a great volume of packages are required for a tour operator to be able to participate.

It has been suggested that "Lack of CRS accessibility may be strangling small hoteliers" [3], which are to a great extent holiday hotels. The Internet, partly because of an increasing penetration and partly because it is a relatively cheap medium, is becoming a very interesting supplement or alternative to the travel agency oriented systems. The latter were created as booking systems for air tickets, but efforts have later been made to adapt them for hotel reservations. Unlike the CRSs, WWW is so far mainly an *information* channel and to a less extent a *sales* channel. This has to do with the fact that issues in connection with payment for the services, which are being sold or booked via WWW still have not been solved satisfactorily. WWW can be utilised both by the individual hotels and the other actors in the distribution chain. If, for example, a hotel is a member of the Best Western chain, or one of the about 10 other hotel chains within TravelWeb (http://travelweb.com) with a total of 8,200 hotels, it is possible for the WWW user to book that hotel *on-line* via WWW and 'guarantee the booking' by including a credit card number. *On-line* here means that the Internet user in the course of a few seconds gets a reply to an enquiry about room availability, and next in the course of a few more seconds gets a reply to their request for a booking. In case the guest wishes to cancel the booking this must take place no less than 24 hours before the guest was to show up at the hotel. In case of 'no-show', hotels in the chain reserve the right to payment for the first of the booked hotel nights, by debiting the credit card number which they have been given. However, most hotels will have to settle for offering the Internet users the opportunity to book *off-line*, which here means either via E-mail or if the hotel has no E-mail box, then via fax. This is because of the technical precondition of on-line booking, both via WWW and the travel agency oriented systems (CRSs), that an electronic database containing information about which hotels at any given time (now or at a future date) have rooms *available* for booking must exist.

For hoteliers there are costs involved with hotel bookings received from the travel agency oriented systems (CRSs) for individual room nights, both in the form of an annual subscription fee and a fee per booking received. These costs can, if the hotels join these systems by a hotel representation firm or hotel chain, diminish the gross profit per room night significantly. Since guests at holiday hotels stay for more nights than guests at business hotels, it is interesting for the owners of the former to note whether the hotel representation firm (which sees that hotels get listed in all the CRSs) base the booking fee per *room night* or per *booking*. One hotel representation firm, Utell International which is the largest of its kind in the world representing 6,500 hotels world-wide and is mainly oriented towards the business segment, base their fees towards the hotels per *room-night*. Others may charge on a per *booking* basis (i.e. per stay). This is the case with Fidelio Nordic, which is based in Sweden, representing 1,200 hotels, mainly Scandinavian. Annual subscription fees and fees per booking to the travel agency oriented booking systems are of course avoided altogether, if the hotelier instead goes for WWW.

Holiday hotels may choose *only* to distribute themselves electronically by WWW, while business hotels should *both* distribute themselves through the travel agency oriented systems *and* WWW. Although *off*-line booking to an ordinary E-mail box following the Internet user's search for information on hotels on WWW is a *slow* form of booking WWW is an information medium and increasingly a sales channel which only few holiday and business hotels alike can afford to disregard.

5 Conclusion

Has the Internet really rescued accessibility (visibility and availability) for small, chain independent, holiday hotels - or are business hotels still at an advantage regarding electronic distribution? The short answer is that Internet/WWW has improved accessibility of both business and holiday hotels. When the only or main electronic distribution channel for hotels was through CRS', the small chain independent, holiday hotels were at a disadvantage compared with business hotels. Does this difference still exist 'after WWW'? Yes. The hotels which are members of a hotel chain or which are represented by a hotel representation firm such as Utell International are still at an advantage. Firstly, the members of *the hotel chains* listed in TravelWeb and a few other similar services offer the WWW users *on-line booking,* for example Microsoft Expedia, http://expedia.msn.com/, which is open to US citizens only. Hotel representation companies such as Utell International are also opening Web-sites (http://www.hotelbook.com/). In future it will give their members, exclusively or primarily business hotels, dual distribution through GDS' *and* WWW. Secondly, the level of connectivity which small holiday hotels will be able to offer the Internet user cannot match that of TravelWeb, since there would be no electronic database containing availability information with which to interact by WWW.

Travel/tour packages are generally distributed electronically through 'Local Distribution Systems' (LDS') such as 'Start' in Germany and 'Smart' in Scandinavia, and not through the GDS'. Although packages are currently being *advertised* on the Web, few organisations currently offer real-time on-line availability information and booking facilities for packages on the Web. 'Start' in Germany offers availability information on last minute package tours (http://www.start.de/) to a great number of destinations, listing the offers by day of departure and by price in increasing order. Details of each offer may be found.

Table 2A. Electronic Distribution Channels for Hotels - the privileged few.

	Distribution byTravel agents via CRS			Direct distribution via WWW		
	Information	Availability data	Real-time booking	Information	Availability data	Real-time booking
Individual products	Yes	Yes	Yes	Yes	Yes	Yes
Packages	Yes	Yes	Yes	Yes	Yes	Yes

Table 2B. Electronic Distribution Channels for Hotels - the vast majority.

	Distribution byTravel agents via CRS			Direct distribution via WWW		
	Information	Availability data	Real-time booking	Information	Availability data	Real-time booking
Individual products	Yes	No	No	Yes	No	No
Packages	Yes	No	No	Yes	No	No

Having checked availability the Internet user may proceed with filling in an order form. Travel documents are issued by travel agencies. Apart from this latter fact, the principle used by 'Start' for packages is similar to that of TravelWeb for individual products. The general idea is to use the availability information which is already stored in an electronic database, at make the best use of it including utilising the Web as an additional distribution channel. TravelWeb cuts out the travel agent, which the 'Start' system does not. The holiday hoteliers it is thus possible to distribute their capacity as part of packages through both LDS' and WWW. They 'just' have to enter an agreement with certain major tour operators, and become part of the packages offered by them. But again there will be a difference between those hotels being 'in' and those being 'out' of the programme of major tour operators. Small tour operators who represent small holiday hotels and who do not distribute themselves through LDS' can still show their packages on the Web. In general they will be unable to offer availability data and booking facilities on a real-time basis for the packages including hotels they offer.

In spite of the advent of the Web, both for hotels as individual products and for hotels as part of packages, there is a division between the privileged few which have dual, real-time based, electronic distribution channels, that is both by automated travel agents and directly to travellers by WWW, and the all the rest. The latter may make themselves present on the Web, that is show their products and offer slow off-line booking functions, but they cannot offer real-time availability information and booking facilities.

References

1. Archdale, Gilbert (1994), "Destination databases: issues and priorities" in A.V.Seaton et.al. (eds.), "Tourism - The State of the Art", pp. 246-253.

2. Beaver, Allan (1992), "Hotel CRS - An overview", Tourism Management, March 1992, pp. 15-21.

3. Beaver, Allan (1995), "Lack of CRS accessibility may be strangling small hoteliers, the lifeblood of European tourism", Tourism Economics, Vol. 1, No. 4, pp. 341-355.

4. Erbner, Arno (1994), T.I.S. GmbH, "T.I.S. Tourism Information System for the Tyrol," in Schertler et.al. (1994) pp. 35-42, i.e. proceedings of the ENTER 94 conference.

5. Go, Frank M. (1992), "The role of computerized reservation systems in the hospitality industry," Tourism Management, March, pp. 22-26.

6. Lindsay, Patricia (1992), "CRS supply and demand," Tourism Management, March, pp. 11-14.

7. Marcussen, Carl H. (1996), "Marketing European Tourism Products via Internet / WWW", Journal of Travel & Tourism Marketing, Vol. 6 (3/4).

8. Moon, R. David (1994), "Deployment of computer networks in leisure-packaged tour marketing in the United States", Journal of Vacation Marketing, Vol. 1, No. 1, pp. 43-60.

9. Murphy, Jamie; Forrest, Edward; Wotring, C. Edward (1996), "Restaurant Marketing on the Worldwide Web", Cornell Hotel and Restaurant Administration Quarterly, February, pp. 61-71.

10. Murphy, Jamie, Edward Forrest, C. Edward Wotring & Robert A. Brymer (1996), "Hotel Management and Marketing on the Internet - An Analysis of Sites and Features", Cornell Hotel and Restaurant Administration Quarterly, June, pp. 70-82.

11. Richer, Paul (1996), "Should travel companies be selling on-line?", Journal of Vacation Marketing, Vol. 2, No. 3, pp. 277-285.

12. Sheldon, Pauline J. (1993), "Destination Information Systems," Annals of Tourism Research, Vol. 20, pp. 633-649.

13. Sussmann, S. (1992), "Destination Management Systems: The Challenge of the 1990s", in Cooper, C.P & Lockwood, A., (eds.), "Progress in Tourism, Recreation and Hospitality Management," Vol. 4, Belhaven Press, London, pp. 209-215.

14. Sussmann, S. (1994), "The impact of new technological developments on destination management systems", in Cooper, C.P. & Lockwood, A., (eds), "Progress in Tourism, Recreation and Hospitality Management," Vol. 5, Belhaven Press, London, pp. 289-296.

15. Walle, A.H. (1996), "Tourism and the Internet: Opportunities for direct marketing", Journal of Travel Research, Vol. 35, No. 1, Summer, pp. 72-77.

Marketing information and decision support on the Internet - New opportunities for national, regional and city tourist offices

Karl W. Wöber (karl.woeber@wu-wien.ac.at)
Institute of Tourism and Leisure Studies
Vienna University of Economics and Business Administration, Austria

Abstract

Since reliable information is the key issue in marketing many tourist offices have used the enormous advances in computing to create systems that more or less extensively record tourism data. Nevertheless, some organizations have encountered difficulties in efficiently using this data for advertising purposes or to plan their marketing strategies in conjunction with various decision support techniques. For this reason alone, it has become essential to create a system that gives users efficient access to all opportunities that computer technology can offer.

In this article the author describes recent experiences made during the development and implementation of a tourism marketing-information system (TourMIS), which has proved to be useful in research and consultation for more than 10 years. Its nationwide acceptance turned the system from a local to a distributed decision support system, which caused new problems in data maintenance and the model's interface.

1 Introduction

The marketing function of tourism organizations includes advertising a country, region or city nationally and internationally as a tourist destination. Further marketing activities undertaken by tourist offices include the provision of appropriate marketing research studies and the maintenance of information centers. Irrespective of the size of tourism organization, management needs data on which to base decisions, plan, organize and control the various activities. Meijs and Dijks investigated the influence of different information system architectures on benchmarks for local tourist offices [8]. The figures of their survey make clear that tourist offices should pay more attention to information facilities and services.

Since 1984, the Austrian National Tourist Office has been using a decision support instrument called TourMIS. TourMIS was fundamentally updated at the beginning of 1990, and an extensive database was available to tourism managers not only on national but also on regional level [10] [12]. This computer-aided decision support system was set up by the Austrian Society for Applied Research in Tourism (ASART) together with the Institute for Tourism and Leisure Studies at Vienna's University of Economics and Business Administration. The project was realized in close collaboration with the Market Research Department of the National Tourist Office.

Table 1: Market research sources in TourMIS

data source	type of information and major crosstabulations	surveys and update interval	data format
ÖSTAT Austrian statistical office (Vienna)	overnights, arrivals, capacities 25 European countries of origin, 18 different capacity classes, Austria and its 9 federal states	1960-, every month	time series
OECD and WTO statistics (Paris/Madrid)	overnights, arrivals 21 countries of origin, 9 destinations (Austria and its main competitors)	1970-, annual	time series
ASART Austrian Guest Survey (Vienna)	proportion of repeat visitors, travel purpose, age, transportation, organization, length of stay, travel motives, expenditures, sort of accommodation, activities, income, profession, education, Eurostyles-type 16 countries of origin, Austria and its 9 states as destination	1988-, all 3 years	original
ETM European travel monitor (Munich)	trips, age, travel purpose of guests in Austria 25 European countries of origin	1989-, annual	time series
ETM European travel monitor (Munich)	e.g. trips, travel purpose, organization, transportation, overnights, expenditures 25 European countries of origin/destination	1989-, annual	time series
FECTO European city tourism survey (Paris)	e.g. overnights, arrivals, capacities, expenditures, travel motives, attractions, ... 61 European cities, 21 countries of origin	1983-, annual	time series
Stft German Travel Survey (Starnberg)	age, transportation, travel purpose, sort of accommodation, organization, length of stay, activities, travel motives, expenditures, ..., Germany (east and west) as country of origin, 10 destinations	1975-1990, annual	original
OeNB Austrian National Bank (Vienna)	population, economic growth, rate of unemployment, change in consumer price index, foreign exchange rate, intern. tourist receipts/expend., imports/exports, ... 21 countries of origin	1980-, annual	time series
ASART Austrian hotel and restaurant panel study (Vienna)	60 variables like: turnover, cash flow, return on investment, ... location, size, category, form of a restaurant or hotel	1982-, annual	original

TourMIS stores data of international travel statistics, the results of travel and guest surveys and selected economic indicators for the most important countries of origin. It gives tourism managers access to a large number of information sources that were previously unknown or inaccessible to them via a computer system. According to the survey techniques TourMIS distinguishes between two different storage formats. Data from regular statistical publications providing tourism volume figures (e.g. bednight and capacity statistics) are stored in time series format, whereas data from guest or travel surveys are stored in the original data format allowing full flexibility to the users' queries. The program provides access to more than 10,000 time series and results of 25 European guest and travel surveys representing more than 40,000 interviews (Table 1).

The data maintenance is carried out by the Austrian Society of Applied Tourism Research. This association is represented by the most important tourism organizations and is located at the Institute of Tourism and Leisure Studies at the Vienna University of Economics. There, distorted and incomplete data from sources with different surveying methodologies are carefully rectified and forecasted. In general, all data are updated annually, however, more crucial time series are recorded monthly. The maintenance of data from visitor or travel surveys is carried out as soon as the results have been published.

2 Experiences with the off-line maintained PC-version of TourMIS

Originally TourMIS was designed as a single user system for one company. It consisted of a retrieval system for storage and access to key marketing variables by a user friendly interface. Over the years more sophisticated strategic planning tools have been added. Several procedures for numerical operations and more complex marketing models, like a portfolio analyzer, a program to solve multi-attribute-problems and a budget optimization program where attached to the basic retrieval system [5] [6] [7] [11]. These extensions to the models and methods interface caused an exponential increase in the database.

In 1991 the Austrian National Tourist Office, as a non-profit organization with the target to support Austrian's tourism economy, decided to open the system to all branch offices. After a two-year trial period, TourMIS was installed at seven branch offices in Austria. At the end of 1994 about 50 tourism managers used TourMIS to learn about the tourism environment and to take action with respect to it. The maintenance of the database was organized off-line by sending a set of diskettes every month to all participating offices. Similar to the years before, the increase in managers involved caused new requirements to the database and the analytical part of the decision support system.

The data monitored by the regional and city tourist offices are completely different from the NTO's focus. For instance, the evaluation of the main competitors and the resulting market volume and market share calculations differ tremendously. As further enhancements to the database system became necessary, the maintenance of the individual databases got very difficult and cost intensive. Furthermore, the regional organizations with limited resources to purchase powerful computer hard- and software, were not able to fully utilize all decision support tools offered by TourMIS.

Figure 1: TourMIS system design

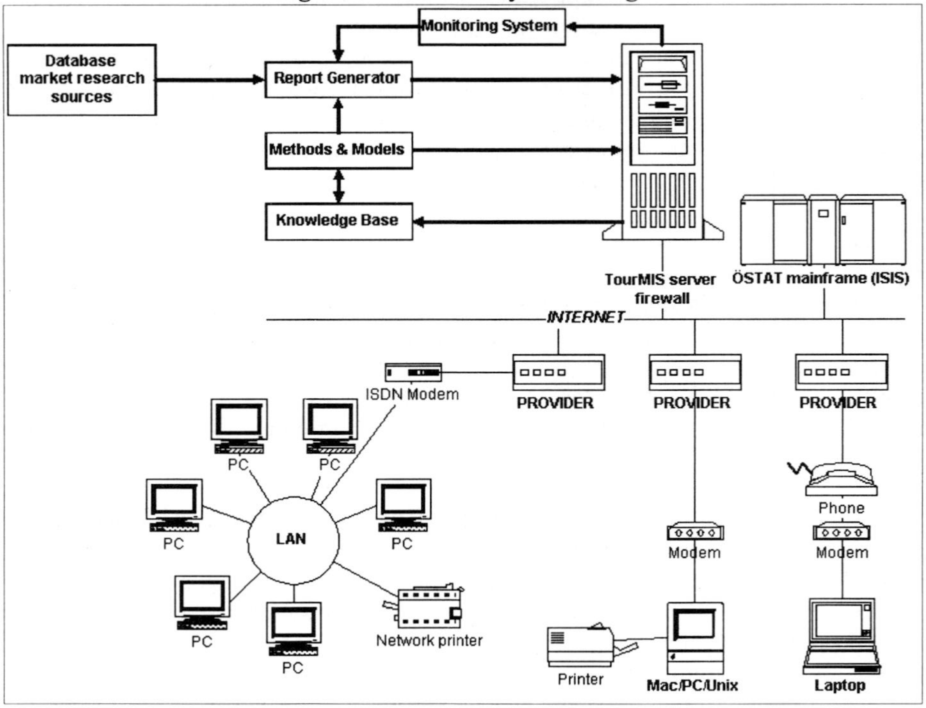

3 Advantages of the Internet for a Marketing Information System

Determinants of success for marketing decision support systems were evaluated by Oude Ophuis et al. [9]. The first and in many ways the most critical elements is the user interface. The manager and his stuff must communicate with the system. The design of the software must allow rapid retrieval of data and comfortable production of print-outs of tables and graphs, even if the operator is inexperienced. According to these recommendations and considering the experiences made with the PC-version of TourMIS the Austrian National Tourist Office decided to transfer TourMIS to an on-line distributed marketing-information system in 1993. As the organization's branch offices are located all over the world, different forms of wide area network facilities were evaluated.

The employment of the World Wide Web's (WWW's) hypermedia functionality for supporting tourism information transfer was originally suggested by Clyde and Landfried [2]. Electronic delivery of hypermedia market research information via international computer networks, such as the Internet, is an interesting alternative to the traditional distribution platforms. Finally the Austrian National Tourist Office decided in favor of Internet, as this distribution channel offers easy access to large quantities of information by powerful browsers and file transfer protocols (Figure 1). This decision was also strongly supported by a survey among 51 European city tourism offices in which the author had evaluated the accessibility to the Internet. In 1995, from all interviewed tourism manager 53% already had Internet access, whereas 40% were planning to install it in the near future [14].

Additional arguments supporting this strategy refer to the users' utilization behavior of the existing stationary PC-system. Hence, experience showed that tourism managers were primarily using the system's retrieval functions and their perceptions of information quality were strongly related to actuality and availability of the required data.

Figure 2: Example of data representation

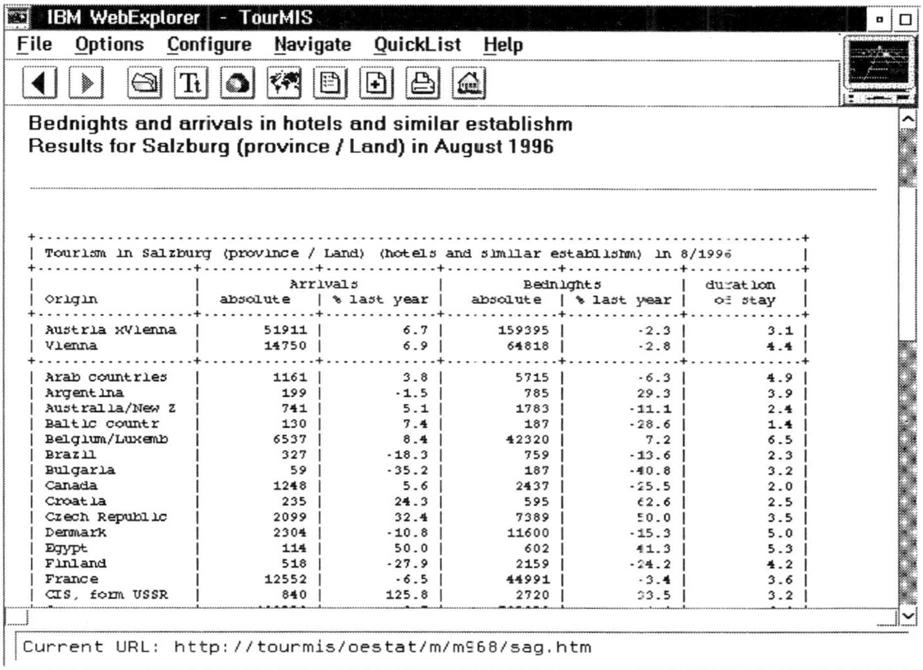

Tourism in Salzburg (province / Land) (hotels and similar establishm) in 8/1996

Origin	Arrivals absolute	Arrivals % last year	Bednights absolute	Bednights % last year	duration of stay
Austria xVienna	51911	6.7	159395	-2.3	3.1
Vienna	14750	6.9	64818	-2.8	4.4
Arab countries	1161	3.8	5715	-6.3	4.9
Argentina	199	-1.5	785	29.3	3.9
Australia/New Z	741	5.1	1783	-11.1	2.4
Baltic countr	130	7.4	187	-28.6	1.4
Belgium/Luxemb	6537	8.4	42320	7.2	6.5
Brazil	327	-18.3	759	-13.6	2.3
Bulgaria	59	-35.2	187	-40.8	3.2
Canada	1248	5.6	2437	-25.5	2.0
Croatia	235	24.3	595	62.6	2.5
Czech Republic	2099	32.4	7389	50.0	3.5
Denmark	2304	-10.8	11600	-15.3	5.0
Egypt	114	50.0	602	41.3	5.3
Finland	518	-27.9	2159	-24.2	4.2
France	12552	-6.5	44991	-3.4	3.6
CIS, form USSR	840	125.8	2720	33.5	3.2

The hard- and software conditions in the various branch offices were completely different in terms of central processing unit, peripheral equipment and operating system, which made the development of a widely used front-end computer program very difficult. One of the best parts of the WWW, and arguably the reason it has become so popular over other Internet services, is its capability to display both text and graphics in full color on the same page. If someone can access the Internet he can access the WWW. It doesn't matter what operating system he is running and the WWW is not limited to any kind of hardware.

4 Structure of market research data and its possible hypermedia presentation

Crockett describes information selection and information structure as the most crucial factors for the success of an executive information system [3]. Also Königer and Janowitz see two reasons for the growing inability to cope with information today [4]. First the quality, speed and efficiency of work suffer because relevant information is not accessible where and when it is needed, and on the other hand, people find themselves incapable of taking in all the information that is targeted toward them. The moderate success of many computer-based marketing information systems is due to the fact that their developers have paid too much attention to the technical details of information processing and not enough attention to the definition of the information that should flow into the system. Therefore, the development of a decision support system for strategic marketing planning needs a clear and precise structuring of the available marketing research information. In general survey data can be classified by:

1. definition of the market (depending on market segmentation approach)
2. the strategic business unit (e.g. certain product)
3. sort of information to be monitored (e.g. turnover)
4. period of time (e.g. last month)

If there are multiple data sources for one and the same sort of information (e.g. average length of stay can be evaluated by sample surveys or calculated by number of bednights divided by arrivals), an additional fifth dimension *data source* has to be added. For the development of a tourism marketing information system the adaptation of the terms *market-definition* by *guest-segmentation* and *strategic business unit* by *destination* have proved to be advantageous. For instance, a typical data query in TourMIS is illustrated in Table 2.

Table 2: Structure of market research data

general structure of market research data	adaptation to tourism information systems	TourMIS example
market-definition	guest segmentation	all Italian guests
strategic business unit	destination	Austria
sort of information	sort of information	number of bednights
time	time	1995
source	source	OECD report

Given millions of alternative combination possibilities, as mentioned in the above sections, it is necessary to build an easy to find structure for data retrieval and maintenance. Probably the easiest and most logical way to structure market research information on the WWW is in a hierarchical or menu fashion, illustrated in Figure 2. In hierarchies, the home page provides the most general overview to the content below and it defines the main links for the pages further down in the hierarchy. The user starts with a list or menu of major topics (e.g. data sources); selecting one leads him to a list of subparagraphs (e.g. destinations); which then leads him to a list of further subparagraphs (e.g. sorts of information), and so on. Finally he will find himself on the last level of this information pyramid where he will find the actual figures and tables (Figure 2; an example of an input screen shows Figure 3).

In a hierarchical organization, it is easy for the users to know their position in the structure; choices are to move up for more general information, or down for more specific information. Providing a link back to the top level enables the user to get back to some known position quickly and easily. A popular form of document organization on the WWW is a combination of a hierarchical structure with a linear one. The combination of linear and hierarchical documents works well as long as there are appropriate clues regarding context. Because the user can either move up and down or forward and back, it's easy to lose one's mental positioning in the hierarchy when one crosses hierarchical boundaries by moving forward or back.

Given the above system description it seems to be obvious, that the generation of the hypermedia pages cannot be produced manually. Indeed, none of the present TourMIS Internet pages are written with an hypertext markup language- or other text editor. All WWW pages are generated by special interface programs (WWW *report generator module* in Figure 1). Although this causes incredible savings in reporting costs, it captures a considerable amount of the server's storage capacity. As it is unlikely that all possible query combinations will be asked by the tourism managers, only the most frequently requested evaluations are presented in the system. The less usual data inquiries are processed on e-mail request.

5 Conclusions and further research

A disadvantage of information distribution via WWW compared with traditional database systems is the constraint of the predetermined hierarchical structure. The researcher who sets up the information system has to decide which dimension of the data structure will appear on which level of the hierarchical system. All information on the top level will be easy to find and should therefore lead to the most common information queries. The tourism manager's perceptions of service quality will strongly depend on the clear hierarchical structure, and on the fact, whether the existing hyperlinks meet his expectations.

As market research information is changing fast this optimization problem can be best treated with an adaptive system approach (*monitoring system* in Figure 1). An important issue therefore is to gather and subsequently analyze information on the user behavior under different technical, organizational and economic conditions. The marketing information system TourMIS has built in statistical user tracking applications to do detailed behavior studies. A database stores historical case-based information about the users, like date and time of the users' access, their path through the marketing information system, the employed type of WWW browser and the standard user identification number (IP-address). Regular reports about the most requested information sources will also offer insights to the managerial needs in tourism marketing research, and therefore could be used as guidelines for research investments in the future.

Figure 3: Example for data input possibilities

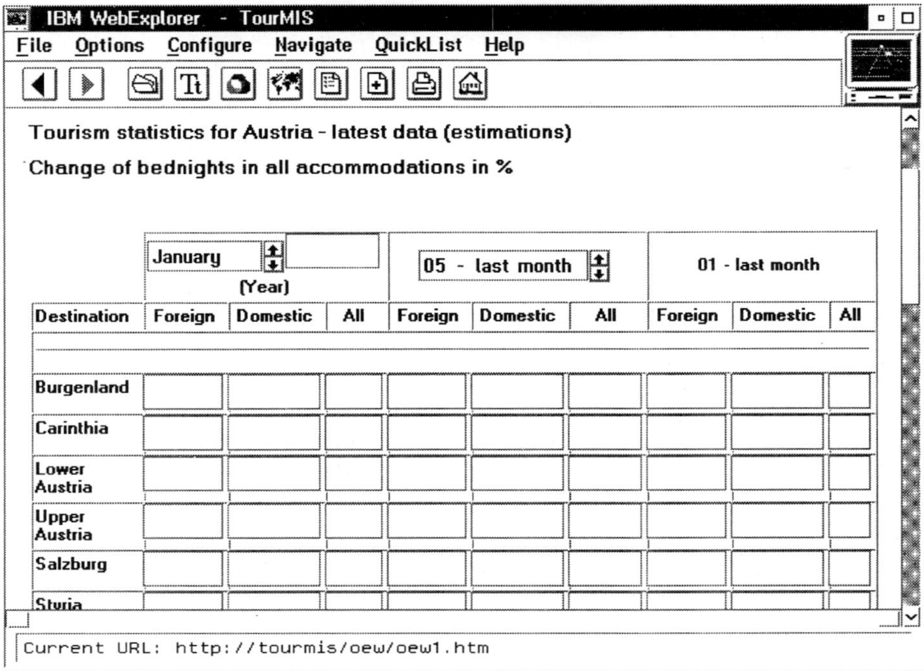

A marketing information system can monitor and report data, but the system can also perform calculations for the purpose of comparing different resource allocation alternatives. At the moment these models are developed according to the researchers' perceptions of the managerial needs. These models can be downloaded and applied always with the most up-to-date data offered by the marketing information system. Future extensions in browser technology will offer interactive capabilities (e.g. Java) which will make on-line market simulations possible, user handling easier and eventually managerial use more frequent.

Notes

The study was carried out with financial support from the Austrian National Tourist Office, the regional tourist offices in Tyrol, Salzburg, Vienna, Burgenland, Styria, Upper and Lower Austria, the Federation of European Tourist Offices', the Ministry of Economic Affairs and the Austrian hotel and restaurant communities. Preliminary results of the study have been reported in [10] [11] [12] [13].

References

[1] Calantone, R.J., Mazanec, J.A. (1991), Marketing Management and Tourism, *Annals of Tourism Research*, 18 (1), pp. 101-119.

[2] Clyde, S., Landfried, T., OuR-TOWN: A Distributed, Multi-media System for Travel and Tourism, in: Schertler et al. (eds.): Information and Communication Technologies in Tourism, Springer, Wien-New York 1995.

[3] Crockett, F. (1992), Revitalizing executive information systems, *Sloan Management Review*, Summer, pp. 29-47.

[4] Königer, P., Janowitz, K. (1995), Drowning in Information, but Thirsty for Knowledge, *International Journal of Information Management*, Vol. 15 (1), pp. 5-16.

[5] Mazanec, J.A. (1986), Allocating an advertising budget to international travel markets, Annals of Tourism Research, 13 (4), pp. 609-634.

[6] Mazanec, J.A. (1986), A decision support system for optimizing advertising policy of a national tourist office. Model outline and case study, *International Journal of Research in Marketing*, Vol. 3, pp. 63-77.

[7] Mazanec, J., How to evaluate a travel market, econometric modeling vs. multiattribute decision making with management estimates, Aix-en-Provence: Les Cahiers du Tourisme (Centre des Hautes Etudes Touristiques) 1986.

[8] Meijs, C., Dijks, G., Information system architectures and benchmarks for local tourist offices, in: Schertler et al. (eds.): Information and Communication Technologies in Tourism, Springer, Wien-New York 1995.

[9] Oude Ophuis, P.A.M., Wierenga, B., Huizingh, E.K.R. and de Jong, C M., Determinants of Success for Marketing Decision Support Systems, in: Bergadaà, M. (ed.): Proceedings of the 24th EMAC conference, Vol. II, Essec, Paris 1995.

[10] Wöber, K.W. (1994), Tourism Marketing Information System, *Annals of Tourism Research*, Vol. 21 (2), pp. 396-398.

[11] Wöber, K.W., Expertenschätzungen in touristischen Entscheidungsunterstützungs-systemen, Service Fachverlag, Wien 1994.

[12] Wöber, K.W., Strategic Planning Tools Inside the Marketing-Information-System in Use by the Austrian National Tourist Office, in: Schertler et al. (eds.): Information and Communication Technologies in Tourism, Springer, Wien-New York 1994.

[13] Wöber, K.W., Entwicklung von Wettbewerbsstrategien mittels verschiedener Portfolioansätze im Verbund, in: Zins, A. (Ed.): Strategisches Management im Tourismus, Springer Wien-New York 1993.

[14] Wöber, K.W., Services and Functions of European City Tourist Offices, Report of the Research and Statistics Working Group of the Federation of European Cities' Tourist Offices (FECTO), Administration Board Meeting, Dublin, March 22nd 1996.

How to assess WWW-applications for tourism information systems from the end-user perspective. Methodical design and empirical evidence

by

Christoph Passmann

Wolfgang Pipperger

Prof. Dr. Walter Schertler

Department of Organisation and Strategic Management,

University of Trier, D-54286 Trier, Germany

Abstract: In the spring of 1996, an empirical research project was conducted at the University of Trier which aimed at evaluating an Internet-based electronic information and booking system for the tourism industry. During the development of the new „TIScover" system by the TIS Corp. in Innsbruck, Austria, an early prototype was tested to assess its user-friendliness and to learn more about customer behaviour in online-booking systems. The empirical data was used to determine strengths and weaknesses of the „TIScover"-prototype and to further improve its enduser performance.

In this paper the authors propose an empirical design to test WWW-based travel information systems and present empirical results of how to design enduser friendly Web-pages for tourism information and booking systems.

1. Introduction

The evolving technologies in telecommunications and particularly the World Wide Web (WWW) have created the opportunity for both touristic sites and intermediaries to present themselves in a completely different and customer-oriented way. However,

making good use of these newly emerging technologies is a demanding task (Schertler, pp.17 1994).
The WWW enables touristic content providers (CPs) to present information in an attractive way using multimedia applications. Traditional catalogues are bound to text and pictures. WWW-presentations, however, are able to combine text, pictures, sounds and videos to generate a more comprehensive multimedia focused impression.

At the same time, WWW-based tourism information systems enable CPs to update their contents at every time to meet customer demands for quick and accurate information. Whereas traditional brochures lose accuracy over time, WWW-presentations solely depend on update-frequencies to remain accurate and useful. The skyrocketing „last-minute" business shows that today's travellers ask for quick information and fast booking capabilities. Tourism information systems thereby represent a strategic response to changing customer demands.

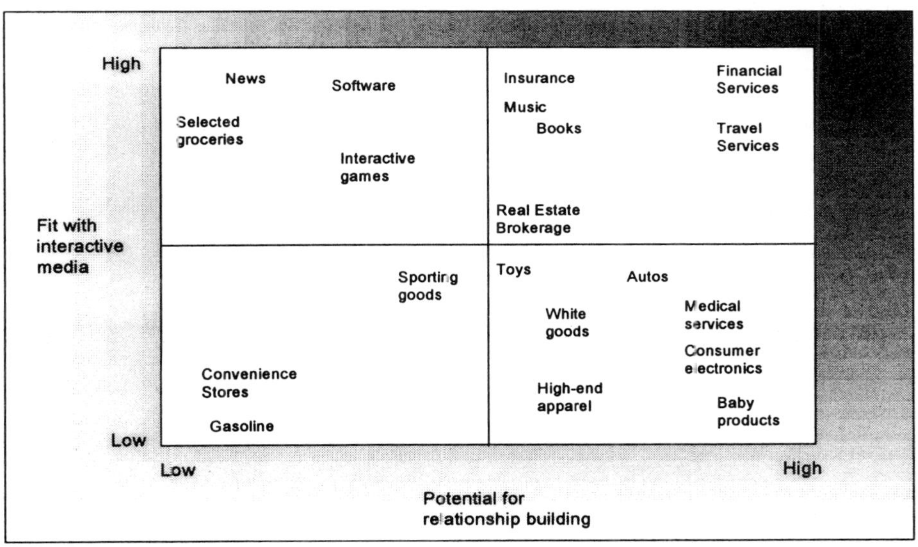

Figure 1: Categories suitable for digital marketing (The McKinsey Quarterly, p 12.)

As figure 1 shows, tourism information systems are likely to become a powerful marketing tool in the future. Hitherto, online-transactions were limited to one-way information retrieval. However, the broad opportunities of the Internet pave the way for interactive transactions such as online-purchasing of touristic services. However, user acceptance will be **the** critical factor for content providers using new technologies such as the WWW. Based on the studies conducted at the University of Trier, one of the most influential obstacles towards acceptance of tourism information systems is the lack of personal assistance and advise. Users tend to feel uncomfortable with an impersonal assistant. Partly this impression will vanish as tourism information systems become more common. But the critical task will still be to create user-friendly and self-explaining systems that provide personalised assistance to the greatest possible degree.

Empirical evidence on user-behaviour within tourism information systems is needed to enhance user-friendliness and - eventually - acceptance. The study at the University of Trier provides some valuable insight and aims to be a part in this process.

2. Objectives

The empirical tests of the „TIScover“ prototype were designed to serve the following purposes:

- To evaluate the prototype in terms of its user-friendliness,
- to generate the empirical data needed to further improvement of the system,
- to give some general insight into end-user behaviour in online-booking systems.
- to compare the old vs. the new tourism information system („TIS@WEB“ vs. „TIScover“)

The underlying questions on which the study was based can be summarized as follows:

- Do customers generally accept electronic booking systems?
- Do customers understand and value the advantages of these systems?
- Is it possible to construct complex interactive procedures (e.g. booking procedures for touristic services) in such a way that even those customers without profound experience with PC and Internet applications are able to understand them intuitively?

- How can simple navigation in the system be ensured?
- What are common mistakes using the „TIScover“ prototype?
- Which features of the system are important for end-users?
- What is the trade-off between useful features of the system and its corresponding complexity?

We will not be able to elaborate on each of these questions in further depth since this would not serve the purpose of this paper. However, we feel that these questions generally constitute the framework for evaluating touristic WWW-applications.

3. Design of the study

In the course of the study a group of forty people (subjects) were probed on their perception of the „TIScover“ prototype. Unfortunately, it was not possible to conduct the study with a larger sample due to the complexity of the tests. In an earlier test, the simulation of booking procedures turned out to be extremely time-consuming, particularly since the subjects did not know the prototype beforehand. Furthermore, the tests required the presence of at least one interviewer to maintain the system and to record completion times and typical mistakes. The sample was evenly grouped into male and female as well as young professionals and students. This grouping was set by the TIS Corp. in order to evaluate the ability of less-experienced subjects to work with the prototype rather than to reflect reality (by far the largest part of Internet users are male students which are the most experienced as well). The average age of the subjects was 29.3 years. The sample was therefore composed of young and well-educated subjects representing potential online-bookers.

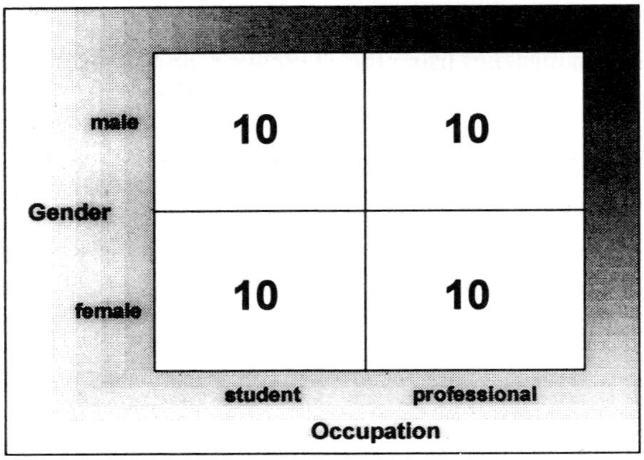

Figure 2: Sample Grouping

The test environment was defined to approximate typical home-based Internet transactions as closely as possible. The interviewers did not interfere with the actions of the subjects. Their only task was to ensure the proper operation of the system (some functions of the prototype proved to be unstable) and to keep record of the actions and mistakes of the subjects. The subjects worked on 2 ordinary Windows95-PCs with modem and Netscape 2.01.

System Specifications:

Computers	2 Windows95 PCs
Processors	Intel Pentium 100
Monitors	Siemens-Nixdorf 15'' / Sony 15''
Modems	28.800 bps
Internet Browser	Netscape Navigator 2.01

The prototype was installed on a WindowsNT server which was connected to the intranet of the University of Trier.

The subjects were selected according to the specifications set by the sample grouping (male, female, student, professional). All subjects were required to have at least basic knowledge of PCs and the Internet. They were informed about the test by either e-mail or by word-of-mouth at the University of Trier.

The structure of the empirical study at the University of Trier can be summarised as follows:

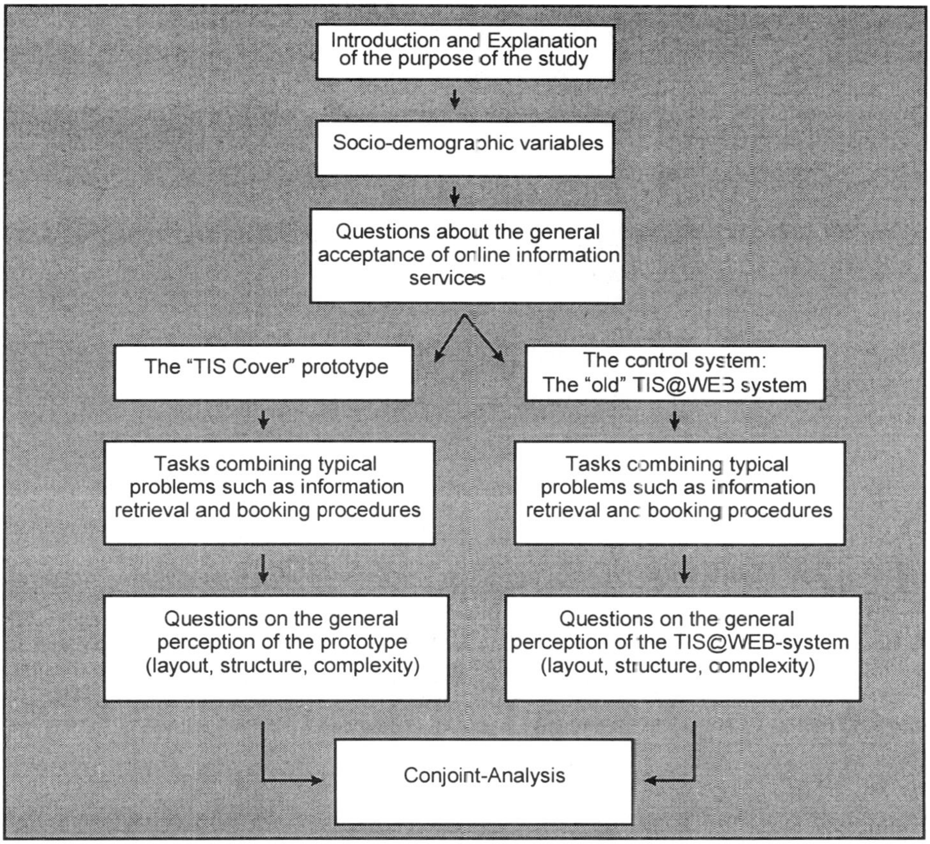

Figure 3: The Trier Model

On average, the subjects needed about 2 hours to complete the test. The questions and tasks in the questionnaire referred to both the „TIScover" prototype and the „old" TIS@WEB information system to ensure comparability. Corresponding questions were identical to the maximum extent possible. To avoid distorting effects half of the subjects had to first work with the old TIS@WEB system and then with the new „TIScover" system while the other half proceeded vice versa. All subjects were given 20-30 minutes to get acquainted with both systems and to test their features before starting with the questions and tasks. The questionnaire was divided into three separate parts. In the first part the subjects were asked about their general perception of newly emerging technologies such as the Internet or Online-Media.

In the second part of the study they had the opportunity to examine both systems in depth. This part of the test combined questions about the general perception of both systems such as layout and structure and questions which confronted the subjects with typical tasks such as retrieving information about a hotel or area and booking accommodation. The subjects were given specific criteria to use the search-functions of both systems and to get appropriate results. Furthermore, they were also given information about their personal situation and preferences („Suppose, you have a family with two children and you would like to go skiing. Your children do not know skiing yet and you would like to send them to basic ski-training. You need accommodation in a quiet mountain village with care centres for your children and guided mountain tours for you and your wife...“). There were five tasks including all relevant booking problems such as retrieving information and finding appropriate villages, hotels, and package journeys. While the subjects were working on the tasks completion times as well as typical mistakes were recorded by the interviewers.

In the third part a conjoint analysis was conducted to probe the subjects' preferences on different ways to gather information and to book. From the viewpoint of the TIS Corp. it was most interesting to learn about the utility of electronic information and booking systems. The process of purchasing touristic services can be split up into three basic factors: Information and assistance, the quality of information, and the process of booking. This constitutes a 3x3x3 matrix with 27 different purchase alternatives. With a statistical method these 27 alternatives can be reduced to 9 basic designs. All subjects had to rank these designs. By statistical means, the individual contribution to utility can be calculated for each component.

The three factors and their alternatives:

1) Information and assistance

- personal information and assistance in travel agencies,
- information retrieval through electronic media, then personal assistance in travel agencies,
- information retrieval and assistance through electronic media.

2) The quality of information (speed, accuracy, opening hours)

- hourly updated online-information, booking confirmation at once, 24 hours access,
- printed catalogues, booking confirmation within a week, normal opening hours,
- both personal advice through travel agency, printed catalogues, booking confirmation within one day, normal opening hours.

3) Booking alternatives

- online-booking with electronic media,
- booking in travel agencies,
- booking by telephone

As mentioned above, these nine alternatives can be combined to nine specific designs which were ranked by the subjects. By statistical analysis, we were able to calculate the subjects preferences with respect to booking procedures. This data could be used to generate further insight into strengths and weaknesses of electronic information and booking systems. Furthermore, it could be used to separate those features of added-value from those without utility to customers. Since not all features of online information and booking systems are of equal utility to customers, conjoint analysis is an appropriate way to show valuable features to concentrate on. **We learned that there is an important trade-off between the features of an electronic system (features increase the system's capabilities) and its corresponding complexity (which decrease ease of handling).** Conjoint analyses are useful means to maximise the system's overall utility by providing the information about relevant factors.

See the results of the conjoint analysis in the next part.

4. Results

The study produced important evidence about strengths and weaknesses of the „TIScover" prototype. Most of the evidence referred specifically to the prototype and will not be presented in this paper. There are, however, some important general implications which determine the quality of such systems considerably. We developed some general recommendations which will be explained below:

- Homepage design and structure:
  The system homepage is the central starting point for any user. During the study we found out that most problems occurred on the starting pages of the system. Once the subjects had found the right path to complete the task, only minor problems arose. Homepages **must** be designed and structured in such a way that their logic can easily be understood.

Figure 4: The „TIScover" Homepage (prototype)

Figure 4 shows the „TIScover" Homepage that we tested. Most subjects had severe problems to understand the meaning of the four central buttons („news", „scout", „navigator", and „gazette"). Each of the four buttons lead to an independent set of options to retrieve information, search or book. The graphical design of the buttons could not be intuitively interpreted by the subjects. Since no further information on their meaning was given, the subjects resorted to „try and error"-systems to complete the tasks. The empirical data of the study showed that

graphical attractiveness does not necessarily help user-friendliness. As mentioned above, the „TIScover" prototype was tested in comparison with the „old" TIS@WEB system. Though TIS@WEB was graphically much less attractive, 85% of the subjects valued the structure of its homepage „very good" or „good" compared with only 20% for the „TIScover" prototype.

- Standardised procedures:
  Selection and booking procedures should be standardised as far as possible to take advantage of learning effects. The „TIScover" prototype combined completely different procedures. The subjects had to click hyperlinks, press buttons or check boxes to activate similar functions of the prototype. **Most subjects had severe problems with ever-changing procedures**.

- Language:
  Especially buttons should not be labelled in foreign languages. The „TIScover" prototype used a lot of English expressions which were not understood by the subjects (even Austrian-German words caused problems). **The use of foreign language expressions does not contribute to the user's comfort**.

- Personal address:
  According to earlier studies at the University of Trier, most customers feel uncomfortable with impersonal electronic assistants. They prefer personal expertise in travel agencies. Addressing users personally can diminish the feeling of mistrust and make the system's usage more comfortable. The electronic system becomes less technical and more personal. **Instead of impersonal directions like „date of arrival" users should be addressed personally: „When do you want to arrive?"**.

- Frames:
  Frames are useful means to enhance user-friendliness. However, they should not be used extensively and their contents should be structured logically. **Users have to be able to understand the meaning of frames**.

- No „dead-ends“:
  There should be no pages with „dead-ends“, that is, **users should never be forced to use the „back“-button in their Internet browser**. The „TIScover“ prototype had numerous pages with no further links which proved to cause severe problems.

During the test, a conjoint analysis was conducted to probe the subjects' booking preferences. According to our analysis of this sample customers prefer to

- **be informed** online by PC,
- **get assistance** and personal expertise in travel agencies with normal opening hours,
- **book** in travel agencies.

On average, this design can best satisfy customer needs. This means that online information and booking systems' core value is its information quality. **Information quality assurance is therefore paramount!** To change these preferences, online systems will have to concentrate on personalised assistance and emphasise security. Mistrust seems to be the major factor which hinders online booking.

5. Methodical Criticism

The design of the study entailed some important problems which can be organised into three categories:
a) Limited approximation to reality
b) Limited operation of the prototype
c) Limited scope of testing

Ad a): Limited approximation to reality
It was not possible to create a perfectly realistic test environment. The subjects knew that they were **not connected to the Internet** and not actually booking. In reality, the need for booking-security would be extremely high, since customers do not want to book anything they are not completely sure about. In online booking systems mistakes cannot be attributed to someone else because customers are responsible for their

actions. During the test, this lack of uncertainty may have lead to quick actions without the fear of mistakes. Probably some subjects who pressed the final button „Book Now“ in the prototype would never have done so under realistic circumstances.

Ad b): Limited operation of the prototype
Since the prototype was not connected to the TIS database, it could not come up with realistic query-results. Instead, standardised results were programmed into the prototype. The questions and tasks were organised to match the programmed query-results. However, these results were not always logical (a complete match would have been impossible given the immense number of different functions). Thus, depending on the way the subjects navigated through the system the prototype did not always function properly. Whenever this happened, the interviewers explained the problem and asked the subjects to continue without regard to „strange“ results. Interviewer interference was also necessary because the prototype sometimes terminated due to programming errors. These problems might have had a distorting impact on the subjects' behaviour in the system.

Ad c): Limited scope of testing
The study could only test individual behaviour within the system. **All subjects started on the prototype's homepage**. There are, however, various external factors which could not be tested. In reality, customers must have proper information about the Internet address of the system or at least about ways to get that address. That is, successful working with a touristic WWW-system not only requires knowledge about the system itself but also about the complete WWW network and its features.

6. Conclusion

The empirical study at the University of Trier aimed at developing a design to broadly assess touristic WWW applications. Since user-acceptance will be the crucial factor within the industry, further empirical studies will be needed to enhance the abilities of online-systems and increase their orientation to customer needs. The design which we presented here will be further developed in future studies.

7. References

The McKinsey Quarterly, 3/1996.

Schertler, Walter (ed.), Tourismus als Informationsgeschäft - Strategische Bedeutung neuer Informations- und Kommunikationstechnologien im Tourismus, Wien 1994.

Schertler, W. / Schmid, B. / Tjoa, A.M. / Werthner, H. (eds.), Information and Communication Technologies in Tourism, Wien 1995.

A Qualitative Analysis of Hotel List Providers on the World Wide Web

Claire L Dorren , Dr Andrew J Frew
Napier University, Edinburgh
c.dorren@napier.ac.uk; a.frew@napier.ac.uk

Abstract This paper provides an analysis of initial investigations into Hotel List Providers (HLP) who distribute accommodation (and destination) information using the WWW as the enabling technology. A representative sample of internationally-focused sites as well as those which specialise in the United Kingdom as a region was investigated and a methodology is presented for the ongoing analysis of such Web-based sites. A current, generic profile of WWW users is established and this data is used to provide a platform on which to build profiling of Web based tourism consumers. Using empirically derived assessment criteria, this sample of Hotel List Providers was examined and the work placed in the context of larger scale studies by the authors into both international reservations systems and the role of intelligent agents concluding with suggestions for progressing the work in the area of site assessment.

1. Introduction

New technology has been frequently cited as one of the most profound influences impacting upon the UK tourism industry and among these technologies, the emergence of the World Wide Web (WWW) has been remarkable, particularly in its potential for opening up new distribution channels. The WWW is increasingly being used by *mainstream* consumers and it is therefore pertinent to assess how these new and evolving systems may be best aligned to meet not only the needs of tourism businesses themselves but also those perceived by the end consumer. Key participants in this scenario are the Hotel List Provider (vide infra) who are utilising these new communications media to facilitate on-line distribution.

These developments are significant for tourism service providers who sell a *perishable, intangible, heterogeneous and high risk* product. Multimedia features incorporated within the WWW reduce the risk factor involved in this purchase decision and enable greater dissemination of the vast quantity of information often associated with travel packages, at a single contact point. The result of a recent survey (Majewski 1996) indicates that the majority of WWW users access tourism based sites simply to gather and retrieve information. The WWW, therefore, has the potential to assume a large proportion of future on-line commercial activity of tourism services on the Internet.

The unique selling proposition of the WWW for tourism services lies in its capacity to market directly to the end consumer, eliminating many of the layers previously associated with electronic reservation processing, to this end the new technology and opportunities facilitated by the WWW may be regarded as a threat by established network service providers. Within this arena the WWW will affect the reengineering of electronic advertising and distribution. The unique characteristics of the WWW addresses the prerequisites of any distribution channel, *viz*

"Creating access for the *consumer* in one or more places, *convenient* for him, but *remote* from the place and time of delivery/consumption of the services being bought." (Middleton 1988)

Consumers are the key to the potential of this approach especially with the continuing trend towards independent customised travel packages (Cockerall 1991). The Internet and in particular the WWW is a *convenient* source where tourism transactions may be handled from beginning to end from a single point of contact. There is usually 24hr all-year access with independent users having no subscription or membership costs (no switching costs). As for *remote* accessibility, the WWW facilitates browsing in an on-line media-rich environment where the consumer may interact with local, regional, national or global databases pertaining to tourism services without actually having to visit. The benefits, therefore, of this communications standard to the small and/or unaffiliated tourism service provider is twofold; firstly it reduces the entry barriers to global marketing and distribution by minimising the effect of product branding, all businesses may be more or less equally well represented and secondly, cost-effective options are available for getting information on-line, especially with the development of approaches such as the Internet/Web Gateway model (Dorren and Slater 1996).

2. Emergence of the Hotel List Provider

Hotel List Providers, as defined for the purpose of this research, are concerned with adopting a WWW based direct marketing approach to the end consumer of travel related services, and in particular serviced accommodation. Traditionally electronic reservations networks have serviced the information and decision making requirements of many participants including third party information brokers such as Travel Agencies and Representation Companies and, in the guise of Global Distribution Systems and Centralised Computerised Reservations Systems, they have enjoyed an exclusive and secure trading environment. The status quo, however, has been challenged by the unprecedented success and global acceptance of a 'new' de facto open standards communications network - the Internet. Several of the main players in the traditional electronic markets e.g. THISCo, SABRE, Utell, are themselves establishing a significant presence on the WWW (Marcussen 1996) and additionally, most large hotel groups and consortia are utilising the WWW as an extension of their marketing reach although not all have applied a rigorous marketing strategy to their approach (Murphy 1996). A recent report (Activemedia 1995) also suggests that travel related services are one of the most likely niche markets to achieve commercial success on the WWW.

The nature of the Internet challenges the established electronic trading procedures and distribution channels, by facilitating direct access between tourism service providers and consumers (these issues are discussed in greater detail elsewhere (Dorren & Frew 1996)). The Internet facilitates information sourcing and gathering, and is increasingly becoming accepted as a mechanism for digital commerce. This interactive, direct access point of information and point of sale technology combines to establish the potential for a new generation of electronic reservations network for our industry. Given that the emerging Hotel List providers aim to offer a commercially viable point of information and sale to the end consumer, it is appropriate to assess the relative effectiveness of these sites, in terms of their marketing, distribution and reservations effectiveness, as perceived by tourism service consumers.

3. Generic Web Use

The WWW has been investigated almost since its inception, and consequently, there are various sources of research which aim to profile the characteristics of WWW users. An investigation into the current status of Hotel List Providers on the WWW firstly requires to establish the generic trends of the Internet/WWW and for this reason and to set the Hotel List study within a reliable framework analysis of existing one-off and longitudinal Internet User studies are used to place the narrower field in context. There is no shortage of statistics available relating to WWW users, unfortunately not all of them are reliable, or even *realistic*. Some Internet user surveys do not distinguish between the Internet and the WWW for user demographics, but perhaps more alarming is the range of statistics quoted for hosts/users; 60 million - 2.5 million (CNIDS, FIND/SVP, O'Reilly, Times Mirror, Economist, Net User) These differences are difficult to reconcile for use as baseline data for further, industry specific research, however what is comparable from all Internet user surveys is that the medium is becoming more reflective of the general population at large.

The following WWW user profile has been generated from a series of longitudinal surveys of WWW use conducted by Georgia Technology Research Corporation fifth WWW user survey. The *typical* WWW user is a 33 year old male, married with a moderate to high income, is most likely to be a college educated and working in educational or computer related employment. The average user realises most of the data security issues posed by the Web, but still feel strongly that they should be able to control their personal details as tracked on the Web. The predominant problem with using the Web is that of speed of access to load and view Web pages.

4. Research Methods and Approach

Assessment criteria were established as a framework for examining the effectiveness of HLP sites and involved the elicitation and subsequent analysis of a range of value judgements, concerning user perceptions. As no definitive directory of HLP sites exists, a sample was generated using on-line Search Engines and keywords. This method was adopted to replicate the searching technique which may be used by a reasonably experienced WWW user, but one also with little prior knowledge of Hotel List sites. Table 1 details the HLP sites located using these standard Search Engines.

Table 1 Identification of HLP population, by standard Search Engine

Site Name	Site URL http://www.	Scope	I	Y	L	AV
AA Hotel Guide On-line	.theaa.co.uk/hotels	**	✗	✓	✗	✗
AccomoDATA	.accomodata.co.uk	**	✓	✓	✓	✓
Accommodation Search Engine	.netxtra.co.uk/accom	*	✓	✓	✓	✓
All the Hotels on the Web	all-hotels.com	*	✓	✓	✓	✓
AutoEurope	.wrld.com/ae/index.html	*	✓	✓	✗	✗
B & B in the UK	.visitus.co.uk	**	✗	✓	✗	✗
Britain B & B Budget Guides	.usbusiness.com/elstead/brit1.htm	**	✓	✓	✓	✗
British Travel Information	.inect.co.uk/vacation	**	✓	✓	✓	✗
Expotel Hotel Reservations	.netxtra.co.uk/netxtra/expotel	*	✓	✓	✓	✓
Hotels & Travel in the UK	.webscope.com/travel.uk.html	**	✓	✗	✗	✗
Hotel-UK	demon.co.uk/hotel-uk	**	✓	✓	✓	✗
HotelNet	.theshoppes.com/~ala/hotelnet.htm	*	✓	✓	✓	✓
Hotels UK Ltd	hotels.co.uk	**	✓	✓	✓	✓
Stay UK	.stayuk.co.uk	**	✗	✓	✗	✗
The Hotel Guide	.hotelguide.ch	*	✗	✓	✓	✗
The Lodging Guide Worldwide	.marriott.com.lodging/lodging.htm	*	✓	✓	✓	✓
The Virtual-Inn	.virtual-inn.co.uk	**	✓	✓	✓	✓
TravelNow	.travelnow.com	*	✓	✓	✓	✓
TravelWeb	travelweb.com	*	✓	✓	✓	✗
UK Hotel & Guest House Directory	s-h-systems.co.uk/shs.html	**	✗	✓	✓	✓
Worldhotel	worldhotel.com	*	✓	✓	✓	✓

* Sites with an International Geographic Scope** National Geographic Scope
I=Infoseek; Y=Yahoo; L=Lycos; AV=Alta Vista

The initial population of 21 sites was further stratified, those with <75% success rate using keywords were removed and the sample of 15 HLP sites (shaded) settled. A rigorous assessment framework was devised after considered evaluation of numerous generic and travel specific WWW sites. An iterative process, in the form of peer review was applied with several recognised experts in the field of Travel and Tourism IT offering comment on the criteria and approach. The assessment framework was categorised into 10 sections, each containing a range of features and their respective values and all sites were accessed and assessed using this specifically defined and rated set of criteria.

The most appropriate research technique for a study of this nature is that of a ***Panel Survey***, whereby the panel participants are issued with the assessment framework and asked to score each site, independently, using stated criteria. The following hardware and software specifications applied; Netscape Navigator (Browser) version 2, 486DX Processor, 8Mb RAM, SVGA Monitor, 640/480 pixels, 256 Colours, 128kB/s leased line connection *(No sound cards or multimedia equipment was used).* A population of undergraduate and postgraduate Hospitality and Tourism management students was established with members having experience of using the Internet, little or no prior knowledge of Hotel List provider Sites and in general terms reflected the stereotypical characteristics mentioned above. Site assessment was scheduled to be completed within one week (28-31 Oct 96), a narrow time band limits the differences between sites and none of the Hotel List Providers under investigation were alerted concerning the assessment. Each panel member was provided with the assessment framework and given a set of Hotel List Site Uniform Resource Locators (URL) to access and assess.

Each participant received instructions concerning their input including basic information concerning *why* this study was being undertaken, *who* is likely to benefit from the overall survey results. *what* the output of the study may be and *how* their participation will strengthen the outcome. They were provided with the assessment framework, section and marking criteria were explained prior to commencement and any queries settled and issued with 2 site URL's in an arbitrary manner. Each participant had one hour in which to complete site analysis. Survey participants were issued with a scoring sheet, as detailed in which was completed during analysis of the site, date and time of analysis were specified, as were additional qualitative comments generated. A total of 70 individual users participated in the survey giving 107 site accesses. HLP sites identified in Table 1, were contacted. These results are now presented in correlation to scope and origin of site. Specific site features under analysis in each section are *italicised.*

1. General *Name of site (1), Accuracy of URL (1), Location identification (1), Awareness of geographic scope (1), Currency used (1).*
2. Site Objective. *Destination information only, Accommodation information only, All travel associated information*

	1.General **Score (0 - 5)**		2. Site Objective **Score (0 - 2)**	
	Accumulated	*Mean*	*Accumulated*	*Mean*
Geographic Focus				
International (n=6)	79%	4.0	73%	1.5
National (n=3)	75%	3.8	71%	1.4
Domicile Country				
USA Based (n=5)	76%	3.8	72%	1.4
UK Based (n=4)	78%	3.9	74%	1.4

3. Overall Perception This site gauged nine generic features: *Time to load pages (4), Graphics quality (2), Photographic quality (1), Effective use of tables (1),*

Effective use of frames (1), Effective use of bitmaps (1), Ease of navigation (4), Quality of information (4), Ease of contact (4).

4. Additional Features These features were primarily concerned with; *Relevant links (4), Links to additional information (4), Notification of updates (1), Uniqueness of information (2), e-mail to maintainer (1).*

	3. Overall Perception **Score (0 - 22)**		4. Additional Features **Score (0 - 12)**	
	Accumulated	*Mean*	*Accumulated*	*Mean*
Geographic Focus				
International (n=6)	68%	15	56%	6.8
National (n=3)	57%	12.4	48%	5.7
Domicile Country				
USA Based (n=5)	68%	15	57%	6.9
UK Based (n=4)	56%	12.3	60%	7.2

5. Specialist Features Significant credit was given for implementation of various technologies; *Multimedia applications (4), Independent search engine (4), Site security (4), Databased accommodation (4).*

	5. Specialist Features **Score (0 - 16)**		6a Accommodation Database; Absolute Numbers **Score (0 - 10)**	
	Accumulated	*Mean*	*Accumulated*	*Mean*
Geographic Focus				
International (n=6)	45%	7.2	73%	7.3
National (n=3)	37%	5.9	57%	5.7
Domicile Country				
USA Based (n=5)	49%	7.8	75%	7.5
UK Based (n=4)	39%	6.3	59%	5.9

6. Accommodation Database

(a) Absolute Numbers: Here size of site was scored in terms of the approximate number of accommodation providers featured on the site, indicating scope of coverage. *Less than 30 (0), 30-100 (2), 101-300 (4), 301-3,000 (6), 3,001-30,000 (8), 30,001+ (10)*

(b) Business categories: Sites were assessed here relative to the proportionate 'mix' of accommodation providers distributed on their site, in terms of ownership, grade and location. *Predominantly chain/consortia located in major city centres (1), Balanced mix of all accommodation types and locations (4).*

(c) Concentration in Home Market: The home market concentration was introduced to indicate the degree of specialisation within the domicile country, this section was included to recognise site providers who specialised on a national level. *Coverage from 0-19% (0), 20-30% (1), 31-39% (2), 40-59% (3), 60-79% (4), 80+% (5).*

	6b. Accommodation Database: Business Categories **Score (0 - 4)**		6c. Accommodation Database: Concentration on Home Market **Score (0 - 5)**	
	Accumulated	*Mean*	*Accumulated*	*Mean*
Geographic Focus				
International (n=6)	62%	2.5	55%	2.8
National (n=3)	70%	2.8	53%	2.7
Domicile Country				
USA Based (n=5)	59%	2.4	57%	2.8
UK Based (n=4)	69%	2.7	55%	2.7

7. Customer Focus End users were asked to indicate which group of travellers they 'believed' were being targeted through this site, for example independent, group, business and leisure. Sites were scored for each appropriate group. *Caters for Business travellers (1), Leisure travellers (1), Independent travellers (1), Corporate business and travel agencies (1).*

8. Rates Information Availability of rates is considered essential information for any end user wishing to book accommodation. Sites were scored in relation to the degree of accuracy of displayed rates information. *No rate information given (0), Price ranges indicated (2), Exact prices quoted (4).*

	7. Customer Focus **Score (0 - 4)**		8. Rates Information **Score (0 - 4)**	
	Accumulated	*Mean*	*Accumulated*	*Mean*
Geographic Focus				
International (n=6)	71%	2.9	55%	2.2
National (n=3)	69%	2.8	51%	2.0
Domicile Country				
USA Based (n=5)	73%	2.9	66%	2.6
UK Based (n=4)	70%	2.8	54%	2.2

9a.Reservations: Information with end users specified as being important in the accommodation querying and booking stages were assessed. *Ability to Check availability (4), Make 'best' suggestion (2), Response to queries (4), Ease of reservations (4), Ease of cancellation (4), ease of payment (4).*

9b.Reservations The actual procedures available for making on-line reservations were scored depending upon the degree of real time reservations ability. *Ability to Book (accommodation) on-line (4), Book with e-mail (3), Book with fax (2), Cannot book accommodation on site (0).*

	9a. Reservations **Score (0 - 21)**		9b. Reservations **Score (0 - 4)**	
	Accumulated	*Mean*	*Accumulated*	*Mean*
Geographic Focus				
International (n=6)	61%	12.7	76%	3.1
National (n=3)	40%	8.4	72%	2.9
Domicile Country				
USA Based (n=5)	61%	12.8	78%	3.1
UK Based (n=4)	43%	9.0	67%	2.7

10. Evaluation End users were asked to state overall if they believed that the site under investigation was, in their opinion, providing value for money and if the time they had spent browsing the site for information was well spent

	10. Evaluation **Score (0 - 5)**	
	Accumulated	*Mean*
Geographic Focus		
International (n=6)	58%	2.9
National (n=3)	46%	2.3
Domicile Country		
USA Based (n=5)	58%	2.9
UK Based (n=4)	48%	2.4

Table 2 Comparative Table of Mean and Accumulated Totals

Assessment Category	**International Focus**		**National Focus**		**USA Based**		**UK Based**	
	x	**%**	**x**	**%**	**x**	**%**	**x**	**%**
General	4	79	3.8	75	3.8	76	3.9	78
Site Objective	1.5	73	1.4	71	1.4	72	1.4	74
Overall Perception	15	68	12.4	57	15	68	12.3	56
Additional Features	6.8	56	5.7	48	6.9	57	7.2	60
Special Features	7.2	45	5.9	37	7.8	49	6.3	39
Absolute Numbers	7.3	73	5.7	57	7.5	75	5.9	59
Business Category	2.5	62	2.8	70	2.4	59	2.7	69
Home Market	2.8	55	2.7	53	2.8	57	2.7	55
Customer Focus	2.9	71	2.8	69	2.9	73	2.8	70
Rates Information	2.2	55	2	51	2.6	66	2.2	54
Reservations (A)	12.7	61	8.4	40	12.8	61	9	43
Reservations (B)	3.1	76	2.9	72	3.1	78	2.7	67
Evaluation	2.9	58	2.3	46	2.9	58	2.4	48

% Total Accumulated score expressed as a percentage. **x** Mean Score

5. Conclusions and Future Investigations

Longitudinal profiling of WWW users, especially those predisposed towards purchasing on the WWW is a prerequisite to effective target marketing and business success on the WWW. Specific tracking of tourism service consumers in this medium is necessary, to distinguish user requirements and relative significance placed on a range of inter related site features. This study illustrated that opportunities remain to improve and differentiate HLP sites to achieve competitive advantage within this market place. Site maintainers may well wish to address these issues in future development strategies.

The assessment framework used was designed to elicit subjective responses concerning perception of successful (or otherwise) HLP performance and there are clear indications of significant differences on almost all criteria between sites, especially those with different geographic focus and location. Further investigation concerning this group is underway and seeks to assess HLP performance against their *own* perceptions of priorities. This second phase will extend to focus group discussion and assessment of the same group of HLPs to ascertain and to rate site effectiveness. This work will coincide will a self assessment questionnaire issued to HLP sites, the views of users and site maintainers shall be compared and contrasted.

Research by the authors into the changing nature of the international reservations networks makes it clear that there will be a major metamorphosis in these over the coming years, catalysed by WWW representation and thus the nature of the networks per se is worthy of ongoing investigation. How users navigate these electronic networks, filter, report and action information is increasingly likely to be through direct and indirect agent technology and therefore inter-related research into agent/network interfaces, particularly from the end user perspective is timely and finally, given that increasingly information and reservations facilities are being presented on the Web, it is becoming apparent that there is a need for an independent assessment mechanism which can rate providers and thus offer a service to the consumer of tourism products.

It is felt that the above research may help form the basis for such a rating system; the initial research indicated that the assessment criteria applied above could be suitably scaled up with only minor modifications, the HLP sites subject to assessment may now be broadened in both scope and in geographical location and a much larger participant population used to more fully establish the reliability of such assessments. It is the authors' opinion that ultimately such ratings will become integral to this area with assessment being achieved through a combination of 'expert assessors' and on-line surveys of users. In an ideal world such on-line assessment could be standardised and incorporated into each HLP site!

References

Activemedia (1996), http://www.activemedia.com/TrendsExec.html

Anderson, C. (1995) 'The Accidental Superhighway' Economist 1st July Pp 5.

Cockerall, N. (1991),Outbound Markets/Market Segment Studies 'European Independent Travel' EIU Travel & Tourism Analyst No. 4 pp. 38-48.

CNIDS (1995), CommerceNet/Neilsen Report 1995,http://www.nielsenmedia.com

Dorren, C., Frew, A.J.(1996), Intelligent Agents and the UK Hotel Sector,.Journal of Hospitality Information Technology, Vol 1, No 1, paper 4.

Dorren, K., Slater, A. (1996), Leveraging Tourism on the Web Hospitality Information Technology Association , Conference Proceedings, Edinburgh

FIND/SVP (1996) Find/SVP survey 1996 , http://etrg.findsvp.com/features/newnet.html

Georgia Technology Research Corporation (1995) fifth WWW user survey. (http://www.cc.gatech.edu/gvu/user_surveys/).

Irvine, D. (1995) 'The Internet and how to Swing it' ,NetUser Spring Issue 1 Pp. 10.

Majewski, R. (1996), User Report 1 & 2 , http://www.demon.co.uk/travmag/

Marcussen,C.(1996), 'Hotel representation Companies and the Internet', Hospitality Information Technology Association, Conference Proceedings, Edinburgh.

Middleton, V. (1988), Marketing in Travel & Tourism , Butterworth and Heinemann UK Pp. 186.

Murphy, T .,Forrest,E., Wotring, C., Brymer, R. (1996), 'Hotel Management and Marketing on the Internet' Cornell Hotel, Restaurant Administration Quarterly, Vol. 37 No. 3 pp. 70-82.

O' Reilly/Trish (1996), http://www.cyberatlas.com/www.ora.com/survey

Times Mirror (1995),Internet Use in the United States: 1995 Baseline Estimates and Preliminary Market Segments. http://www2000.ogsm.vanderbilt.edu/baseline/1995.Internet.estimates.html

Tourism Management 'Travel Agents and the Information Highway' Tourism Management Vol.12 No.2 Pg.154 1995.

Developing Tourism in the Lunigiana

E. Bracci[x], P. Coltelli[x], G. Cresci[x], C. Magnarapa[x]
A. Montanari[+]

(x) CNUCE - Institute of CNR, Via S. Maria 36, 56126 PISA
(+) Consorzio Pisa Ricerche, P.zza D'Ancona, PISA

Abstract

The rapid evolution of hypermedia technology together with the increasingly wide use of the Internet can be effectively exploited to publicise the great artistic and historic heritage of Italy and its regions. Aiming to increase tourism, many public administrations have been looking with great interest to projects enhancing exposure of their territories and permitting virtual visits to museums, towns, archaeological sites and other sights.
This paper describes a project, partially funded by the European Community, whose main goal is to produce a tool to allow tourists with different interests to cover virtual routes in a historical area named the "Lunigiana".
From a technical point of view, key aspects discussed are related to integrating Geographic Information Systems (GIS) with Database, Multimedia, Information Retrieval (IR) and Internet technologies.

1. Project aims

The application of multimedia technologies represents an attractive means to develop tourism. Integration of GIS features in multimedia tourist systems is however quite a difficult problem that has yet to be completely solved by todays technology. The CNUCE, Institute of the Italian National Research Council, has been carrying out research in this field [Mogo92]. Current multimedia tourist systems do not support geographic features such as on-the-fly computation (i. e. distances, shortest routes, etc.) or display functions (i. e. pan, zoom, etc.), which are critical when the system must cover a wide area (a town, region, or larger). Such limitations can be overcome only by means of GIS tools.
Over the last two to three years, the growth of the Internet and its user-friendliness has made these systems accessible to a very large number of users, thus rendering the joint approach (Multimedia + Internet) an extraordinary tool for tourist marketing. Many examples of multimedia tourist systems are viewable on the Internet [Wand96] [City96].
The project described in this article intends to implement a tourist system building upon CNUCE experience and exploiting today's powerful software tools. The system integrates features of database, multimedia, GIS, IR and Internet systems, and focuses on an area in central Italy called the Lunigiana.
Although the system has been implemented for a specific geographic area, the approach is absolutely generalisable, so that the tool can be applied to different areas and subject data as described in the following.

2. Funding and partners

The project has been funded by:

- European Community;
- Regione Toscana (regional authority);
- Comune di Massa (local municipality).

for a total amount of 250 KECU over a period of three years (96- 98).
European Community funding was awarded within the framework of structural funds for areas in industrial decline and is aimed at stimulating the creation of new employment. Funding by local authorities (region and municipality) imposes the targeting of areas under their jurisdiction.
The following partners are involved in the project:

- CNUCE - CNR is responsible for system design and supervision of technical activities.
- Consorzio Pisa Ricerche, an industrial consortium devoted to applied research, has been charged with implementation.
- Soprintendenza ai Beni Architettonici di Pisa, the local branch of the Ministry for Cultural Heritage, is responsible for cultural data collection and the user interface.
- Comune di Massa has provided tourist data (i. e. on hotels, restaurants, events, etc.)

3. Why the Lunigiana?

Rather than the smaller modern area of the same name, the project actually deals with the ancient area of the Lunigiana, roughly corresponding to the Province of Massa (Fig. 1). This area of about 1000 Km^2 and 100,000 inhabitants possesses many characteristics that make it ideal for tourist development:

Fig. 1. The Lunigiana Area

- It is an area with few economic resources, only marginally involved in industrial activities, and where the natural environment as well as traditional crafts and customs have been preserved.
- It has a long and interesting history beginning in prehistory and developing throughout Roman times and the Middle Ages, as testified to by abundant archaeological findings, villages, castles, churches and museums. Lunigiana is located on the "Via Francigena", a road used since medieval times by pilgrims on their way to Rome from the North and West of Europe.
- There is no single dominant cultural or tourist attraction, but rather many sights which reveal the history, culture and evolution of this land. The situation is common to many other Italian regions: historic sites spread throughout the countryside. Thus, a geographic approach is best adopted in order to visit cultural places and link them to historic events and traditions.
- The Lunigiana is quite near and easily reached from the well-known sea-side resort area, the "Versilia". Tourism in Versilia has a long tradition (more than a century) and remains today one of Italy's most popular bathing resorts. The chance to divert some tourism from the often overwhelmed Versilia to the Lunigiana would be welcomed by the local administrators of both areas, and for many tourists complementing a sea-side holiday with cultural (as well as natural and culinary) excursions could represent a pleasant change of pace.

4. The tool's features

The main goal of the project is to promote the artistic, historic and natural resources of the Lunigiana by making the region known to a much wider population of potential users. In order to achieve this, the tool to be implemented must address the varying needs of different types of users: from leisurely browsers, to occasional tourists and students to experts with specific interests.
Satisfying such a wide range of interests requires:

- A simple and attractive user interface; this is a crucial aspect as it affects the way users will perceive the system and its data [Salm89] [Leon89].
- Inclusion of a large number of data types (texts, graphics and maps), most of them with quality requirements (i. e. high-quality images) to satisfy expert requisites. Defining the data and their connections is an essential activity in information systems design involving both developers and experts in the subject. In this case, the problems are attributable to data complexity, rather than quantity.
- Functionality for both tourists and experts: simple, pre-defined analyses must be implemented for general users (occasional tourists), while experts must be allowed to conduct more in-depth analyses (for instance, dynamically building queries on a database or creating new routes based on individual requirements). The major difficulty here consists of making sure users do not get lost while navigating the system. Moreover, basic tools in support of data analyses include subject-specific documentation (called a 'subject book'), in which topics are thoroughly described in simple, yet scientifically correct terms, as well as a dictionary of technical terminology.

Both cultural and tourist data must be considered. Not all cultural data regarding the Lunigiana have been included; at this stage interest has been focused on the following themes:

- Historic villages;
- Castles;
- Medieval churches;
- The "Statue-stele" museum.

The first three items are closely linked, as they represent the civil, military and religious expressions of the art and science of architecture, and through them tourists can come to understand the main aspects of the social and economic life of this region during the Middle Ages. The fourth item provides a glimpse of the area's pre-historic development through the sculptures typical of that era.

Tourist data include all the information needed to make a real visit, that is, road maps, means of transportation, hotels, restaurants, camp sites and so on. They also include information on events, shows and curiosities, and provide suggestions on how to meld the entertaining with the cultural.

Suggestions are not related to a site but rather to site neighbourhood. To implement this, landscape has been partitioned by means of a grid and suggestions are connected to grid elements. The sizes of grid elements are not regular; they are inversely proportional to the concentration of cultural sights.

From a technical point of view, the tool must allow:

- Different navigation modes: on thematic maps, on hypertexts, and on database. Users can switch from one mode to the other and the system will manage the change: some functions available in one navigation mode are inactivated in others (e.g. suggestions are only available in geographic navigation mode).
- Integration of alpha-numeric and geographic queries. All queries return lists of places located on maps. Both lists and maps must be available for further selections.
- Creation of dynamic routes taking into account cultural interests, roads and distances, hotel and restaurant locations, etc. Pre-defined routes must also be supplied to meet needs of the hurried tourist.
- Route display by means of vector elements on a geo-referenced raster background.
- Differential local and remote use. When used in a local (LAN) environment, presentations make use of large, high-quality images and maps. When used in a remote (WAN) environment, some displays could be less detailed and some geographic functions disactivated in order to reduce network traffic and wait times.
- General services such as print-outs, bookmarks, etc.

Tourist systems currently available on the Internet do not include all the features described. One can find systems of differing complexity: from simple advertising tools, to complex teaching ones [Pers96]. Many allow geographic navigation and virtual visits, but do not support user geographic interaction or route definition. Furthermore, geographic and theme components are generally kept separate; results of queries on *objects* having geographic attributes are generally returned as lists or tables, not maps. Selections are made on tables [Pers96] or on pre-defined elements of a map. [Cali96] supports some geographic functions operating on a single theme only.

5. Data organisation

A suitable, well-organised data structure is a basic requirement for any information system. The system must operate on different data types (alpha-numeric, geographic, images, etc.), many of which are difficult to classify and structure. Different types of operations must be performed with various levels of complexity (e.g. tourists vs. experts). Operations include:

- structured queries on alpha-numeric data;
- retrieval of words in free text searches;
- geographic navigation and enquiry;
- hypertext navigation.

Each type of operation can be performed easily by means of specialised software environments: Data Base Management System (DBMS), Information Retrieval (IR), Geographic Information Systems (GIS), HyperText/HyperMedia (HT/HM). Unfortunately, integration of so many software environments into a single computer system entails problems of compatibility, performance and cost. In practice, a compromise between functionality and complexity must be found.

HT/HM is an intuitive and very attractive approach. However it has several shortcomings related mainly to data complexity and quantity. Large amounts of data can require long navigation paths. Complexity generally implies many connections (relationships) between data; in a hypertext this means a great number of links. Long navigation paths and many links increase the risks of winding up confused or lost in the system [Sign95]. Finally, some kinds of analyses (i. e. all the *objects* with a specified property) are difficult to implement in a hypertext.

Data with such characteristics could be managed well through a DBMS approach. However, the use of such systems is limited by their main defect - a complex user interface unsuitable for users with limited computer skills (in practice, knowledge of a specialised language such as SQL is required).

Short descriptions are always required because the properties of cultural objects cannot be completely described by data structures. An IR tool is then needed to manage free (unstructured) texts and complement DBMS queries on structured data.

The integration of specialised environments is obtained by associating the following to each *object* described in the system:

- Its own series of attributes whose values can be part of a continuous range or a number of discrete pre-defined alpha-numeric strings. SQL queries operate on these attributes and return lists of selected *objects*.
- A brief description in free text: a search engine operates on these descriptions to select all the documents where a keyword (or a combination of keywords) occurs. The engine returns a list of selected documents.
- A hypertext with sensitive elements or anchor points (so called "hot-spots" - words, images or map elements) activating links to the subject book and a dictionary. Fig. 2 presents a schematic illustration of a hypertext structure with

four levels (home page, *object* description, subject book and dictionary) and their connections.

- Geographic co-ordinates of the site where the *object* resides. Such data are used to place *objects* on maps and allow geographic operations.

This integrated approach allows combining the simple and intuitive interface of HT/HM systems with the powerful analytic functions of DBMS, IR and GIS. The major emphasis is on detailed data analysis and modelling in order structure as much information as possible. This reduces the hypertext complexity and amount of free text descriptions as well. Fig. 3 shows a simplified conceptual scheme of DBMS data.

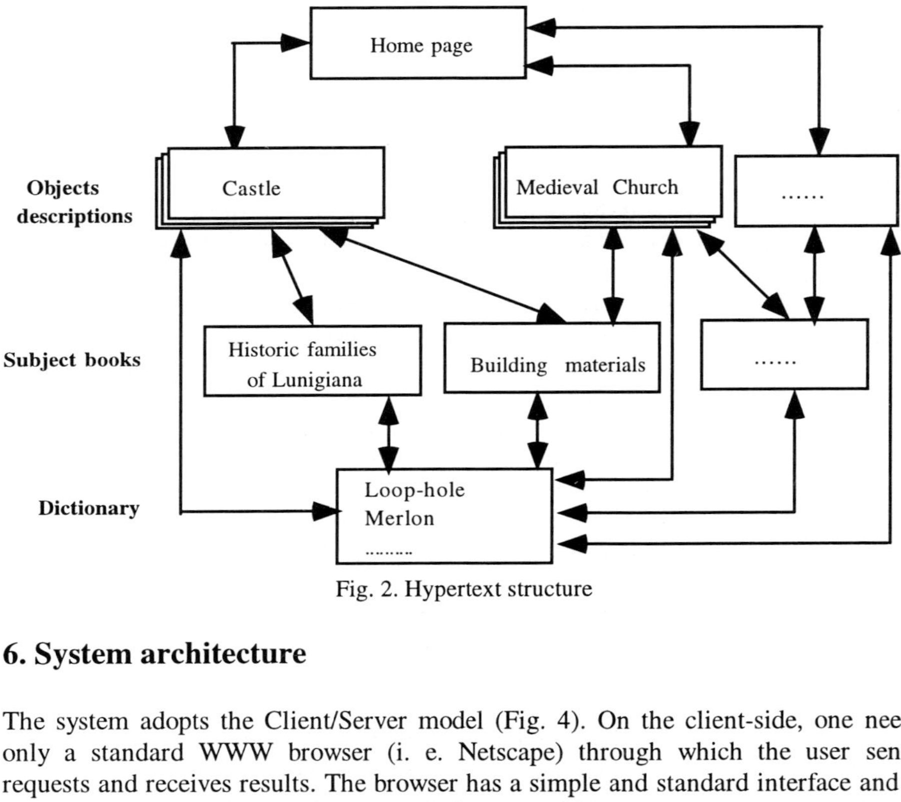

Fig. 2. Hypertext structure

6. System architecture

The system adopts the Client/Server model (Fig. 4). On the client-side, one needs only a standard WWW browser (i. e. Netscape) through which the user sends requests and receives results. The browser has a simple and standard interface and is the only system component the user must interact with.

The browser can operate in two modes:

1. With the HTTP server alone: requests can be satisfied directly by this server or redirected to the DBMS or GIS server. In the latter cases, the server acts as a gateway by translating a HTML request into an SQL or geographic query. The HTTP server also must convert the replies returning from DBMS and GIS into HTML format.
2. If the HTTP, DBMS and GIS servers operate on the same hardware platform, the browser can interact directly with DBMS and GIS. This can be made transparent for users using a Java compatible browser. Such a solution requires more

complex design and implementation, but can be useful in reducing network traffic and response times.

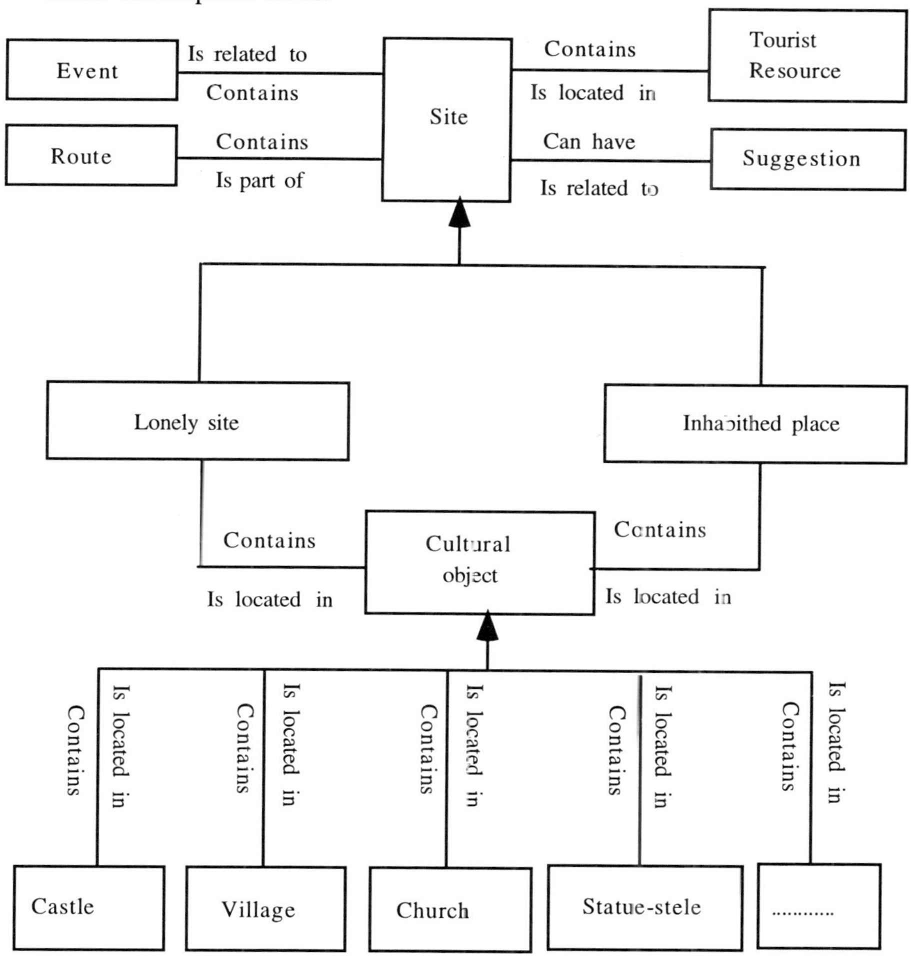

Fig. 3. Simplified database schema

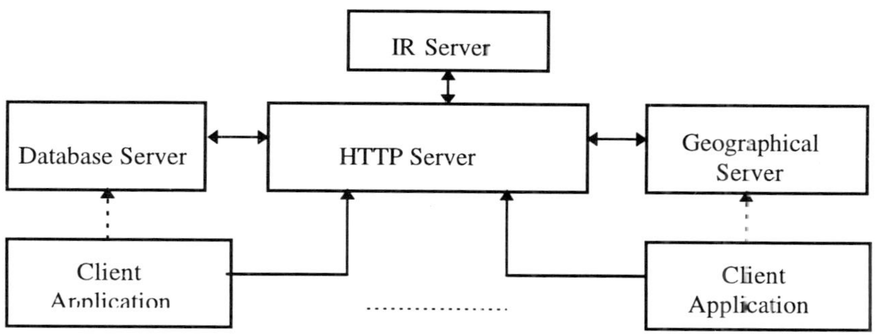

Fig.4. System architecture

Whichever solution is adopted for client application, four logical servers are required. Choices in this area are strictly determined by the hardware platform and operating system. There is no specific constraint for the HTTP server; these are available for UNIX and PC platforms with DOS, Windows, or Windows NT operating systems. In no case are costs an issue.
DBMS functions are also available for many environments. Although, powerful platforms are generally required to manage large amounts of data, this is not the case here, and a simple (and inexpensive) product can be used (i. e. MS Access).
IR functions are supplied by many products (low cost or even freeware) that can be integrated well with HTTP servers.
The GIS server could be a major problem; many packages entail high costs and complexity, and often require powerful platforms. In our case, a product such as ArcView, a low-cost product available for most common platforms, seems to provide all the required functions.
All four servers can be hosted on a single hardware platform: a high-end Intel workstation will likely be used for development and testing. Because of the speed of technological advancement, the system to support the tool's operational life will be configured at project's end in order to achieve the most cost-effective solution. Windows NT, by virtue of its compatibility with clients, the large base of applications and its reduced cost, is likely to be the operating system.
Application programs are required both on the server side, in order to transmit queries from HTTP to the DBMS and GIS servers, and the client side, in order to complement standard browser functionality. On the first side, CGI (Common Gateway Interface) scripts can be used, while on the latter, the Java programming language seems to be the best choice because of its simplicity and platform independence.
The architectural and software choices made for the project allow production of a general-use tool where new data and functions can be added without requiring global architecture modifications. The tool can support geographic and tourist data and functions for any country. It can also be used to support subject data related to other castles, villages and churches; but in this case modifications may be required to support new features not encountered in the Lunigiana. Such features will probably be related more to data properties than relationships, though they will likely also affect functions. Finally, addition of new topics requires a review of the data model and specific action on implementation, but general architecture, data organisation and many functions can be preserved.

7. Project timing and resources

The project, started in late spring of this year, will finish in the autumn of 1998; at that time the software tools and data must be ready for use. Following the principles and rules of information systems design, the project plan includes:

- design;
- data acquisition;
- prototyping;
- implementation;

- testing.

At the time of writing this, the design phase is near completion. Data analysis has been completed, function analysis is near completion and system architecture definition is in progress. Full design will be accomplished by January 97.

Data acquisition has just started and will continue for the whole of next year, and a first prototype will be available by next spring.

Initial planning allocated 8 person-months to design (mainly by the CNUCE and Soprintendenza), 24 person-months to prototyping and implementation (Consorzio Pisa Ricerche) and 14 person-months to data acquisition (the Soprintendenza and Massa Municipality). It now seems that we had slightly underestimated the resources necessary for system design - unforeseen difficulties were encountered and are described below.

8. Conclusions and future developments

The tool, whose development is now well under way, can effectively enhance tourism in an area with poor economic resources. Thus, the project conforms to the aims of European Community funding.

During the design phase a few difficulties have been experienced, mainly related to data complexity and the large amount of specialised knowledge involved. Some data are intrinsically difficult to structure, and experience in this field is scarce. Historians, archaeologists, architects, art experts are involved in data and function analysis: definition of a common language for both software developers and cultural experts has proved to be another quite difficult task. Also, many computer experts are required to integrate HT/HM, IR, GIS and DBMS and select appropriate development environments and tools. Rapid technological evolution has the potential to create further problems: incorrect design choices can easily lead to the creation of quickly obsolete tools.

Despite such design and implementation difficulties, early presentations and technical documents have been well-received by local administrations, which are thus preparing for a second phase of the project. This new phase will include:

- Collection of new data mainly on Roman times (the name Lunigiana comes from Luni, an important Roman port).
- Addition of another region, the area of the Apuan Alps, with its great natural resources and prevalent economic asset, the so-called white "Carrara" marble, used since the time of ancient Rome.
- New applications: 3-D display would be very effective in landscape representation and allowing now totally or partially destroyed castles to be reconstructed on the basis of historic documents and images; thus, giving the virtual tourist the chance to roam the Lunigiana, not only as it is today, but as it was in bygone times as well.

9. Acknowledgements

The project involves the work of many people; the authors wish to thank all those who contributed to the partial results described here. A special thanks goes to Olga

Renda for her contribution to data analysis and Giuseppe Muccioli for his technical support in defining the system architecture.

10. Bibliography

[Cali96] http://city.net/countries/united_states/california/san_francisco/maps/mapquest.html

[City96] http://www.city.net/

[Leon89] M. K. Leong & others: *Towards a visual language for an object-oriented multi-media data base system* - Visual Database System - North Holland, pp. 465-495, 1989

[Mogo92]P. Mogorovich, C. Magnarapa, M.V. Masserotti, S. Mazzotta - *Merging GIS with multimedia technologies: the case study of an information system for tourist applications.*
EGIS 92 Conference Proceedings, Vol. II, pp. 1085-1094, March 1992

[Pers96] Perseus Project is available at the following Internet address:
http://www.perseus.tufts.edu/

[Salm89] Salmon, Slater - *Some aspects of the human - computer interaction*
Computer Graphics: systems and concepts.- Addison Wesley, pp. 455-483, 1989

[Sign95] O. Signore - *Issues on Hypertext Design*
DEXA 95: Database and Expert Systems Application - Proceedings of the international conference - Springer Verlag, pp. 283-292, 1995

[Wand96] http://www.spottedantelope.com/wander.htm

Dynamic Web Links for Tourism Information Services

Kanchana Kanchanasut, Pensri Charoenchai
CSIM, Asian Institute of Technology,
P.O. Box 2754, Bangkok 10501, Thailand

Abstract

A tourism information homepage is often found to serve many purposes or many different categories of users ranging from curious browsers, potential travellers to virtual tourists. With HTML's static links, in order to make one set of information shareable among different target audiences, a fixed set of guided searches may be provided through a set of fixed menus from which a user can make selections appropriate for their session mannually. In this paper, we propose to incooperate dynamic hyperlinks to a Tourism Information Service which is being provided distributedly on the WWW. At the beginning of a session, a user can specify what kind of session he/she is planning to do. From this piece of information, the user will be provided with more relevant pages and lankages dynamically.

1 Introduction

Today an increasing number of tourism information services have adopted the World Wide Web (WWW) technology to provide information for the global community. With the exponential growth of the WWW, the Internet is getting very crowded making bandwidth a serious issue of concern which needs to be economized in Web page design. The concern is far more greater when the tourism information service has to be provided on a low speed connectivity network. Apart from optimizing file sizes by reducing unnecessary graphics, the number of times for downloading a new page within a user's session (a session starts from when the user enters a given homepage) should also be optimized. If downloading occur too frequently either due to the way the web page or the hyperlinks were designed or due to mistakes during navigations, users could easily decide to leave their sessions because more time had to be spent waiting, getting disconnected or discovering unavailable links.

Providing a tourism information service by a government tourism promotion organization means that one has to satisfy heterogeneous user groups which is quite unlike designing a commercial promotional service for a supplier or a tour operator who have specific target audiences in mind. To satisfy all types of users would be a formidable task however efforts to reduce wasteful searches or the so-called loss in space syndromes should be one of the major design criteriors. The following obervations were made through our experience when designing *Sawaddee Thailand* [5].

- Information provided may be given different values by different groups of Website visitors at different points in time, eg. information about a destination with historical information may be uninteresting for people looking for practical information only. Hence a hyperlink should be made to different places depending on the users' requirements. This means that the session's objectives should be taken into account when chosing appropriate information for the users. We refer to this as *session type.*
- While during some sessions, a user may enjoy browsing or visiting various pages with no specific purposes, but in some other times the same user may prefer to get down to what is being sought after as fast as possible. *Access modes* or speed required should be made available as an option for the users. While no particular care is needed for normal access mode, the number of links along a search path must be minimized for a speedy access mode and detours may be encouraged for casual browsers.
- Different *categories of tourists* must be catered for thus requiring different presentations of information for the same topic, e.g. for a given destination, points of attractions for different groups are quite different and hence should be presented appropriately. For example, Bangkok for a culturally oriented tourist would mean temples or musuems should be made more prominent than shopping centers or other sport facilities.

From the above observations, it is felt that whenever there is a hyperlink several choices based on session types or access modes or categories of tourists should be taken into account before a link is established. Not all factors may be taken into account at the same time because not all of them are relevant for the same link during a navigation process. So when a destination is mentioned on a practical information page, it is no longer necessary to check for session type but access modes and tourist categories remain important.

With HTML citehtml, only static links can be specified where only one link is allowed at a time. Thus in trying to provide information for heterogeneous groups of users, whenever a choice is to be provided the use of menus has to be adopted. Menus could be provided on a separate page which requires an extra download or as a pop-up menu. Users of the information service could navigate through the vast amount of information provided on the homepage making selections whenever choices occur and backtracking whenever mistakes are made. Very often these mistakes are caused by misinterpreting the choices provided by the menus or by users getting disoriented after a long period of navigating. A common hypertext lost in cyberspace syndrome could become much more severe because of the non-homoginity of the WWW environment where different styles, bandwidths or speeds exist.

2 WWW and dynamic hyperlinks

To decide on behalf of a user which link to be made at any point in time we must obtain sufficient information about how the user sees the Web session being conducted. Users are asked to supply this information when the session begins and this information must be maintained throughout the session whenever the user uses any Web pages belonging to the same information service.
The WWW is based on a client and server architecture based on a communication protocol, HTTP [2]. The connection between the client and server is stateless which is perfectly fine when we are only concerned with static hyperlinks as provided in HTML. To provide dynamic links, decisions as to which links to be made have to be done during the session which depends on the information about that particular instant of the session. This means that we have to overcome the problem of statelessness of the HTTP where information about the user's intent must be maintained throughout the user's session.

We implement our link manager as an applet in Java [3] which gets initiated when the top-level page is entered and remains to be used over multiple pages which follow on. The user is asked to enter information about the session which are session types (general browsing, trip preparation, surveying destinations and information gathering), access modes (express, normal and slow) and tourist categories (business, nature and culture etc). The applet is used to generate links for subsequent pages belonging to the same information provider or a group of collaborative providers forming a single tourism information service. The session is controlled by the broswer running the applet initially started by entering top-level page from any of the servers within the collaborative group as shown below.

For example, a user describing himself as a low-budget traveller requring speedy access to a railway time-table may specify that the session is for a trip preparation with express access is required. Any click on "railway' should immediately brings out a time-table and fares instead of general railway information or a history of the railway lines.

A single static HTML link may now be replaced by an applet where a number of dynamic links get generated. As for the above user, bus and train time-tables together with the fares, a city map and a list of guest houses should be given at the same time if the user clicks at a tourist destination.

3 Related work

Dynamic links scheme proposed in this paper is not based on learning or user statistics which is where most of the work in hypertext dynamic configurations seem to be concentrated on. Here pre-defined choices are provided by page designers but the choice of which path to be taken depends on the dynamic environment of the users' sessions. Our approach of having a link generator which dynamically establish a link for the WWW is close to those proposed in citeStan. In their system, however, the link generators are parts of the servers instead of the clients while in ours we assume that a tourist information service can be distributed over several servers hence link generators are parts of the clients.

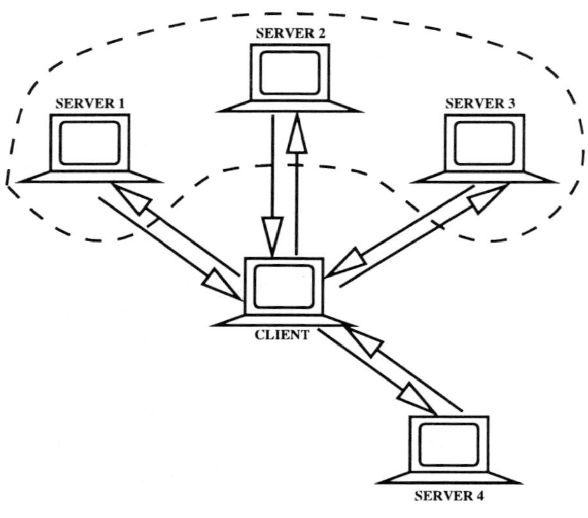

Fig.1. A Tourism Information can be distributed over many servers (within dotted line) while a session is controlled from the browser.

4 Conclusion and future work

We have proposed a tourist information system with dynamic links without making use of any learning or access patterns. In this scheme the number of page accesses can be reduced at the expense of design complexities where for each hyperlink the designer must take care in pre-defining appropriate links. The idea can be extended to include some inference mechanism for appropriate links from past access patterns as in citeStan or navigation profiles.

References

[1] Berners-Lee, T. and Connolly, D., Hypertext Markup Language - 2.0. IETF, RFC: 1866. Nov 1995. http://www.cis.ohio-state.edu/htnin/rfc/rfc1866.html.

[2] Berners-Lee, T., Fielding, R. and Frystyk, H., Hypertext Transfer Protocol-HTTP/1.0. IETF Internet-draft (1995). http://wwww.w3.org/pub/WWW/Protocols/HTTP1.0/draft-ietf-http-spec.html.

[3] James Gosling and Henry McGilton, The Java Language Environment, White Paper, Sun Microsystems (1995).

[4] Tak Woon Yan, Mathew Jacobson, Hector Garcia-Molina and Umeshwar Dayal, *From user access patterns to dynamic hypertext linking*, Computer Networks and ISDN Systems, 28(1996) 1007-1014.

[5] Tourism Authority of Thailand,*Sawaddee Thailand*, http://www.tat.or.th/
pages 253-263,

The Role of Intranet in the Tourism Domain
(Extended Abstract)

Roland R. Wagner
FAW, Johannes Kepler University Linz, Austria
wagner@ifs.uni-linz.ac.at

1. Introduction

The Intranet, the internal Internet, is perhaps the most important innovation in the last year. There are estimations that three quarters of the Web servers going up today are for internal corporate use. An Intranet is an internal corporate or organizational framework that uses Internet technologies to let employees browse and share electronic information remarkably easily. An Intranet is a private version of the Internet's World Wide Web , but it is available only to people inside the organization. Over the next 12 or 18 months the real pay-back of typical business will come from using Intranet web sites to give employees the information they need to be more effective. [Gates 96]
The Intranet portends a revolution in information processing. There are several distinguishing features of the Intranet:
- Uses TCP/IP (transmission control protocol/Internet protocol) for both wide-area and local-area transport of information,
- Uses HTML (hypertext mark-up language), SMTP (simple mail transport protocol) and other open Internet-based standards as the means of moving information from client to servers,
- Is completely owned by the corporation and not accessible from the Internet-at-large by the general public, and
- Is managed by IS with a similar set of tools, attitudes and procedures as they currently manage their legacy networks [Strom 95]

2. Evaluating Web Technology on Its Own Merits

To understand the true power of Web technology we should follow Ryan Bernard's suggestion to stop thinking about it as strictly an Internet tool and start evaluating it on its own rather impressive merits. We should look at on-line information system that has every feature we could possibly desire, then look at what Web technology already offers [Bernard 96]:
- Inexpensive client and server software
- Intuitive, document based GUI interface
- Requires no training
- Retrieves virtually any document type on-line
- Supports multimedia
- Supports imbedded hypertext links to local and remote documents

- Supports SQL queries or other interactive retrieval, display, and updating of database information
- Supports corporate-wide standardization of on-line interfaces
- Supports on-line forms, data-entry and other two-way interactive communication between users and computers
- Supports e-mail applications
- etc. etc.

An evaluation of these features shows us the full power of the Web technology as a basic IS-technology for future applications. (see [Bernard 96])

3. Executive Information Systems on the Fingertips in the Tourism Domain

For the author's opinion one of the first applications of the Intranet in the tourism domain will be the comfortable access to decision support relevant information for the executives. For this reason research and development efforts is taken for the user friendly access of extracted data from various distributed databases containing relevant information of the tourism. Data which are preferably accessible by virtual OLAP-engines will be used by the executives without the help of intermediate administrators of executive information systems who prepare some 'generic' tables. Furthermore for the tourism domain access with metaphors which are more suitable for this area, such as maps etc. can be created to allow a more natural view of executive data in this area. The use of Java/Active X technologies allow a local computation of special applets which are streamlined for management applications in the tourism area (e.g. geographical drill downs of management information, integration of intelligent agents for different complex analyses)

4. Intranet Use for the Tourist

One of the first and successful implementations of the Intranet for the customer is the FedEx Advanced Tracking applications. The FedEx Intranet application was designed for the internal use. With this Intranet tool the employees are given a simple user friendly tool for detecting the status of every package at any time of the day using the data stored in the distributed databases. Because of the successful internal use of this software FedEx decided to open parts of this Intranet application for its customers. The customer is now able to track the status of his/her package at any time of the day, anywhere in the world. The customer is able to follow the journey of his/her package even it is still in transit.([FedEx 96])

In the tourism area these kind of applications (quite complex Intranet applications opened for the customers) are still very rare. The next step in the tourism area could be expected in the area of reservations, where pending flight reservations could be tracked by the customers. Tourist packages where different components with incomplete information (respectively not yet confirmed information) can be continuously monitored by the customer. It is clear that simple static information such as precise description of sport equipment, rooms etc. can be easily attached via the Intranet. The tourist has the capability to browse through his/her individual details of an arrangement. A very successful implementation of such an Intranet application from another domain is the PartNet implementation which is an Internet-based part information service that allows engineers and designers to search part supplier's product information databases with very detailed and useful multimedia information. ([Partnet 96])

5. Conclusion

This paper should give an overview on the possibilities of the Intranet in the tourism domain. We focused on two applications of the Intranet namely the use of the Intranet for Executive Information Systems and its use as an Internet application for the tourist.

6. References

[Bernard 96] Ryan Bernard
Building a Corporate Intranet
http://webcom.com/wordmark/sem_1.html

[FedEx 96] FedEx Tracking
http://www.fedex.com/cgi-bin/track_it

[Gates 96] Bill Gates
Intranets help solve the puzzle of sharing information
http://www.microsoft.com/corpinf-g/column/1996essay/intra124.htm

[Partnet 96] About PartNet
http://PART.NET/about.htm

[Strom 95] David Strom
Creating Private Intranets: Challenges and Prospects for IS
http://www.strom.com/pubwork/intranetp.html

Electronic marketing and business reengineering -with examples from the tourism domain

Janusz Wielki
Wroclaw University of Economics
jwielki@ss5.wsi.opole.pl

1. Introduction

A large number of contemporary companies all over the world have to cope with difficulties. Not only are their competitive capabilities deficient in low efficiency but also in the organizational structure and work style. Many tenets and techniques that were very expedient some time ago and brought companies to success had to be changed. The organizational structure of most contemporary companies is based on Adam Smith's division of labor which means dividing work into repeatable tasks. Using this philosophy of work, companies created **pyramidal organizational structure** with a great number of controllers, planners and managers of all level. With the passage of time such a hierarchical structure became ineffective. The distance between the management and product users dilated. Companies turned out to be too slow and not enough flexible. Quick and rapid upheavals were the most important reasons that took place in their business environment. Customers expectations increased violently. For many companies their organizational structure and the way they used to do their work began to constitute a very big problem. Some managers noticed that if companies want to meet the requirements of the contemporary market and their customers, they must focus on **processes** instead of tasks. And because of this the whole company should be seen as a **set of processes.**

2. Process description

How the **process** could be defined? According to [HARR91] **process** could be described as *an activity or set of activities that takes an input, adds value to it and provides an out put to an internal or external customers by using organization resources.* Definition of [ÖSTE95] underlines the role of information technology. It says that **process** *is a set of activities to be performed in a specified sequence, supported by IT applications which value creation is composed of the outputs to process customers.* Because of the process complexity we can describe a process hierarchy as follows [HARR91]: *macroprocess, subprocess, activities, tasks.*

In many organizations the most important processes are the production and business processes. Within this elaboration, attention will be concentrated on business processes.[HARR91] defines **business processes** as "*all of service processes that support production processes*" .

According to him there are five important distinguishing features of any business processes: *flow, effectiveness, efficiency, cycle time, cost.*

Flow should be understood as the methods used for transforming input into output. **Effectiveness** of the process, measures the quality by checking how well the process meets expectations of its end customer. **Efficiency** tells how well resources were used to

provide an output. The fourth characteristic of business process is **cycle - time**. It informs how much time is needed for transformation, from input to final output. The last characteristic of the business process is its **cost**. It is a very important aspect of the process and it means the expenses of the entire process.
Probably the most important aspect of every business process is its **customer** or **customers**. As stated by [HARR91] **process customer** is a *person or company which receives, directly or indirectly output from the process*. Process customers can exist inside the organization and in this case they will be called **internal customers**. Customers outside the organization are called **external customers**.
There can be distinguished five types of customers[HARR91]:

- **primary customer**
- **secondary customer**
- **indirect customer**
- **external customer**
- **consumer**

3. Changes to be introduced

In the sequence we will introduce the notions of **change** that should be led in companies cognizant of its necessity. Furthermore an important issue is understanding of what sort of changes should be introduced; as well as the way they should be performed and by whom.
In the 80's the panacea for companies' drawbacks seemed to have been the total quality management (TQM) but results realized, far less dramatic than expected [HARW95].The reason of the negative reputation of TQM was one-dimensional solution approach. Organizations also wanted to improve other aspects of their performance like service, costs, innovation [ZOHO95].
H.J.Harrington suggests another way of introducing changes in organizations. To describe process of changes he introduces the definition of **business process improvement** (BPI). According to him " *business process improvement is a systematic methodology developed to assist an organization in making significant advances in the way its business processes operate*" [HARR91].
M.Hammer together with J.Champy goes further than Harrington and introduces the conception of **business reengineering** . This conception has become very quickly the most popular one in the business world of the 1990s. Money that American companies have spent on business reengineering is very big. According to [ZOHO95] by the end 1997 the business process and information systems reengineering is expected to achieve the value of $ 52 billion dollars. Not everybody is so enthusiastic about business reengineering. Some, like Paul A. Strassman, considers business reengineering as nothing new, just "*a clever buzz word*" [STRA94]. He puts forward an allegation against Michael Hammer of his using violence and revolutionary methods in rhetoric and practice [STRA94], [STRA95]. What really does this emotive conception called business reengineering mean ? The core of business reengineering is **discontinuous thinking.** It means "*identifying and abandoning the obsolete rules and fundamental assumptions that underline current business operations*" [CAHA95]. What is the definition of **business reengineering**? According to [CAHA95] **business reengineering** is "

the fundamental rethinking and radical redesign of business process to achieve dramatic improvements in critical, contemporary measures of performance, such as cost, quality, service, and speed".
There are four key words in this definition: *fundamental, radical, dramatic, process.* First of those words means that business reengineering is the process where there is no assumptions; nothing is given. The most important is what *should be,* completely unimportant stays what is. **Radical** redesign means paying no attention to all structures and procedures being used so far and preparing brand new ways to achieve the goals of the company. The word **dramatic** means that business reengineering should be introduced only when there is a real need for a crucial change. The last important words from the definition of business reengineering is **process** [CAHA95].The characteristic of the processes was given before, so there is no need to reiterate it.
As stated by [CAHA95] there are several characteristics of almost every reengineered business process:

I) Several jobs are combined into one .
People cannot and shouldn't be specialized only in performing single tasks. Where it is possible, tasks should be combined into one to make one person to be responsible for the whole process. Such a person is called **a case worker**.

II) Workers make decisions
Implementing business reengineering into a company, causes compression not only horizontally but also vertically. It means that in many cases workers will have to make decisions by themselves.

III) The steps in the process are performed in a natural order

Business reengineering introduces delinearisation to all processes where it is possible; what means that very often, jobs can be done simultaneously .

IV) Process has multiple versions
To meet the requirements of all kind of customers it is necessary for the company to have multiple versions of the processes. It is also important to standardize all work procedures of every version of the process. Standardization guarantees that people performing work will use the best ways to accomplish goals [HARR91].

V) Work is performed when it makes the most sense
In reengineered company work is shifted across organizational boundaries to improve process performance.

VI) Checks and controls are reduced
In reengineered processes controls are reduced and they exist only in these points of process where are really necessary and make economic sense.
VII) Reconciliation is minimized
Processes reconciliation is minimized by cutting the process number of external points of contact.

VIII) A case manager provides a single point of contact

Case managers act as if they were responsible for the whole process. They behave like intermediary between the process and customer.

IX) Hybrid centralized/decentralized operations are prevalent

Fast development of **information technology** (IT) enables companies to operate as if their units were completely autonomous. Using suitable equipment workers have quick and from every point, access to necessary information and they also can send informa tion to headquarters. It gives vital advantage and gives possibility for a quick response to the market requirements.

4. The role of IT in business reengineering

As it was mentioned above there is a significant role of information technology in changes introduced into companies. M.Hammer and J.Champy indicating its very important role in reengineering call it "*essential enabler* " [CAHA95]. But role of IT is very frequently misunderstood by companies and especially their executives. Managers often consider that putting, for example, a new quick computer into existing processes, will speed them enormously and make those processes extremely efficient. Computers very often will accelerate old processes but in a small range, much smaller than expected. Another problem which happens in companies is a lack of integration of hardware and software being used. It causes the real problems for the data flow and exchange within the organizations. Such a situation is often caused because of unintelligent introducing of so called "*computerization*". According to [CAHA95] the true and powerful role of information technology lies in breaking obsolete, existing sometimes for a long time rules and making completely new ways of performing work. So, using of IT is not an automation of existing processes but arranging them quite differently by means of latest capabilities of technology.

5. Change performance

Before changes will be introduced it is very important to select proper process or processes for reengineering because the real object of reengineering are the processes, not the whole company. Choosing processes it is necessary to remember that improving the individual processes is not a goal of reengineering. The true aim of business reengineering is enhancing customer satisfaction and company profitability, so it is important to have a global view of all changes being done [HARW95]. According to [CAHA95] to choose which process or processes require reengineering and what is the order in which they should be tackled, three criteria should be used:

- *dysfunction* - one should look for processes which are in the deepest trouble.
- *importance* - it is crucial to point out processes which have the biggest impact on functioning the whole company
- feasibility - it means that for reengineering those processes should be chosen which are most susceptible to successful redesign at the moment.

As stated by [ZOHO95] the major functional areas of business reengineering nowadays are: finance/accounting (28%), customer service (16%), sales and marketing (12%), in-

ventory distribution (9%). Managers expect that the targets of reengineering in the close future will be such processes as: financial (59 per cent),sales and marketing (58 per cent) and supply chain management (39 per cent) [CANW95].
After choosing the proper processes for reengineering, the next step is to understand the functioning of them. People who will reengineer must be cognizant of the process working, how the flow looks inside them. Without deep and meticulous understanding of the process functioning, it is impossible to introduce the real reengineering. The simplest way for understanding the process is its observation and participation in it for some period of time. Also, very useful can be preparing flowcharts of the process. There are many types of flowcharts but for understanding business processes four of them can be very expedient: *block diagram, ANSI standard flowchart, functional chart, geographic flowchart* [HARR 91].
Proper choosing of processes to be reengineered and good understanding of them is really very important thing. But without suitable groups of people, who will lead and control the whole process, nothing will be changed in the company. As stated by [CAHA95] in business process reengineering the following participants should be involved:
-**Leader** is a person, mainly a senior executive, who has a clear vision of the kind of company he or she wants to create in the future.
-**Process owner** - senior level manager who is responsible for reengineering a process .
-**Reengineering team** - a group of people between five and ten in number, who perform the real reengineering work. Every reengineering team should consist of two members: *insiders* and *outsiders*. *Insiders* can be described as people who currently work inside the process which will be reengineered. *Outsiders* are people who don't work in the process, so they present a higher level of objectivity.
-**Steering committee** -a group consisting of senior managers, including process owners, whose role is to prepare a strategy for reengineering.
-**Reengineering czar** - a person who coordinates the whole reengineering activities inside the organization, and who supports each process owner and reengineering team.
It is necessary to underline the need of the top management's support for the changes. It seems to be obvious that without this sort of support, any kind of changes appeared to be impossible to be introduced. Many trials of process reengineering failed because of the lack of top management's support [HARW95].
Introducing reengineering into the processes is a very essential issue to be solved. The problem lies in measuring the improvement progress. The goal is the deep redesign, but how good should the process be? To find out what company's goal should be, is often very helpful to look outside the organizations in order to observe, how processes are well performed by other companies. A process of searching for organizations that do something in the best way, and learning from them how they perform task in order to emulate them, is called **benchmarking** [CAHA95]. More formal definition of benchmarking is given by Harrington. As he states "*act of systematically defining the best systems, processes, procedures and practice is called benchmarking* ". In fact it is the only method for measuring how deep changes are. Generally, however, it is difficult to find consistent, reliable method of measuring the costs and benefits of business reengineering. A lot of managers use conventional financial measurements like: return on investment, increased profitability, and reduced costs to determine how effective reengi-

neering is [CANW95]. Thus, there is a real need for developing completely new tools which could be expedient in business reengineering. There is one more important question at the end. In what way fundamental changes should be introduced ? Should the companies choose revolutionary way, as proposed in [CAHA95] or rather, the way of evolution, proposed by [HARR91] ? H.Österle states that both ways should be combined, because business reengineering requires revolution at the beginning and in the next years continuous refinement (i.e. evolution) of processes [ÖSTE95].

6. Electronic market - advantages and disadvantages

As it was mentioned earlier, market and customers play crucial role in strategies of most contemporary companies. The process of sales and marketing is becoming increasingly, the targets of business reengineering. The most important reason is in huge development of electronic market in the last few years caused by matured technology. A great number of companies has noticed quite new possibilities for their expansion, growth and functioning. Thus, they have to reengineer their old processes in order to adjust them to a new situation.

Generally, the history of electronic market is not very long. In the early 1970s some companies, mainly in the USA, created single-source electronic sales channels hoping that it would be a source of their strategical advantage. But those first attempts were not successful. In the next years there were many other endeavors to create electronic links between suppliers and customers, but the real explosion began in 1990s with the rapid growth of the Internet and particularly its part called World Wide Web (WWW). Increasing, companies' interest in the Internet has made much more probable vision of using it for creating real electronic market. Since each transaction comprises three phases: *information gathering, negotiation, settlement*, **electronic market** can be defined as " *market places put into action by means of telematics which support all phases of market inter-transaction, including the formulation of prices for goods and services*" [SCHM94]. The definition given above describes electronic market in a narrow sense. In fact, another more liberal definition is more suitable for describing current state of electronic market. It says that it should be understood as " *information systems supporting one or more phase and/or function of coordination within market systems*" [SCHM94]. As it was mentioned earlier, the Internet is the most important area of electronic market creation and development.

According to [HOKR93], the **Internet** can be described as:

1) a network of networks based on TCP/IP
2) a community of people who use and develop those networks
3) a collection of resources that can be reached from those networks
"

The Internet's history began in 1970's with created by US Department of Defense experimental wide area network called **ARPAnet**. Introduced in 1973 and 1974 TCP/IP protocols enabled all ARPAnet computers to communicate with each other regardless their operating system. This was the beginning of the Internet. In the next years new services appeared [SHAH94]. The Internet's growth is really impressive. On the Janu-

ary 1st 1996, the number of hosts was 9,472,000. The number of Internet users is being assessed for 40 mln. Nowadays the most important part of Internet is **World Wide Web**, initiated by Tim Berners Lee from the European Laboratory of Particle Physics (CERN) in Geneva. He proposed a conception of WWW in 1989 and two years later it started to function. Web is an information service based on hypertext and it provides an access to multimedia complex documents, databases. The World Wide Web is increasing with a very big speed as the whole Internet. It is the first example of **hypermedia computer-mediated environment** (CME). **CME** can be defined as: "*a dynamic distributed network, potentially global in scope, together with associated hardware and software for accessing the network, which allows consumers and firms to:*
1. provide and interactively access hypermedia content (i.e. 'machine interaction') and
2. communicate through the medium (i.e. 'person interaction')" [HONO95].
In this new environment, a marketing communication model changes. Traditional marketing communications model is **one-to-many** process where company transmits a content *through a medium* to a big group of consumers . In this model, there is no interaction present between consumers and companies. The new communication model in a hypermedia CME is **many-to-many** process in which two types of interactivity are possible: *through the medium* and *with a medium.* Not only can firms provide a content (*hypermedia)* to the medium *(distributed computer network)* and interact with it, but also consumers themselves. In **many-to-many** communication model, primary relationships are not between sender and receiver but primary becomes interaction with CME.[HONO96]. Not only does the communication model change in hypermedia computer-mediated environments like Web. Since Web is a quite a new environment for leading marketing activities, conventional marketing activities have to be transformed; it is difficult to implement them in the current form [HONO96]. It seems to be particularly important issue because it is predicted that within 6 years there will be more than $600 billion dollars over the Internet[KILL95].
There are already some suggestions about reconstruction marketing activities in this information intensive environment. For example in [GLAZ91] there are several proposals referring to: market attractiveness, competitive position and organizational structure. There are such corollaries, among others, as: "*shorter product life cycles*", "*breakdown in the formal distinction between producer and consumer*", "*buyer/supplier power shift in favor of buyers* ", "*breakdown in distinction between marketing and other departments in firm*", "*focus on cooperative as opposed to competitive strategy* ".
There are also other attempts in this area. Generally there is an urgent need for creating new marketing paradigms including cooperation and competition with strong focus on active consumers role in a new communication model.
Internet and particularly Web can be used for many business purposes as [SHAH94]:

- *Creating a Client Base*
- *Product Analysis*
- *Market Analysis*
- *Expert Advice and Help*
- *Recruit New Employees*
- *Rapid Information Access*
- *Wide Scale Information Dissemination*
- *Rapid Communications*

- *New Business Opportunities*

Nowadays six categories of commercial Web pages can be defined [HONC95] :
On-line Storefront - combines elements of direct marketing with in-store shopping, and there is also potential for them to be extremely efficient. But there are also challenges for such shops. The technology of the on-line security payment is not yet ready. Some problems stay unknown as : *a consumer behavior issue, segmentation problem, impact of customer designed incentives.* Especially important seems also to be offering something that would stimulate customer's *" trial and repeat",* some kind of *„cyberhook"*[CHAS95].

Internet Presence-gives opportunities for firms to present themselves and present their offer to current and future customers. The Internet presence can be divided into :
* *Flat Ad* which means single electronic page without hypermedia links.
* *Image* this kind of page is used when information content is of lower importance than picture
* *Information* which means detailed and rational description of the company and its offer.

Content- this kind of Web site can be divided into three categories:

- *Fee-Based*
- *Sponsored*
- *Searchable database*

Mall - site which consists of collection of on-line storefronts.

Incentive Site- is a unique form of advertising which should attract potential customers to a site. The real goal is to pull customer to the commercial site behind it.

Search Agents - are Web sites used to find other Web sites through keyword search. The meaning of such Web sites is particularly important because of poliferation of Web sites and increasing difficulty of finding anything.

7. Security

The description given above points out many new business possibilities which appeared for companies together with the development of the Internet. However, there are also problems slowing its commercialization. Undoubtedly considered as two most important of them are *security aspects* and *lack of implemented payment methods*. Both of them will be widely discussed now. It seems that most important aspect which is necessary to be solved for the quick development of electronic commerce is providing the Internet users with such security system or systems which will guarantee that exchanged information will not be read or changed by anybody who is not its recipient. It became particularly important when companies started to treat Web and the whole Internet as a marketing channel or a channel for information exchange between companies subsidiaries, it became to be obvious that at that moment secure channels had to be established.

Security is necessary to guarantee privacy very widely, from e-mail massages to information important for electronic transactions. So its significance is really difficult to be overestimated. Generally, to encrypt messages, two cryptosystems are used [ZIMR93]. The first one is called **secret key cryptography** or **symmetric cryptography.** In this system a single key is used for the massage encryption and for decryption. Thus, the sender and recipient of the message have to use the same key, which means that a key must be transmitted through the secure channel before encryption. Because of the necessity of secret key exchanging through secure channel, sometimes with many recipients, this cryptosystem seems to be a very impractical one for the Internet use. The second system, much more feasible, is called **public key cryptosystem** (PKCS) or **asymmetric cryptography.** In this system each user has two related complementary keys: *public key* and *secret key (private key).* Each public key can be distributed via communication networks without any protection. Everybody can use someone's public key to encrypt a message but only a recipient can decrypt it, using its secret key.

There is also a possibility for a message authentication. Authentication can be achieved by means of **digital signature.** Signed message can be verified by anyone. The first step in creating digital signature is preparing **message digest.** Message digest is a message of any length reduced to a string of a fixed length. It "represents" the whole message and is used to detect potential changes which could be made. To create digital signature, message digest is finally encrypted with sender's private key and is added to the encrypted message.

Public key cryptography is also used to create **digital envelope.**
In this method, the message to be sent, is encrypted with randomly generated secret key and then the secret key is encrypted with recipient public key. Recipient receives encrypted message and encrypted secret key [KALI93]. There are also some problems with cryptosystems usage. In some countries, like Iran and France, encryption usage is illegal or restricted as in USA [MCEL96].

The best known example of the secret key cryptoysystem is US Federal **Data Encryption Standard** (DES). DES has fixed key size of 56 bits and is used mainly by financial institutions [BABR5].

As to public key cryptography, the best known standards are offered to developers of computer systems by **RSA Laboratories.** Probably the widely known implementation of public-key technology, is created by Philip Zimmermann, **Pretty Good Privacy** (PGP). This system can be used to protect e-mails and data files with no need for secure channels for prior key exchange. Since public key encryption algorithm is comfortable for use, but much slower than single key encryption, PGP combines both of them using mentioned earlier *digital envelope.* As a conventional key PGP does not use DES but different encryption algorithm called **International Data Encryption Standard** (IDEA) with longer, 128 bits key size. Very important feature of Pretty Good Privacy is data compression which is made before encryption [ZIMR93]. There are also some protocols using public key cryptography for providing privacy in the Internet. One of them is Netscape's **Secure Sockets Layer** (SSL) protocol. This is a security protocol providing privacy for client/server applications, communicating over the Internet. SSL provides a layered implementation of encryption. It provides additional layers between application and TCP/IP connection layers in the network protocol stack. SSL provides security by encrypting the communication channel between a client (e.g. Navigator) and a server. In SSL protocol application data messages are carried by fragmented, compressed and encrypted. [FRKK96].

Another protocol being used for providing security in the Internet is **Secure Hypertext Transfer Protocol** (S-HTTP). S-HTTP is a security-enhanced version of HTTP. S-HTTP protocol and earlier mentioned SSL are not mutually exclusive. Both of them can co-exist in a complementary fashion by layering S-HTTP on top of SSL [FUNB95]. S-HTTP enables incorporation of various cryptographic message format standards into Web clients and servers.
Message to be sent can be protected in three ways : *by signing, encryption, both of them.* [RESC94].
Another security protocol being introduced by Visa and MasterCard is **Secure Electronic Transaction** protocol (SET). The SET protocol is limited to financial part of shopping. Its goal is providing a secure payment processing using bankcard products over open networks (like the Internet) and the private ones. Generally, the Secure Electronic Transaction protocol uses cryptography to: *provide information security, ensure payment integrity, authenticate cardholder and merchants.* SET protocol uses both encryption methods: secret key cryptography and public key cryptography. In SET protocol there is also introduced a new application of digital signature called **dual signature.** A dual signature is generate by creating the message digest of two messages. In the next step two digests are linked together. Finally, the message digest of the result is computed and encrypted with the signer's private key [VISA96].

8. Payment systems

Apart from security there is one more element which should be urgently established to turn World Wide Web and the whole Internet into a highly efficient electronic market for goods and services. This element is reliable payment system or systems. Such a system should be [[OMPE95] : *scaleable, anonymous, secure.*
The need for the system to be highly *scaleable* is very important, taking in consideration the fact that there are million users of the Internet and its number is growing quickly. *Anonymity* of the system is also a very important feature. The system must offer customers some protection against attempts of information collection. *Security* of the payment system is its most important feature but this aspect was previously discussed.
Nowadays there are about 50 proposals of Internet payment systems [PEIR96]. Some of them are being implemented, others are only projects. On the whole, the Internet needs quite new payment methods because it is obvious that the traditional ones are not suitable for the use over the Internet and possess many deficiencies.

All currently proposed Internet payment systems can be divided into four groups [CRED96]:

- ***Smart Card Based Systems***
- ***Credit Card-based Systems***
- ***Digitized "e-cash" Systems***
- ***Payment Clearing Systems***

The system called **Mondex** that has been developed in England is a very good example of the *smart card based system*. Mondex card is a usual plastic card with built-in micro-processor which serve as an " electronic purse ". It can be loaded with money which is stored until the payment is made. Since Mondex card allows to send and receive money instantly through phone lines it can be easily used for person-to-person payments across the world. Transactions made by means of Mondex card do not need any authentications and signatures. There is also a security system on Mondex card. It enables its user to lock the card, using a personal code so that only the card's owner can use the money on it or check the transaction log [MOND95]. Prepared by **Visa** and **MasterCard** payment system, based on discussed earlier protocol **SET** (*Secure Electronic Transaction)*, can serve as a good example of a *credit card-based system*. The SET system is expected to be fully ready in the fourth quarter of 1996. There were also earlier attempts of usage credit cards for payment making over the Internet. Such systems were based on transmission of credit card details by means of phone or fax, but sending such information had caused additional costs and had not been specially comfortable for customers. The SET system is expected to change this situation and create good, secure credit card-based payment system. There are several steps in SET payment system. Before the initial purchase can be made, every cardholder must register its card, on-line with its bank. In the next step issuer checks the account's authenticity and then sends electronic certificate to customer. This certificate which is stored in customer's computer, proves that its card is valid. Having digital certificate, the customer can begin shopping. In the same way merchants, who want to conduct electronic commerce, have to register. Before the customer's purchase, the merchant needs to provide him or her its electronic certificate. It can be done in two ways. After checking validity of merchant's digital certificate, the customer sends its order electronically to the merchant. In this moment merchant seeks authorization for the amount of the purchase. After approval completing, the order is processed and confirmed. The whole transaction takes only seconds to be completed[VISA96] .

As the *digitized "e- cash" systems* are concerned, one of the most widely known is prepared by **DigiCash** system called **Ecash.** This system is composed of three elements:

- ***Banks*** which mint coins, validate existing ones and exchange real money for Ecash
- ***Buyers*** who have accounts with a bank and can withdraw or deposit Ecash coins there.
- ***Merchants*** who accept Ecash coins as a payment for goods or services.

There is a special software required for buyers and merchants who want to use Ecash system. The buyers' one is called a **cyberwallet** and is necessary for withdrawing and depositing coins from a bank and for paying or receiving coins from merchant. To guar-

antee security of transactions, Ecash uses RSA public-key cryptography. Very important feature of Ecash is providing full anonymity and transaction security. The only problem can be in the future size of the database of spent coins, especially when a large number of people will start using this system.
The most commonly quoted example of payment clearing system is **The First Virtual Internet Payment System.** This system requires no special hardware, software or encryption. To be able to make shopping buyer must register its credit card with First Virtual to receive individual **VirtualPin** (account ID). Process of registering is made over the telephone. When the customer wants to buy an item he or she sends VirtualPin instead of a credit card number to the seller. Seller sends a message to First Virtual containing the buyer's VirtualPin and the amount of the sale. First Virtual sends customer e-mail to confirm purchase. After receiving confirmation , customer's credit card is charged by First Virtual and the buyer is notified via e-mail [FIRS95].

9. Electronic marketing examples in tourism and other domains

Description given above showed quite new possibilities, challenges and also limitations which appear for companies in accordance with dynamic development of the electronic market. In fact the growing number of companies try to involve the Internet into their business strategy. Not all of these attempts can be considered as successful ones but there are many good and innovative examples of the Internet usage in commerce. Now it is a good moment to present some of them.

One of the first attempts of electronic market usage is connected with United Airlines and their reservation system Apollo. Current version of this system is still in use. It is a system led by American Travel Corporation, which service called **PCTravel** allows making the air reservation and ticketing service through the World Wide Web. There is no fees for using the system. Using this system customer, from USA or outside, has an access in real-time to schedules, fares and tickets for more than 700 airlines worldwide by means Apollo Reservation System. This reservation system is protected by SSL protocol. There are two airline tickets payment options: credit cards and money transfer. There are several different methods of entering credit card information : through the on-line form, directly to PCTravel by modem, by fax, by phone [PCTR96].

One of industries which quickly noticed huge potential laying in the Internet is tourism. As an example from this industry can serve **Internet Travel Mall**.
The goal of Internet Travel Mall is to provide Internet users a centralized source of amusement, entertainment, recreation and other travel related information, products and services. It also provides links and databases to other Internet travel related services and sites.
Through the Internet Travel Mall customer can view entire brochure, review rates and book information and also contact via e-mail to book reservations or ask questions about products or services. Internet Travel Mall does not provide on-line reservations or ordering but some of their advertisers do this. Owners of travel - related business are offered full range of advertising packages, using text, color images, graphics, audio or video. They are also offered quarterly summary of the number of travel shoppers who

access advertisements each month. Each company's Web site is listed with at least 16 Internet Directory services so more customers can find it [ITM96].

One of the companies which established its business in the Internet is famous **Harley-Davidson.** It is a quite small company employing fifteen people with gross annual sale between $3million and $4 million. Harley-Davidson decided to establish Web site to create international presence which would lead directly to concrete sales. It was specially important because of their lack of distribution channels and large advertising budgets. Not less important motivation for establishing Web site was attempt to create a dialog between a company and Harley enthusiasts worldwide. So site features a calendar of upcoming Harley-Davidson events, service tips, Racer's Corner as well as T-shirt contest that rewards riders who send the best pictures of themselves riding Harleys. Commercial part of the Web site promotes Harley-Davidson and Buell motorcycles, offers direct sales of spare parts, collectibles, clothing and accessories. The credit card is a basic payment method. Although Harley-Davidson site is on insecure server, customers have shown no hesitation in using credit cards. To protect credit card users it is also possible for customers to register their address and billing information. Orders can be placed by e-mail, authenticated with a password. Customers calling from USA can also place orders via a toll-free number. Harley-Davidson's example shows that Web marketing can be specially useful for small companies, enabling them to compete with big corporations even with much smaller budgets [LISH95], [HARL95].

Quite a different kind of company is the **Internet Tradeline** It is a company whose target is to provide with global trading service to bring buyers and sellers together in a low-cost communication and marketing medium. It is designed to create direct communication link between buyers and sellers and to facilitate complete international trade transactions. The International Tradeline, Inc. is concentrated on a business - to - business transactions.

The offer of its services is very wide and comprises: on-line product search, on-line presentation, marketing analysis, assistance with negotiating contracts, preparing import-export documents etc. So far about 30 thousand companies from the whole world have put their offers in the Internet Tradeline. Although ITI is an American company and will be soon quoted on American Stock Market, they open their subsidiaries in Poland, Russia , Turkey, Greece, Korea, France, Germany, Hungary, England and many other countries around the world [INTR95].

The last example presented here is **Bargain Finder**, which is an experimental service introduced by Anderson Consulting as a part of the Smart Store Virtual initiative investigating an electronic commerce. Bargain Finder serves as on-line intelligent agent for finding pop or rock CD records from nine virtual retailers. Most of them offer on-line secure transactions. Bargain Finder helps to find the best price. The only thing a customer has to do is to type an artist's name and album title. Bargain Finder searches databases of all nine virtual shops and present its findings. Next, a customer chooses the most interesting offer and links with a selected shop to make on-line order. Bargain Finder is an interesting concept of comparison shopping which can be implemented for different kind of products and services shopping. Such a concept is particularly important because resources of the Internet are huge and finding the best offer is frequently time consuming [BARF95].

10. Conclusion

This short overview points out the current state of electronic market and electronic marketing with its important chances, as well as problems and limitations. It also presents the essence of business process reengineering, the fundamental issue in the context of the Internet rapid development. The Internet as a new business environment, redefines roles of both market participants - customers and companies. Increasing customers' role is accompanied with companies new possibilities. Companies have to reposition themselves rapidly and rethink the way they lead their business processes in order to involve Web marketing into their general strategy. In this new situation, the role of electronic marketing and its influence on business reengineering seems to be particularly significant. The forthcoming years will show how many companies have noticed these new possibilities and have been able to reengineer their business processes, in order to adjust them to a new situation.

References

[BABR95] Back A., Brooks P., Brown A.; The Simple Key Search Protocol ; A Cyberpunks Working Group, July-August; [Piete Brooks@computer-lab.cambridge.ac.uk]; 1995

[BARF95] Bargain Finder ; Your Intelligent Agent for Comparison Shopping; [http://bf2.cstar.ac.com/bf.html]; 1995

[CANW95] Business expands scale and scope of reengineering; today's objectives stress revenue growth, survey finds; [http://www.newswire.ca/c3094.html];1995

[CAHA95] Champy J., Hammer M.; Reengineering the corporation; Nicholas Brealey Publishing; 1995

[CHAS95] Chase L.; Pro-Active Marketing on the World Wide Web; Marketing Tools Magazine; 1995

[CRED96] Crede A.; Electronic Commerce and the Banking Industry: The Requirement and Opportunities for New Payment Systems Using the Internet;[http://shum.cc.huji.ac.il/jcmc/vol1/issue3/crede.html]; 1995

[FIRS95] First Virtual Internet Payment System; Payment System Summary (Buying and Selling); [http://www.fv.com/intro.html]; 1995

[FUNB95] First Union Corp.; SSL versus S-HTTP; [http://www.firstunion.com/library/securi.html]; 1995

[FRKK96] Freiser A.O., Karlton P., Kocher P.C.; The SSL Protocol, Version 3.0; [http://home.netscape.com/info/security-doc.html]; 1996

[HARW95] Hall G., J.Rosenthal, J.Wade; How t o Make Reengineering Really Work;Harvard Business Review;Nov-Dec 1995

[HARR91] H.J.Harrington; Business Process Improvement; McGraw-Hill, Inc.;1991

[HARL95] Harley-Davidson of Stamford Connecticut Homepage;[http://www.hd-stamford.com/]; 1995

[HOKR93] Hoffman E.,.Kroll E.; FYI on „What is the Internet" Network Working Group Request for comments; [gopher://ds1.internic.net/00/fyi/fyi20.txt];1993
[HONO95] Hoffmann D.L, Novak T.P.; Marketing in Hipermedia Computer-Mediated Environments: Conceptual Foundations;
[http://www2000.ogsm.vanderbilt.edu/ cmepaper.revison.july11.1995/cmepaper.html]; 1995
[HONO96] Hoffman D. L., Novak T. P.; A New Marketing Paradigm for Eletronic Commerce; [http://www2000.ogsm.vanderbilt.edu/new_mark.html]; 1996
[HONC95] Hoffman D.L, Novak T.P., Chattrjee P. ;Commercial Scenarios for the Web: opportunities and challenges; [http://shum.cchuji.acil/jcmc/vol1/issue3/scenario.html]; 1995
[ITM96] Internet Travel Mall Homepage; [http://io.com:8001];1996
[INTR95] Internet Tradeline Homepage; [http://www.trade90.com];1995
[KALI93] Kaliski B.S.; An Overview of the PKCS Standards;
[http://www.rsa.com/pub/pkcs/ovwerview.txt];1993
[KILL95] Killen's 1995 study; Opportunities in electronic commerce; in: [SCHU95]; 1995
[LISH95] Lightman A., Shears N.; Harley-Davidson of Stamford Connecticut;[http://arganet.tenagra.com/V1I2C1.html]; 1995
[MCEL96] McElhearn K.; Ubervista is watching you; TidBITS # 333;
[http://wais.sensei.com.au/macarc/tidbits/searchtidbits.html]; 1996
[MERI96] Merit Network Inc; Internet Growth Statistics;[ftp://nic.merit.edu/nsfnet/statistics];1996
[MOND95] Mondex Homepage; [http://www.mondex.com/mondex/home.html];1995
[OMPE95] O'Mahony D., Peirce M.; Scaleable, Secure Cash Payment for WWW Resources with the PayMe Protocol Set;
[http://www.w3.org/pub/Conferences/WWW4/Papers/228/paper.html]; 1995
[ÖSTE95] Österle H.; Business in the Information Age, Heading for New Processes; Springer; 1995
[PCTR96] PCTravel Homepage; [http://www.pctravel.com]; 1996
[PEIR96] Peirce M.; Payment mechanisms designed for the Internet;
[http://ganges.cs.tcd.ie/mepeirce/onintern.html];1996
[RESC94] Rescorla E., Schiffman A.; The Secure HyperText Transfer Protocol;
[http://www.commerce.net/information/standarts/drafts/shttp-1.0.txt]; 1994
[SCHM94] Schmid B.; Electronic Markets in Tourism;
[http://wwwiwi.unisg.ch/iwi4/cc/genpubs/enter/index.html]; 1994
[SCHU95] Schultzer D.; Payment methods over the Internet;
[http://www.IInl.gov/fstc/projects/commerce/public/payment2.pdf]; 1995
[SHAH94] Shah R.; The business of the Internet: an introduction to the Internet for commercial organizations; [http://www.rtd.com/people/rawn/business.html]; 1994
[STRA94] Paul A. Strassmann; The Hocus-Pocus of Reengineering; Across The Board;
[http://www.strassmann.com/hocuspo.html]; 1994
[STRA95] Paul A. Strassmann; Reengineering; excerpted from The Politics of Management; The Information Economics Press;[http://www.strassmann.com/iep/info-politics.html]; 1995

[VISA96] Visa International; SET specification, Book 1: Business Description; [http://www.visa.com/cgi-bin/vee/sf/set/setbus.html]; 1996

[WILD95] WilderC.;An Electronic Bridge to Customers; [http://techweb.cmp.com/iw/509/10mton.html]; 1995

[ZIMR93] Zimmermann P.; PGP User's Guide, Volume I; [ftp://ftp.demon.co.uk/pub/pgp/pgp26uix.zip];1993

[ZOHO95] Hoffman Z.; Business Process Re-engineering : a new strategic paradigm shift in change management; [http://alpha.bzlogi.hu/stagg/hoffman.html]; 1995

Information Technologies in Hospitality Businesses: The case study of La Plagne, France

Dr Dimitrios Buhalis, Steve Keeling, Adria Lacorte and Naomi Reynolds.

School of Urban Development and Planning, University of Westminster,
35 Marylebone Road, London NW1 5LS
TEL: 0171-911-5000 X 3112, FAX: 0171-9115171, EMAIL:buhalid@wmin.ac.uk.

Abstract

Information Technology prevails the development of contemporary management and marketing transforming the best business practices and redefining the competitive advantage of each actor in the economy. In the tourism and hospitality industries information technology has traditionally been under-utilised due to a wide range of reasons. The majority of the industry is subjected to an inevitable competitive disadvantage which may result on the one hand in the expansion of multinational corporations which have both the resources and the financial backing for investing in technology, and on the other hand in the development of technology oriented enterprise which would serve or even replace members of the tourism industry and achieve competitive advantages within the digital economy context. This paper concentrates on ITs penetration in the hospitality organisations in the Alpine resort of La Plagne, France. Hospitality is provided by hotels, self-catering accommodation and tourist residence and there is a high presence by leading French Companies which operate horizontally integrated facilities throughout the country. Although the presence of ITs at the moment is limited, hoteliers appreciate the technological developments and gradually invest towards the improvement of their system. The paper examines the level of technological penetration by reporting on the existing facilities as well as analysing the future plans for technology incorporation. In particular it highlights that attitudes and beliefs of decision makers are critical for ITs adaptation and that education and training are crucial for the future competitiveness of hospitality organisation.

1 Information technologies in accommodation establishments

The arrival of the Information Superhighway in the early 1990's coupled with ongoing developments in Information Technology (IT) heralds the dawn of an electronic age in which a whole system of ITs is being rapidly diffused throughout the tourism industry [29]. The revolutionary capabilities that ITs present, provides innovative hospitality managers with a strategic tool with which to re-engineer established business practices in order to achieve competitive advantages and enhance competitiveness in the global tourism arena. Previous research carried out in SMTEs of Wales and the Greek Aegean Islands [18,19], revealed the acutely low level of ITs application and strategic weaknesses of small hospitality establishments. A comparable approach is utilised at the French ski resort of La Plagne to test whether peripherality and similar vulnerabilities undermine the ability of SMTEs to take advantage of ITs and identify factors which facilitate ITs adoption.

1.1 The ITs promise to accommodation establishments

Many of the large hotels are at a mature stage in computer literacy and ITs penetration. They have been quick to capitalise on the capabilities of ITs by reshaping intra- and inter- organisational practices to suit the contemporary business environment and consumer [8]. Property Management Systems (PMSs) integrate the entire range of hospitality functions and departments and hence increase efficiency and a more profitable use of resources. Opportunities for product differentiation also arise whereby ITs initiatives can be employed to add value to the tourism product. More critically, however, ITs provide tools for real globalisation of the industry and empowers organisations to reach their new and existing markets effectively. However, smaller properties have consistently failed to incorporate ITs in their operational and strategic management as a result of their deficient economic resources, lack of expertise, functional disadvantages and management structure. As a consequence they suffer from marginalisation and find it increasingly difficult to communicate with consumers in the marketplace. The new generation of travellers have emerged as computer literate, information hungry and ever more demanding consumers, requiring the highest level of technology and service in their decision making and purchasing processes [33]. As tourism is an industry based almost entirely on information, the decision by the consumer to accept or reject the elusive tourism product will increasingly be subject to a compelling description, representation and the successful dissemination of data. ITs will also allow customers to respond by reserving and purchasing tourism products in an equally sophisticated manner.

1.2 Taking advantage of the Internet and World Wide Web (WWW)

The rapid expansion of the Internet and WWW provides a unique mechanism for organisations' marketing and distribution as it holds the unprecedented ability to disseminate in depth information and receive direct feedback from consumers in an open environment [32]. Accommodation establishments have already taken advantage of the new potential in order to strengthen their distribution function and reduce their costs. Inter-Continental hotels quote costs of $15 for a telephone booking, $4 for a CRS booking and $2 for and Internet booking [18] highlighting its cost effectiveness as an instrument for electronic commerce. Innovative small hospitality operators also discover that the Internet can provide an international representation at very reasonable cost and thus restructure their marketing function to reap these benefits. Hence the Internet is also emerging to become a fundamental facility supporting all the functions of everyday personal and professional life. Undoubtedly, therefore the provision of Internet and other electronic services will become an integral part of the customer service and value- added product of accommodation providers in the new millennium, regardless of whether they are business or leisure orientated. For example, the newly opened Lindner Congress hotel in Frankfurt boasts product differentiation through the provision of Internet, E-mail, fax and multi-functional ISDN phone with voice-mail in all guest rooms [10].

2 SMTEs and ITs: opportunities and challenges

SMTEs within the hospitality sector have been quantitatively defined as hotels with less than 50 beds or who employ less than 10 employees. Approximately 90% of hotels in the industry are family managed, independent, seasonal and flexible establishments and comprise the bulk of accommodation world-wide. The need to embrace ITs and adopt strategies suited to the rapidly and constantly changing socio-economic and technological environment will be imperative to their prosperity and long term survival [30,15,31,2,7,14].

As tourism demand trends demonstrate that authenticity and differentiation are instrumental for tourism satisfaction, the local character, uniqueness and tertiary location of the typical SMTE is the ideal prerequisite for the new tourist in there search for satisfaction. While the independent property paradoxically may continue to offer the level of personalised service, they may at the same time find it more difficult to remain in contact with their markets unless they exploit modern techniques for interactivity and partnership marketing [29,15,6]. In order for SMTEs to stand on an equal footing to their larger counterparts, they need to overcome a number of both inter- and intra- organisational obstacles which are currently obstructing the path to competitiveness and allowing them to become marginalised from their markets. The vast majority of SMTEs are non-computerised, not well organised with less trained personnel and frequently managed on irrational criteria. Therefore, GDSs provide an incompatible framework for smaller members to distribute their heterogeneous leisure product with an "off the shelf" solution and thus exclude SMTEs from their databases due to their lack of economies of scale [27,2,6]. In addition, the long established dependence of peripheral SMTEs upon tour operators and major travel agencies who provide an 'all in one' solution to the advertising, marketing and reservation duties also presents itself as a strategic weakness. In areas where supply exceeds demand, tour operators promote themselves and in fact are regarded by SMTEs as the only way to fill empty rooms. In return, SMTEs' profit margins may be squeezed as tour operators gain discount of up to 70% off the official rates. Thus SMTEs fail to realise that alternative distribution methods should be exploited through marketing in order to enhance their competitiveness [8,30]. The failure of SMTE decision makers to capitalise on ITs opportunities as a complementary and/or alternative distributor stems mainly from internal constraints, attitudes, and misconceptions. Entrepreneurs are sceptical about allowing external agencies to control their enterprise. At present, managers of small hotels can control and alter the traditional reservation and distribution channels according to their personal and social needs. The misconception that cost will hinder action has been highlighted in past research as one of the top defences which SMTE owners used in order to exempt themselves from the world of ITs [8,19]. However, Chervenak, [9], argues that the consistent reduction in ITs prices will make basic equipment affordable even to tiny organisations. This re-enforces the fact that SMTEs' ignorance and fear of change lies at the forefront of their resistance.

Another handicap which many SMTEs are facing and which also stems from internal attitudes is the reluctance to fully integrate with other destination tourism bodies for their own greater mutual benefit and for the enhancement of the destination as a whole. Destination Management Systems (DMSs) and Regional Integrated Computer Information Reservation Management Systems (RICIRMSs) have been proposed as an ideal distribution and marketing medium for SMTEs [27,7]. However, the protective and individualistic mentality of SMTE owners and managers leads them to renounce any kind of co-operation or assistance which could threaten their independence. In prevailing economic conditions, and certainly for the foreseeable future, it seems highly likely that DMSs and RICIRMSs will only succeed in becoming viable if SMTEs show the ability to break down traditional barriers and view themselves not as a stand-alone unit, but as a node in a web in which to catch the new/sophisticated/demanding tourist. Poon [28] suggests that "there will be no place for small, stand-alone participants, but the world can become the oyster for the small, innovative, flexible and networked enterprise", whilst Buhalis and Fletcher [8] emphasise the importance of increasing the "size of the pie" and hence the benefits for the destination and its enterprises, as opposed to increasing the "size of the slice" for each individual establishment. In this sense, collaborating towards maximising the value added at the destination level would enable small enterprises not only to improve their competitiveness, but also to contribute significantly to the welfare of the indigenous people and increase the satisfaction of consumers.

3 The case study: The La Plagne Ski Resort in France

3.1 La Plagne as a tourism destination

The alpine ski resort of La Plagne belongs to the French departement of La Savoie. This territory is considered the primary French departement both in number of tourists (over 20 million visitors per year), and also in terms of quality tourism (31% of beds at a 3/4 stars' standard) [20]. Tourism represents the main economic activity of the departement as it generates 40% of the GDP and 51% of the local employment [1]. La Plagne is described as the largest ski destination in the world in terms of skiing land [210 Km] with an accommodation capacity of around 45,000 beds [21]. The destination is divided into 10 ski resorts which have adopted a common brand (La Plagne) to promote the resort. French domestic visitors account for 75% of the total *demand.* The rest of the visitors are mainly from Europe and arrive almost exclusively through TOs. Tourism in La Plagne attracts visitors with a medium-high level of disposable income, especially in winter. Consumers look for a particular activity and they seem fairly uninterested in the community life. L'Observatoire Regional du Tourisme Rhone-Alpes [26] observed that business and incentive tourism is growing. The latter has mainly been attracted by the chains of Residences de Tourisme and has already proved very beneficial due to its high level of expenditure and its benefits to the seasonality problem. Although some accommodation establishments offer an excellent customer service, they lack essential business services such as in-room ITs' facilities or Minitel guest services.

La Plagne's *tourism supply* is very diverse. Although skiing is the main motivation for visitation, the resort has developed other types of tourism, adapting their service and marketing to new targets. Regular customers are the backbone of the industry, enabling the area to offer a more personal-service approach and improvements in customer care. La Plagne's leading accommodation firms (85% of the total beds) are self-catering Residences de Tourisme which are controlled by large, horizontally-integrated domestic hospitality chains, offering a more standardised service. Seasonality characterises the operations as 95% of the businesses operate for an average of 6/8 months per year, leaving the resort deserted during the inter-season periods. Ski tourism, has attracted external businesses to the area in order to benefit from this high revenue type of tourism and its offsetting seasonal character. TOs and large domestic hospitality chains have, therefore, purchased the majority of the accommodation establishments. About 1/3 of the businesses receive more than 50% of their customers from TOs and 85% of the beds are managed by national enterprises which usually attract their customers through their own Computer Reservation Systems. This reveals that a high percentage of the businesses are externally controlled and thus decisions, for example on IT investment, are not usually made locally. L'Office du Tourisme de La Plagne (OTP) co-ordinates the resort and, apart from the resort's external promotion, they have developed a Regional Computer Reservation System (RCRS). The RCRS distributes a stock of rooms/apartments to consumers and the travel trade charging a low booking fee to the local hospitality enterprises (4 to 6%). Businesses linked to the system receive bookings through the Minitel service.

3.2. The research methodology

La Plagne was selected as a tourism destination because it fulfils a number of criteria, as a highly seasonal ski resort in the European periphery. Both secondary and primary research were used in order to examine the utilisation of ITs for the operation of accommodation establishments and the resort. Information about the resort was collected from *secondary sources*, whilst local and regional bodies were contacted in order to provide statistics as well as information on tourism planning and management. Secondary research focused on identifying the characteristics of supply and demand in La Savoie and La Plagne as well as exploring penetration of ITs within the region. The results of the secondary research provided a useful basis for conducting primary research with individual establishments. Two main answers were sought by the *primary research*: the degree of ITs' penetration in La Plagne's hospitality sector, and the managerial attitude towards the ITs issue. The latter provides the research with a more dynamic character in the long-term period. Primary research involved 30 personal interviews with hospitality proprietors and managers. Due to the size of the resort, it was decided that a census should be carried out, rather than a sample survey. About 2/3 of the total number of accommodation businesses in La Plagne were interviewed. Interviews were based on a semi-structured questionnaire similar to the ones used by Buhalis [7] and Main [19], which were also used as an interview structure in order to allow interviewers to collect qualitative information. Throughout the discussion, interviewees were able to divert into other interesting subjects which established the general managerial attitude and actions. Both quantitative and qualitative variables were used to provide hard data as well as the reasoning behind the facts, actions and perceptions. Interviewees age mean varied between 41 and 50 years of age and there was an equal distribution between females and males. Enquiries into their level of education indicated that 1/3 of the interviewees had university level education.

3.3. Analysis of the results

The research demonstrates clearly that ITs' penetration has proved to be particularly slow in SMTEs and remote tourism destinations. Although the measurement of ITs' penetration is not an easy task due to the diversity of hospitality businesses, some common aspects have been defined in order to be able to compare and analyse the businesses. For instance, the number of possible computer users within a company or, the different functions performed by computers are two indicators of the degree of ITs' penetration. Quantitative data combined with a qualitative approach identifies SMTEs' perception towards ITs, not only at present but also in the future. The research reveals that ITs' penetration at the destination is in its *infancy* whilst ITs' utilisation by hospitality companies is elementary. Understandably, there is a strong correlation between the number of telephone and fax lines and the capacity of the establishment in terms of number of staff [R= 0.6 Sig= 0.001, R= 0.6 Sig= 0.0009]. Similarly, larger hospitality establishments regarding the number of beds tend to use a higher number of PCs [R= 0.6 Sig= 0.0005]. However, about 1/3 of the interviewees do not have computers and 1/3 have only one Personal Computer as illustrated in Figure 1.

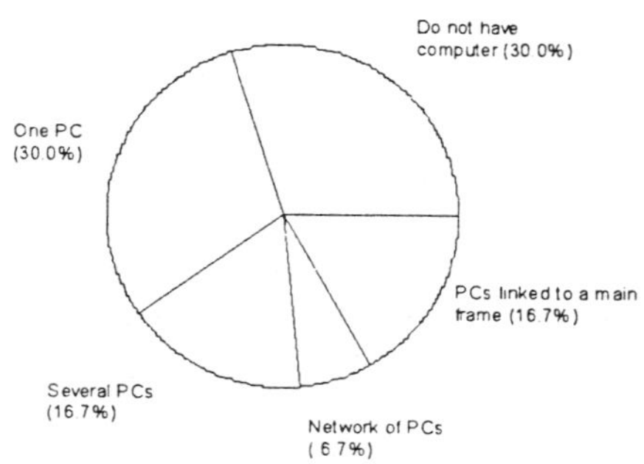

Fig. 1. ITs hardware utilised by hospitality enterprises in La Plagne

Although large companies need more staff who are capable of using computers, 80% of the total number of computers are only operated by the manager/s, rather than other members of staff. Figure 2 demonstrates that computers are almost entirely used for administration and management functions and the main reason for incorporating them is to improve efficiency and eliminate paper work.

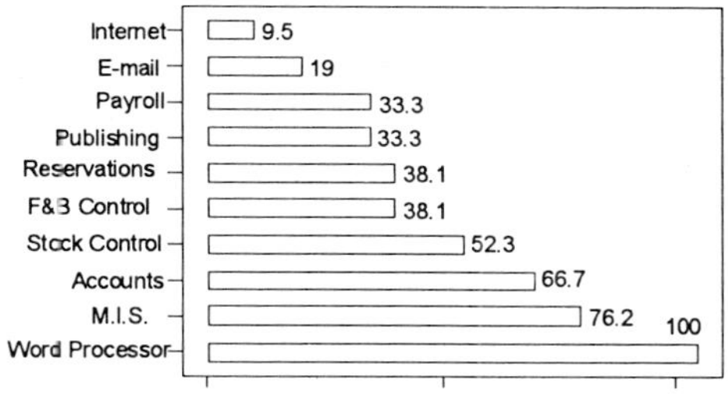

Fig. 2. Functions performed by ITs in La Plagne's hospitality enterprises

Only two interviewees believed they were utilising their computers at full capacity. These were aware of the latest ITs' potential on distribution and promotion, and that was reflected in their ability to reduce seasonality problems, improve their customer service and enhance their position in their distribution channels. In general, the results indicate that the longer the period of operation, the more they tend to utilise ITs as they are able to spread their ITs' investments over more months. There is also some evidence that those establishments which have reduced seasonality problems tend to benefit better from ITs' opportunities such as the Internet. At the same time, whereas RCRS has proved to reduce seasonality, other distribution channels such as Global Distribution Systems (GDS) and Minitel did not formulate statistically significant correlations with the months they operate. The findings also suggest that the RCRS has attracted SMTEs as they have realised the necessity to improve their distribution channels in order to be able to compete with the CRSs of larger hospitality chains. The research indicates that the use of ITs increases the *business satisfaction* from their performance. Thus, those businesses who felt more satisfied with their business performance were those who tended to use the possibilities of their computers fully. Similarly, there is a vague correlation between the number of PCs utilised and the interviewees' level of satisfaction. However, the result is not statistically significant because of the effect caused by external investment decision-making and difficulties of seasonal staff training on ITs.

Qualitative research demonstrates that the under-utilisation of ITs at the resort is firmly related to the marketing function of local tourism. The marketing of the La Plagne's hospitality sector is heavily based on a brochure-based promotion and a distribution system which relies on TOs, chains' reservation systems and local and regional tourism bodies' reservations. Their main objective appears to be the maximisation of customer loyalty and most of their investment is directed towards the satisfaction of their frequent customers. This marketing approach has several consequences. Firstly, some businesses become overdependent on TOs and as a result, they suffer from low profit margins and lack of control over their decision-making. This situation makes them unable to invest in ITs in order to change their marketing strategy. Secondly, there is a remarkable improvement on customer care and partnership marketing. The target market is very well defined and service quality is high, in order to satisfy the customer. However, their customer loyalty strategy prevents the sector from opening their scope and taking advantage of the emerging electronic distribution systems such as the Global Distribution Systems, the Internet and Minitel in order to attract new customers and market segments. This position will pose threats for their future competitiveness when the ability of enterprises to compete will be determined by their position in electronic commerce.

As for the *managerial attitude* towards ITs, the general belief is that new technologies will be indispensable in the near future but, at an individual level, little action is taken to benefit from them. Interviewees at the OTP recognised that ITs are going to play a prominent role in the industry so it is essential to incorporate it in order to remain competitive. The OTP therefore adopts an active role, creating not only the Regional CRS but also a La Plagne Home Page in the WWW (www.skifrance.fr/~laplagne) including a listing of all the local companies. Moreover, they have also organised an Internet course for local managers of the tourism industry. However, the managerial attitude is very reactive and decisions for investment or improvement of ITs are made only in the case of necessity. It is particularly disappointing that most of the interviewees admitted that they do not fully use the opportunities of ITs in order to improve their performance and profitability. The reason for this unconfident and passive attitude is the managers' general ignorance on the benefits to be gained from ITs. This is translated into a limited belief that new technologies are very distant and the result of ITs do not compensate the investment cost and effort of incorporating them. The results demonstrate that the reasons for the non full-utilisation of computers are not economic and they are rather personal as the cost was not identified as a major impediment. On the contrary, the majority of the interviewees argued that their reason for not implementation were lack of knowledge and education of ITs, lack of software and time availability.

4 Conclusions

The under-utilisation of ITs remains a problem in the hospitality industry. This is particularly prevalent in peripheral resorts where SMTEs fail to incorporate new advances in ITs and are increasingly marginalised from the main stream tourism industry. A general ignorance of the benefits to be gained from ITs, inadequate management and training, deficient education and training coupled with external decision-making appear to be the main reasons which prevent SMTEs from utilising new technologies. In addition fear of new methods, lack of education and training as well as different financial resources also contribute to the under-utilisation of ITs. The case study of La Plagne demonstrates a reactive managerial attitude and a limited understanding of the technological changes within the tourism industry. The case is not unique. Similar results are reported by Buhalis and Main [8,19] for SMTEs in Greek and Welsh peripheral resorts. To the degree that SMTEs need to remain competitive, this represents a challenge in terms of tourism planning and management which requires a greater emphasis on training and education in the use of and benefits to be attained from new advances in ITs. This is a necessity if SMTEs are to increase their competitive edge in the new millennium and maintain their contribution to destination's prosperity.

References

1. Agence Touristique de la Savoie, (1994), Chiffres et Contacts, Chambery.
2. Archdale, G., (1993), Computer Reservation Systems and public tourist offices, Tourism Management, Vol.14(1), pp.3-14.
3. ATS, (1996), Observatoire du Tourisme en Savoie: Hiver 95/96, No 15, Chambery.
4. Beaver, A., (1992), Hotel CRS - An overview, Tourism Management, Vol. 13(1), pp.15-21.
5. Beaver, A., (1995), Lack of CRS accessibility may be strangling small hoteliers, the lifeblood of European tourism, Tourism Economics, Vol.1(4), pp.341-355.
6. Buhalis, D., (1993), Regional Integrated Computer Information Reservation Management Systems as a strategic tool for the small and medium tourism enterprises, Tourism Management, Vol.14(5), pp.366-378.
7. Buhalis, D., (1994), Information and Telecommunications Technologies as a strategic tool for small and medium tourism enterprises in the contemporary business environment, in Seaton, A., Wood, R., Dieke, P., and Jenkins, C., (Editors), Tourism-The State of the Art: The Strathclyde Symposium, England: John Wiley, pp.254-275.
8. Buhalis, D., (1995), The impact of information telecommunication technologies on tourism distribution channels: implications for the small and medium sized tourism enterprises' strategic management and marketing, Guildford: University of Surrey Ph.D. Thesis, Department of Management Studies.
9. Chervenak, L., (1993), Hotel Technology at the start of the millennium, Hospitality Research Journal, Vol.17(1), pp 113-120.
10. Churchill, D., (1996), Top hoteld cash in on executive five star fever, Business Travel, The Times, October 15, p.2.
11. Cooper, C., and Buhalis, D., (1992), Strategic management and marketing of small and medium sized tourism enterprises in the Greek Aegean islands, In Teare, R., Adams, D., and Messenger, S., (Editors), Managing projetcs in hospitality organisations, London: Cassell, pp.101-125.
12. Comete, (1996), Bilan d'activite des Hotels, Chambery.
13. EC, (1993), Communication from the Commission: the European observatory for SMEs - Comments by the Commission on the first annual report (1993), European Community COM(93)527, November 5, Brussels.
14. Go, F., (1992), The role of Computerised Reservation Systems in the hospitality industry, Tourism Management, Vol.13(1), pp.22-26.

15. Go, F., and Welch, P., (1991), Competitive strategies for the international hotel industry, The Economist Intelligence Unit, report No.1180, London.
16. Hospitality, (1995), Get on the Internet - or else!, Hospitality, Hotel Catering International Management Association, April/ May, pp.10.
17. Jones, C., (1993), Application of database marketing in the tourism industry, Paper No.1, PATA Occasional Papers Series, San Francisco: Pacific Asia Travel Association.
18. Lyle, C., (1996), The changing economics of travel distribution, Seminar on Tourism and new technologies: Global Distribution Systems, World Tourism Organisation, Cote d'Ivoire,11-12 June.
19. Main, H., (1994), The application of information technology in independently owned hotels, MPhil Thesis, Swansea Institute of Higher Education, University of Wales, Wales.
20. Observatoire Regional du Tourisme Rhone-Alpes (ORTR-A), (1993), "L'emploi touristique 1982-1990", inf'ORT, No 1, October 1993.
21. Office du Tourisme de La Plagne, (1996), Dossier de Presse. Hiver 95/96, OTP, Aime.
22. ORTR-A, (1994), "Saison Hiver 1993-1994", inf'ORT, No 4, December.
23. ORTR-A, (1995a), "Hotellerie de Plein Air 1990-1994", inf'ORT,No.6.
24. ORTR-A, (1995b), "Hotellerie 1992-1994", inf'ORT, No 7, December.
25. ORTR-A, (1995c), "Saison Hiver 1994-1995", inf'ORT, No 8, December.
26. ORTR-A, (1996), "Saison Ete 1995", inf'ORT, No 9, Juin 1996.
27. Peroni, G., (1991), Problems and prospects for the renewal of public tourism marketing in the computer network era, Proceedings of the International Conference on computer networking and the public tourist organisation, Assisi-Perugia, Italy, 18-20 November, 35-59.
28. Poon, A., (1988), Tourism and Information Technologies, Annals of Tourism Research,Vol.15(4), pp.531-549.
29. Poon, A., (1993), Tourism, Technology and Competitive Strategies, Oxford: CAB International.
30. Shaw, G., and Williams, A.M., (1990), Tourism economic development and the role of entrepreneurial activity, in Cooper, C., (ed.), Progress in Tourism Recreation and Hospitality Management, Vol.2, Belhaven Press, London.
31. Sheldon, P., (1993), Destination Information Systems, Annals of Tourism Research, Vol.20(4), pp.633-649.
32. Sun, (1996), Internet essential: The Web open for business, Sun, Surrey.
33. Vogel, H., (1992), Travel and Lodging Industry Marketing Seminar: Managing Marketing and Sales Automation for profit, In conjunction with World Travel Market, 17-18 November, Crafton Hotel, London.
34. Weinstein, J., (1995), More hotels plug into Internet, Hotels, June, pp.15.

Catalysts in introducing information technology in small and medium sized hospitality organisations

Dr Dimitrios Buhalis and Hilary Main

Senior Lecturer in Tourism	Senior Lecturer in Information Technology
University of Westminster	Swansea Business School
Tel: +44 171 9115000 x 3112	Tel:+44 1792 481124
Fax: +44 171 9115171	Fax:+44 1792 481127
Email: buhalid@wmin.ac.uk	Email: HMain@mp.sihe.ac.uk

Abstract

Small hotels are increasingly recognised as pivotal in the ability of destinations to benefit from tourism as well as to satisfy tourism demand. However it is recognised that small hotels are being marginalised from the mainstream tourism industry due to their inability to participate in the transformation of best practices due to their reluctance to utilise information technologies. This paper is based on research undertaken in small and medium hospitality organisation in Wales and the Greek Aegean Islands. It explores the factors determining the adaptation of ITs by examining the stakeholders of small hospitality organisations as well as the push and pull factors they exercise, illustrating a number of catalysts for ITs penetration.

1 Small and Medium-sized Hospitality Organisations (SMHOs)

The vast majority of accommodation establishments around the globe are small and medium-sized, belong to local entrepreneurs are family run, and predominately employ members of the host society. Despite their size, collectively, small and medium-sized hospitality organisations (SMHOs) are extremely important to both national and European economies as they provide stable employment opportunities and support the integration of local economies in peripheral areas, even during recession periods. Typically, SMHOs offer by definition less than 50 rooms, employ less than 10 people, operate in the lower reaches of the market and are often situated in tertiary locations [34,24,24,12].

SMHOs provide a variety of *benefits* for destinations by providing tourists with direct contact with the local character and host population. Therefore they contribute significantly to the authenticity and quality of the "tourism experience", whilst delivering a unique "lodging experience" [11]. Their size provides competitive advantage of flexibility and specialisation and thus they are capable, more so than large, rigid, affiliated hotels, to identify niche markets and promote customised accommodation, achieving differentiation and competitive advantage [26]. SMHOs also have the flexibility to alter their portfolio characteristics, as well as form horizontal and vertical alliances to achieve economies of scope. They attract tourism spending to the host destination and the consequent 'secondary spend' stimulates the multiplier effects generating several rounds of economic activity at destinations [14].

However SMHOs suffer a wide range of *disadvantages* which effectively jeopardise both their profitability and competitiveness, as well as their contribution to regional economies. These disadvantages tend to result from their lack of capital; deficient economies of scale and under-utilised economies of scope; peripherality; insufficient management and marketing skills and expertise; inadequate bargaining power within the distribution channel; and finally lack of representation in the emerging electronic marketplace [1,3,5,6]. To the degree therefore that the public and private sectors need to enhance their *competitiveness and reinforce the prosperity of both enterprises and destinations*, decisions makers should identify methods in order to encourage the utilisation of the strategic tools introduced by Information Technologies (ITs). This paper analyses qualitative research undertaken in Wales and the Aegean Islands in Greece [7,21] and attempts to identify catalysts which will enable SMHOs to incorporate ITs in their strategic and operational management. A total of about 600 hospitality organisations were researched, providing a broad basis for solid inferences. Several push and pull factors are identified, while a stakeholders analysis is undertaken in order to illustrate the decision framework of SMHOs and to facilitate the understanding of critical variables which would enable a higher degree of ITs adaptation in the near future.

2 Technology in hospitality organisations

Technology becomes a main source of sustainable *competitive advantage and a strategic weapon,* especially in the tourism and hospitality industries due to the pivotal role information plays in the description, promotion, distribution, amalgamation, organisation and delivery of tourism products [27,33,8]. Technology can offer significant advantages in operational (eg property management systems), tactical (e.g.financial modelling, yield management) and strategic management (e.g. decision support systems) of SMHOs. Distribution networks and partnerships with other tourism product providers also might provide SMHOs with wider distribution networks and partnerships with complementary products. Increasingly the use of ITs is a major prerequisite in forming strategic alliances; developing innovative distribution methods; communicating with consumers and partners; and satisfying consumer demand. Both customers and partners also tend to place a greater value on organisations which utilise ITs than their competitors [30,14,15,20].

However, despite the technological revolution experienced in the tourism industry, hospitality organisations have traditionally been *reluctant to utilise ITs* [3,4,31,32]. Cho and Connolly note that many hotel managers remain sceptical about the value of investment in ITs and argue that their main failing is in not realising the actual length of time it takes to see actual results [10]. This requires strategic vision and commitment to technology which will secure future diffusion of benefits through the industry and stakeholders. Involving crucial stakeholders in the design process provides the opportunity to match their skills and needs in design by ITs suppliers.
Recent research demonstrates that *technology is under-utilised in SMHOs* in most European destinations, such as the Aegean islands, Wales, and Alpine French resorts

[7,9,21,25]. Less than fifty percent of hotels use ITs at all and those that do tend to concentrate on operational insignificant tasks, such as word processing and accounting. Very few SMHOs use technology for tactical or strategic management decision-making or for linking and/or building innovative information and communication networks. *SMHO managers* who use ITs have tended to be "dabblers", i.e. have bought technology piecemeal with no long-term plan or even specific business use; or "technophiliacs" who have invested, expensively, in technology in the early stages and have lost faith when the technology did not live up to expectations. However, the vast majority of SMHOs failed to incorporate ITs in its strategic business processes re-engineering which would enable them to change their practices in order to take advantage of the new potentials. A gradual, planned, budgeted ITs introduction is crucial to their success, built on a solid foundation, measured against payback, with budgeted growth and upgrading which delivers to business needs. Managerial commitment, continuous training of all members of staff and a paradigm shift to the best practices are required in order to maximise benefits.

There are several identifiable *reasons as to the lack of use of technology* in SMHOs: the lack of training, which is traditional and low-tech; the ill-perceived conception that technology cannot be afforded by small and seasonal hospitality business; the age, educational level, and family arrangements of the SMHOs' proprietors; the deficiency of rational management and marketing functions; and the short-term, operational focus of managers are some of the most critical factors which determine the under-utilisation of ITs in SMHOs. The level of education as well as the training courses undertaken by SMHO managers before or during employment tend not to incorporate an advanced level of technology use and consequently only few managers are competent users of ITs. Thus they remain unable to realise the main potential and fail to take advantage of ITs. As a results SMHOs are increasingly marginalised and suffer significant competitive disadvantages, which will eventually be catastrophic for their ability to attract consumers in the near future. The success of ITs introduction therefore depends on whether technology recipients are willing and capable to initiate a re-engineering of their business processes [1,7,19,21,28].

3 Stakeholder analysis for small and medium hospitality organisations

Stakeholders are the different groups who have a legitimate interest in the activities of an organisation, whilst they are also affected by the organisation's decisions and policies. Often stakeholders can have important influence or exert pressure on an organisation, in order to increase the value offered to them. The stakeholders theory states that, for a firm to remain viable it needs to satisfy the requirements of key stakeholders. This paper identifies key stakeholders in the SMHOs' model and the extent to which they will be catalysts in the introduction of technology by providing push/pull effects. Atkins and Lowe detect increased stakeholder involvement in times of technological turbulence, as the external environment of organisations is changing shape and both interests and benefits of stakeholders need to be readdressed [2].
Owners/proprietors are the most apparent stakeholders of SMHOs as they are often the main investors and managers. They tend to be heavily involved in the operation of

accommodation establishments while they undertake strategic decisions and negotiate with both suppliers and consumers. Owners/proprietors need to utilise technology in order to facilitate both their operational and strategic functions. However, often they tend to lack the expertise in selecting, installing and operating computerised systems and therefore fear that they will lose part of their control, should they allow external ITs experts to undertake these jobs for them. In addition, as ITs do not provide short term return on investment owners/proprietors tend to prefer to invest their capital on more tangible improvements of their facility. As a consequence they tend to take a passive attitude and prefer to use old-fashioned management techniques.

Employees, and in particular managers, also tend to be interested in the prosperity of SMHOs, as they depend their employment and quality of working life. Employees also need interesting, challenging and exciting jobs, while they tend to dislike repetitive, uninteresting tasks. To the extent that ITs can eliminate the repetitive elements of jobs, employees would be interested in utilising more technology and concentrate on the creative elements of their role. However the age and educational background of people often tend to determine their attitude towards new technologies and their ability to comprehend the capabilities of ITs. Frequently young, educated, competent employees play a critical role in introducing ITs in SMHOs, whilst older and less educated employees may fear that technology would jeopardise their job or authority and therefore acquire a negative attitude. It is not unusual for younger and qualified employees to fail to persuade their traditionalist superiors or proprietors, who feel threatened by the new tools which they cannot control and understand. In this situation they tend to feel frustrated and unable to perform their role properly.

Consumers are also stakeholders of SMHOs as they are keen to increase the value for money they get from service providers as well as to maximise their satisfaction. ITs improve the efficiency of SMHOs as well as enable them to differentiate their product through the provision and promotion of specialised products. Consumers increasingly demand a certain level of efficiency and convenience when they deal with hospitality organisations, which cannot be delivered without utilising ITs. In addition, ITs become a core feature of tourism products as business travellers already request in-room ITs facilities, such as several telephone lines, dedicated fax and networked computers. Thus, consumer expectations force the introduction of ITs in SMHOs and their satisfaction would increasingly depend to this provision. *Institutional customers*, such as travel agencies, tour operators and other intermediaries are interested in having easy access to up-to-date information on availability and rates for SMHOs. Unless they can rely on this information, intermediaries would remove principals from their recommended lists and would switch to the ones that offer this type of information, better profit margins and other incentives. Thus the utilisation of ITs for representation in the electronic marketplace, through Global Distribution Systems, the internet, videotex and teletex systems, is pivotal for a harmonious and profitable cooperation between SMHOs and intermediaries.

Moreover, *raw material suppliers* are interested in the well-being of their customers and need to ensure that SMHOs operate efficiently and profitably in order to maintain their custom. Suppliers need to increase their own efficiency in order to enhance their competitiveness. Communication and coordination with their customers is therefore pivotal for their ability to deliver the right product at the right time and price. ITs enables interconnectivity between enterprises and provides opportunities for suppliers to identify and support the needs of individual customers. Thus suppliers can enhance their partnership with SMHOs by utilising technology and hence increasing the value added through the supply chain.

Increasingly tourism regions are managed by *destination managers* who tend to belong to the public sector and aim to co-ordinate and regulate the tourism industry at the destination level. Their personal characteristics, development, and attitudes are critical for the position they adopt towards ITs. However a rational destination manager would probably promote the networking of destinations in order to facilitate interaction between local suppliers and improve the communication with consumers. This will improve the quality of tourists' experience, whilst improving the multiplier effects at the destination level. Hence professional management of destinations would support the utilisation of ITs and the advanced networking of the tourism supply. Finally the *local community* tends to be represented as a stakeholder through owners and employees in SMHOs, while political parties and other lobby groups may influence the direction of the local tourism industry. The higher the competitiveness of tourism enterprises at the destination level, and thus the more they take advantage of the emergent tools, the more prosperous their operation will be and the more contribution they would be offering to the regional economy. Thus local people would gain more benefits if tourism enterprises utilise more advanced management techniques and new technology.

The relationships between stakeholders as well as their interests are dynamic. Hence they change according to the evolution of the external environment, while they also reflect the developments of stakeholders needs. However the above analysis demonstrates that most of the major stakeholders of SMHOs benefit significantly from the incorporation of ITs in the production and management functions. Thus consequently, they are expected to contribute to the paradigm shift experienced in the tourism industry, and demonstrate clearly that ITs should be widely utilised in order to enhance the performance of SMHOs and gain more benefits as stakeholders. This is implemented through the utilisation of a wide range of push and pull factors which aim to influence the decision making process of SMHOs.

4 Catalysts in technology adaptation for SMHOs: Push and Pull factors

Should SMHOs have a choice they would try to maintain a traditional management approach, where ITs would only play a peripheral role. However a wide range of push and pull factors determine that technology adaptation will be critical in their ability to satisfy their stakeholders as well as to improve their performance. **Push factors** are external forces which oblige enterprises to use ITs in order to avoid potential threats or jeopardise some of their functions. Enterprises may not have recognised a need that the technology might fill. If the need has been recognised, enterprises may have not matched a particular technology with the fulfilment of that need. Stakeholders seek to promote, garner, support and push a technology. **Pull factors** provide incentives for enterprises to incorporate ITs in order to gain benefits in their operation. In this scenario a recognised market need is present and enterprises draw on ITs to fulfil it. Hence the *difference between push and pull factors* is the willingness of SMHOs to undertake action. In the former case, action is forced while in the latter action is stimulated and desired. Groth examines the significance of the push-pull factor in harvesting benefit from technology introduction [17]. He defines the pull factor as originating in political, social and economic forces which 'pull' on the technology and the push factor as relating to the efforts of the technology's proponents. The push-pull analysis is critical as it illuminates the attitude of enterprises towards ITs and elaborates on the reasoning for actions undertaken.

Lessons from other industries indicate that the introduction of technology in its early days was predominately a **blend of push and pull factors**. For the *banking industry* the multiplicity of transactions and global expansion pushed the use of computerised systems for the customer. This inter-connectivity also became a pull factor as consumers valued these types of service and provided incentives for banks to utilise more ITs. The advent of technology had visible effects on internal customers, as there was a reduction in human input, leading to structural 'flattening' and a wider degree of information sharing throughout the organisation. ITs also contributed to the enhancement of the banks' marketing activity via extending product range and customer services. Although customers resisted the use of technology at their introduction and preferred to 'interface' with a human clerk, eventually they treasured the wide range of benefits and the more flexible banking methods [22,23]. The introduction of technology to the *retail trade* was 'pushed' via suppliers and scanning/bar coding which increased efficiency in a wide range of functions, such as accounting, stock management, marketing research and reduced both time and cost required for payment.

Pull factors were also utilised in incorporating partnership marketing and increasing consumer loyalty through customer clubs. As in the hospitality industry, affiliated 'chain' supermarkets were quick to adopt the new ITs in order to improve the co-ordination and control of their branches, while independent/smaller retailers were slower to take advantage of technology. *Accountancy firms* had technology 'pushed' to them by complementary suppliers. Spreadsheets and the development of ledger-based accounting packages were a tailor made application for the profession and its requirements to comply with codes of practice and legislation.

Inevitably a wide range of push and pull factors are also exercised on SMHOs towards increasing their ITs utilisation. Figure 1 demonstrates some of these push and pull factors and illustrates how they constrain SMHOs to undertake strategic action towards incorporating ITs in their strategic and operational management.

As far as **push factors** are concerned, several external influences force SMHOs to re-engineer their business processes and utilise ITs. *Education and training* push ITs in

SMHOs, as the incorporation of technology in the hospitality curricula provides a catalyst and promotes ITs to the industry via trained/graduate students who appreciate the emergent benefits and will champion technology in the industry. Green reports that in the USA 51% of hospitality faculties use computers, although only 9% used computers in the classroom [16]. Similar research is not available for Europe but is likely to be considerably lower. To the degree that graduates are competent users of ITs they will be able to recognise needs and undertake suitable action for using ITs in SMHOs, providing evidence to sceptical proprietors that considerable advantages can be gained.

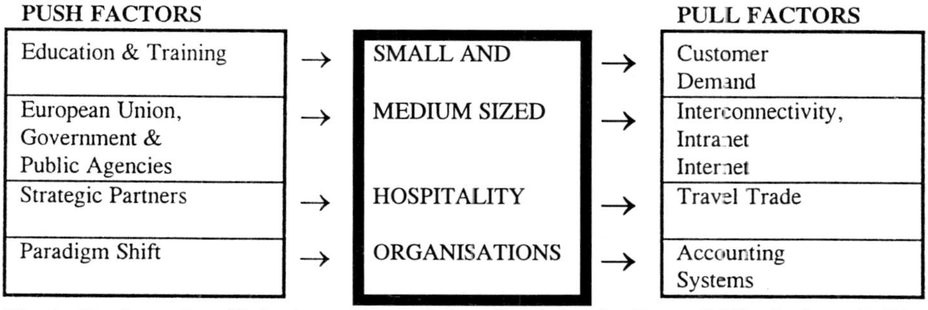

Fig.1. Push and pull factors determining the introduction of ITs in hospitality organisations

Increasingly the public sector recognises that ITs are critical for the competitiveness of private firms. They also recognise that small and medium sized enterprises are instrumental for regional development and have a much more significant contribution to sustainable development than their larger counterparts. As a result the *European Union as well as national governments* concentrate their efforts in providing incentives for enhancing the utilisation of ITs by small firms in order to enhance their market presence and regional integration. SMHOs can benefit from these incentives as they seize the opportunity to gain from discounted equipment, training programmes and help desks [12,13]. *Strategic partners* are other members in the supply chain who are closely interrelated with SMHOs, such as intermediaries and suppliers. Tour Operators have been instrumental in compelling travel agencies and handling agencies to utilise ITs in order to facilitate co-ordination and enhance efficiency. Similarly destination management systems urge SMHOs to utilise ITs in order to be able participate in the networking of destinations and the development of on-line reservations. Hence strategic partners, such as intermediaries and suppliers, can force SMHOs to incorporate ITs and be represented in the electronic commerce.

In general, the *paradigm shift* experienced in the tourism industry illustrates that SMHOs which fail to use ITs will suffer significant competitive disadvantages as they will jeopardise their yield by reducing both market share and profit. As other tourism enterprises increase their ITs usage, SMHOs are forced to incorporate technology in order to cooperate with their partners and maintain contact with consumers. Consumers increasingly expect to be able to access information and make reservations through the Internet. Hence SMHOs start to realise that unless they satisfy this need, they will fail to attract individual and institutional consumers. Finally ITs suppliers emerge to promote applications for SMHOs. They tend to use forceful marketing and often manage to sell their products to SMHOs, regardless whether they understand the entire range of benefits and features emerging through ITs. As the entire industry is transforming itself, SMHOs are forced to follow the flow. A demonstration effect is therefore evident, where proprietors try to "keep up" with competitors and partners.

Several **pull factors** can also be identified as they provide incentives for SMHOs to incorporate technology. Perhaps the most important factor is *customer demand* and the increasing number of computer literate consumers who are empowered by the Internet and tend to use networks for identifying and purchasing various products. Thus, as they attempt to increase their market share they would need to incorporate more technology in order to increase their direct communication with consumers. This also provides an opportunity for disintermediation and empowerment within the distribution channel as it enables innovative SMHOs to distribute their products directly to consumers enhancing their profit margins and their reducing their dependence upon intermediaries [7]. In addition the demand for ITs products during their hotel stay may also increase the awareness of SMHOs of the importance ITs. The electronic superhighway provides unprecedented opportunities for SMHOs for utilising affordable technology in order to improve their global presence and enhance their efficiency. *Interconnectivity within the industry as well as the development of the internet and intranet* enables and empowers SMHOs to distribute and promote their tourism products at an affordable cost. Increasingly SMHOs familiarise themselves with the potential and develop their presence in the networks. A cost and benefit analysis illustrates that they can displace some of their marketing and promotional expenditure to the new media which promises a much wider coverage of the market and a much more efficient reservation mechanism. SMHOs learn gradually how to take advantage of the networks in order to reduce their operational cost. Intercontinental hotels for example reduced their reservation cost from $15 on voice operated systems, to $7.50 for reservations through GDSs and $2 through the World Wide Web [3,18,20].

Similarly with the strategic partners, SMHOs realise that in order to cooperate with the *travel trade* they will need to utilise ITs. To a certain extent this becomes the standard of the tourism industry and whoever will be lagging behind will be unable to attract custom for their products. This is demonstrated clearly as travel agencies refuse to search for accommodation establishments which are not represented electronically, as the cost of locating, and arranging accommodation in hotels not represented in electronic media as well as attempting to collect their commission afterwards, far exceeds the benefit. As a result they channel all their bookings through hotel chains which are represented internationally and have clear and efficient procedures for commission payments. Finally, *accounting systems* is one of the most silent pull factors in the incorporation of ITs in SMHOs. As hotels had no other option but to computerise their accounting systems in order to facilitate their billing systems from various hotel departments, as well as reduce the fees charged by accountants and also comply with regulation, they tend to purchase computerised systems. Once the hardware purchase has been accomplished, proprietors often can identify other applications which can offer direct and tangible benefits. In addition several hotel accounting systems introduce some additional programmes for word processing or room division management. As a result accounting systems have acted as a pull factor for hotel computerisation.

The above analysis demonstrates that several push and pull factors force SMHOs to incorporate ITs within their operational and strategic management. Although these are believed to be the most important, a wide range of additional factors emerge and encourage hospitality organisations to re-engineer their business processes by incorporating technology. Perhaps the most pivotal change is the revolution experienced through the development of the Internet. The internet is gaining commercial viability and is particularly suited to small business where it enables the small business to keep its doors open 24 hours a day, at minimal cost to customers all over the world. Already there are some success stories of small business enlarging their distribution channels on the Web [20]. Software is now designed for those inter- and intra- net connections allowing access to internal databases and applications and secure access from external sources. The Web emerges as a means of achieving true electronic interconnectivity with business partners. SMHOs which are not represented will fail to bridge their distance with consumer and suffer competitive disadvantages.

5 Conclusions

The *stakeholders analysis* illustrates that there several interested parties in the well being of the SMHOs, mainly due to their contribution to the local economies and regional development. Analysis demonstrates that stakeholders adopt a number of *push and pull factors* to force SMHOs to utilise ITs. Recent developments demonstrate that some key stakeholders exercise a more influential role in forcing SMHOs to utilise ITs. Perhaps most importantly the *public sector*, as a stakeholder, increasingly appreciates the benefits introduced by ITs and undertakes initiatives to assist SMHOs to take advantage by improving their equipment and by formulating networks. Hence the European Union has recently offered funding to develop systems to increase the utilisation of ITs by SMHOs and to represent them on the Internet [13]. Moreover there are developments between software houses and government agencies, such as the Tourist Boards, where collaboration is developing an integrated destination system (e.g. Integra & Scottish Tourism Board). These initiatives will be critical in the adaptation process. Moreover *consumers* may be the key stakeholders in the industry, pulling technology through to hotels. They become more and more acclimatised to ordering goods and services on-line, whilst technology connectivity expands to each home and the costs and benefits of on-line reservations supersede traditional booking methods. At present there are many organisations in the process of re-engineering their business and information systems with the burden of integrating crucial historical data and systems with emerging technologies. Unlike them, *SMHOs do not have the problem of "legacy" systems* and thus they can design their systems to take advantage from the emerging capabilities without losing valuable data and recent investments. SMHOs at least have some advantage in that their prior investment in technologies has been minimal. Hence they can learn from others' mistakes and also take the opportunity to form key relationships with ITs suppliers.

Groth concurs that the likelihood of success in introducing technology is enhanced in a pull environment and that it is possible to *convert from a push to a pull scenario* [17], which is what SMHOs have been exposed to over the last years. This is where inter-connectivity as well as customers and the travel trade will provide the necessary marketing pull. Despite the current lack of coordination between stakeholders, it seems that a certain level of co-operation will be essential in order to facilitate the ITs penetration as well as for identifying standards and protocols which will enable SMHOs to achieve a certain level of standardisation and interconnectivity. The key is to align the differing views of the key stakeholder groups, to enable sound communication, interaction and collaboration which will contribute to the success of technology adaptation and proliferation in SMHOs [29].

BIBLIOGRAPHY

1. Archdale, G., 1993, Computer reservation systems and public tourist offices, Tourism Management, Vol.14(1), pp.3-14.
2. Atkins, M., and Lowe, J., 1994, Stakeholders and the strategy formation process in small and medium enterprises, International Small Business Journal, Vol.2/3, pp.12-25.
3. Beaver, A., 1995, Lack of CRS accessibility may be strangling small hoteliers, the lifeblood of European tourism, Tourism Economics, Vol.1(4), pp.341-355.
4. Borsenik, F., 1993, Hospitality technology in the 21st Centry, Hospitality Research Journal, Vol.17(1), pp.259-269.
5. Buhalis, D., 1993, Regional integrated computer information reservation management systems as a strategic tool for the small and medium tourism enterprises, Tourism Management, Vol.14(5), pp.366-378.
6. Buhalis, D., 1994, Information and telecommunications technologies as a strategic tool for small and medium tourism enterprises in the contemporary business environment, in Seaton, A., et al (eds), Tourism-The state of the art: the Strathclyde symposium, J., Wiley and Sons, England, pp.254-275.
7. Buhalis, D., 1995, The impact of information telecommunication technologies on tourism channels: implications for the small and medium sized tourism enterprises, PhD Thesis, University of Surrey, Guildford, UK.
8. Buhalis, D., 1996, The influence of new communications technologies on tourism: seeking competitiveness and competitive advantage, Foro Anual de Turismo, Fundacion Cavanilles de Altos Estudios Turisticos, Benidorm, Spain, November.
9. Buhalis, D., Keeling, S., Lacorte, A., Reynolds, N., 1997, Technology in hospitality organisations: The case study of La Plagne in France, ENTER 1997 Conference, January, Edinburgh, Spinger-Verlag, Wien, (forthcoming).
10. Cho, W & Connolly, DJ., 1996, The impact of information technology as an enabler on the hospitality industry, International Journal of Contemporary Hospitality Management, Vol.8(1), pp 33-36.
11. Cooper, C., and Buhalis, D., 1992, Strategic management and marketing of small and medium sized tourism enterprises in the Greek Aegean islands, in Teare, R., et al, (eds), Managing projects in hospitality organisations, Cassell, London, pp.101-125.
12. EC, 1993, Growth, competitiveness, employment: The challenges and ways forward into the 21st century, White Paper, Bulletin of the European Communities, 6/93, Luxembourg.
13. EC, 1996, Call for proposals for the establishment of a Euroepan co-ordination structure aimed at promoting the usage of electronic commerce through the Internet network amongst

small and medium sized companies operating in the tourism sector and located in the less favoured regions of the Union, Brussels.

14. Edgar, D., 1996, Capacity management: yielding to the short break, Yield Management Conference, Birmingham.
15. Emmer, R., Tauck, C., Wilkinson, S., and Moore, R., 1993, Marketing hotels using Global Distribution Systems, The Cornell HRA Quarterly, Vol.34(6), pp.80-89.
16. Green, C., 1996, Human factors in the implementation of computers in the hospitality education, Hospitality and Tourism Educator, Vol.6(1), pp.63-66.
17. Groth, JC, 1993, Critical factors in exploiting technologies, Management Decision, Vol.31(3), pp.34-48.
18. Hart, P, 1995, Introducing TravelWeb, ABTECH Conference, London.
19. Harris, KJ., 1995, Training and Technology in the hospitality industry: a matter of effectiveness and efficiency, International Journal of Contemporary Hospitality Management, Vol.7(6), pp.24-60.
20. Hewson, D., 1996, To the seaside via hyperspace, The Sunday Times, 26th May, p.10.
21. Main, H., 1994, The Application of Information Technology in the Independent Hotel, M.Phil Thesis, University of Wales.
22. Martell, D., 1988, Marketing and Information Technology, European Journal of Marketing, Vol.22(9), pp.16-23.
23. Mitchell, JW., & Sparks, L., 1988, Marketing Information Systems in the major UK Banks, International Journal of Bank Marketing, Vol.6(5), pp.14-29.
24. Moutinho, L., 1990, Strategies for destination development - the role of small businesses in Goodall, B. and Ashworth, G. (Eds) Marketing Tourism Places, Routledge, London.
25. O'Connor, P., 1995, Using Computers in Hospitality Management, Cassell, London.
26. Poon, A., 1993, Tourism, technology and competitive strategies, CAB, Oxford.
27. Porter, M., 1985, Technology and competitive advantage, The Journal of Business Strategy, Winter, pp.60-70.
28. Price, R., 1992, Technology transfer in the hotel industry, International Journal of Hospitality Management, Vol.11(1), pp.3-23
29. Ruohonen, M., 1991, Stakeholders of strategic information systems planning, The January Strategic Information Systems (UIC) December, Vol.1(1), pp.15-29.
30. Senker, P., & Senker, J., 1992, Gaining competitive advantage from information technology, Journal of General Management, Vol.17(3), pp.31-46.
31. Wardell, D., 1987, Hotel technology and reservation systems: challenges facing the lodging industry, Travel and Tourism Analyst, No.2, June, pp.33-47.
32. Whitaker, M., 1987, Overcoming the barriers to successful implementation of information technology, International Journal of Hospitality Management, Vol.6 pp.229-235.
33. Wong, SY, 1991, Strategic use of IT for small business, Singapore Accountant, September, Vol.7(9), pp.15-21.
34. WTO, 1991, Seminar on small and medium sized enterprises, Report on WTO in Milan, 16-17 December 1991, World Tourism Organisation, Madrid.

The attitude of small and medium-sized tourist enterprises towards Information and Telecommunication Technologies. The case of Italy

Jan van der Borg[+], Valeria Minghetti[*], Laura Riganti[#]

[+]University of Venice, Department of Economics (borg@vega.unive.it)
[*] CISET, University of Venice (ming@vega.unive.it)
[#]Telecom Italia, Centro Studi San Salvador, Venice (riganti@cstudi.telecomitalia.it)

Abstract. In the paper the mechanisms which guide the decision of tourist enterprises to invest or not in ITT are fully analysed. By using a demand-side approach, the different assumptions at the basis of the decision process and the variables that affect the attitude of operators to employ new technologies are investigated. The identification of the main hindrances and opportunities specifically encountered by small and medium-sized tourist firms to the application of ITT provide the basis for a discussion on how a Telecom company can anticipate these developments.

1. Introduction

In recent years Information and Telecommunication Technologies (ITT) have spread with an astonishing speed over the most important sectors of the world economy.
At first sight, tourism is a sector which offers many opportunities for new telematic and informatic tools/services to be applied. In fact, international arrivals are growing steadily as well as turnover and the number of employees and the tourist industry is thought to become the first world industry by 2005. Moreover, tourism is by nature an information-intensive activity. The tourist firms do not sell a ready-to-use product but a package of information on a set of products that together form a trip or a vacation, which may be verified only at the moment of the actual consumption. Last but not least, tourism involves a great number of heterogeneous activities. Hence, the quality of the trip or vacation depends on the ability of all the firms involved in production and marketing processes to communicate rapidly among them and with the final consumer.

But is the tourism industry as a whole really mature for absorbing the powerful stimulus to change coming from technology? Are the tourist firms able to transform their need in a real demand of new technologies?
In spite of the ever increasing enthusiasm among specialists, this contribution intends to show that there still is a long way to go. The travel and tourism industry is

yet distant from the spread of an advanced and fully integrated multimedia system on a large scale.

This paper, based on a research-work carried out in Italy by the International Center of Study on the Tourist Economy (CISET) of the University of Venice in co-operation with the Centro Studi San Salvador (the Research department of the Telecom Italia company), thereto aims at discussing the process which leads tourist enterprises to opt or not for ITT.
By using a demand-side approach, the paper first of all describes the different assumptions at the basis of the decision process and analyses the variables that affect the attitude of operators to invest in Information technology (Section 2). Section 3 shows what the main hindrances to the spread of new instruments and services are for the Italian tourist industry. A short overview of the megatrends which will shape the tourist industry the coming years and may enhance the application of ITT is the subject of Section 4.
In Section 5 some closing notes on the role of the Telecom Italia company in anticipating these developments are included.

2. The adoption of ITT in tourist enterprises. The logical process

Starting point for the construction of the decision process with which tourist firms adopt Information and Telecommunication Technologies is the assumption that such investment derives from the need to manage internal (within the firm) and external (between firms and the market in which they act: suppliers, clients, banks and other firms) information flows more efficiently.
Their perception of the gap that exists between the ITT instruments they possess and those that are required to adequately manage information flows, depend on a variety of factors. Figure 1 represents the investment decision process and allows us to identify the variables that determine adoption of ITT.
Considering the demand side, the variables affecting the quantity and quality of information flows managed by a tourist enterprise are those which concern:

- its *geographical location*;
- its *structure*: sub-sector (accommodation, transport, catering, attractions, TO and travel agencies), typology, size and legal status;
- its *internal organisation*: organisation complexity (number of departments, number of local units), type of contacts (level of standardisation and coding), synergies with existing tools;
- the *management culture*: attitude of ownership/ management towards innovation and specifically towards ITT, evolution of professional skills, etc..

If the importance of some variables is usually highlighted by the management theory, for the other ones their importance arose during some face-to-face interviews we carried out with a small group of Italian tourist enterprises.

Fig. 1. The adoption of Information and Telecommunication Technologies in tourist enterprises. The logical process

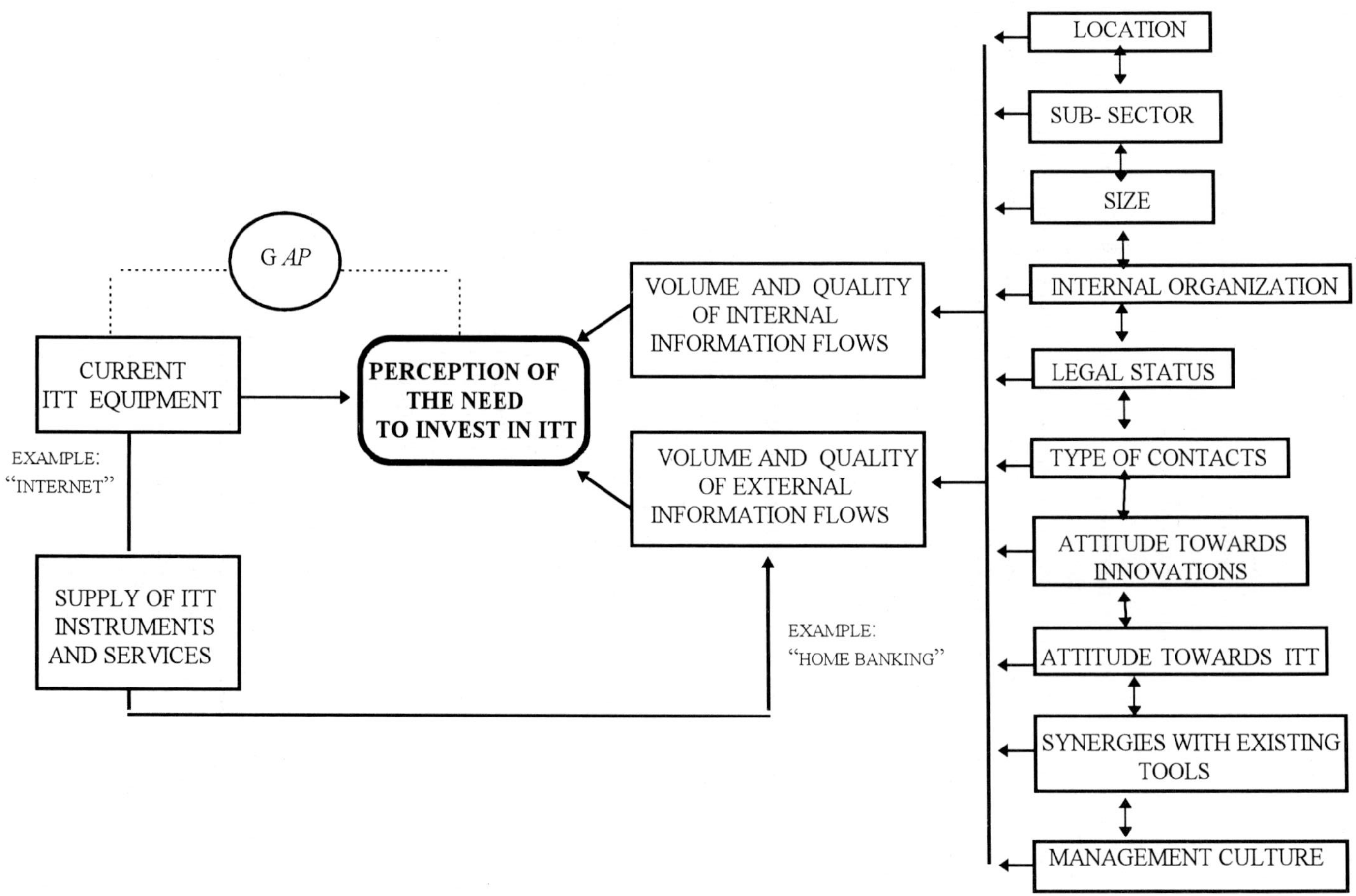

A central *location* has a positive impact on demand for ITT. An enterprise placed in a metropolitan or in an industrialised area usually manages higher and more complex information flows with the surrounding environment than an enterprise located in a suburban or in a less-developed area.
But the volume and quality of information flows also depend on the tourist *sub-sector* the firm belongs to and, given the sub-sector, on the characteristics of the single firm. The spread of new technologies among intermediaries (TO and travel agencies) and means of transport shows a steady, expanding trend. A slow but positive evolution characterises accommodation and attractions, while catering lags behind.
Apart from the sub-sector, the *size* of the enterprise (both in terms of the number of employees as well as the volume of turnover), is one of the variables that has the strongest impact on the attitude towards ITT, along with the *internal organisation* (number of departments and local units). Large-sized tourist enterprises are usually subdivided in many departments and/or local units which communicate with a great number of external actors (suppliers, clients, institutions, etc.). Consequently, these firms develop a major volume of information flows than small-sized enterprises directly managed by the owner. Another element to take into account is the *legal status* (co-operative association, partnership, business company), which is particularly important to determine the level of formality and coding of information flows and so the need to support them with appropriate technology instruments.
The *attitude of management* towards innovations and specifically towards new technologies is the other crucial factor. The more the management is interested in the evolution and the use of technology instruments, the higher the probability that those have a rapid spread among tourist enterprises. More precisely, the firms which are already equipped with some informatic and telematic instruments -- such as personal computers and modems -- or, above all, which possess an access to a digital network have usually a higher stimulus to further invest in ITT than "new-comers".

But the chance that a potential need becomes a real demand, or that the actual demand grows, depends also on the nature of supply, that is on the availability of user-friendly informatic and telematic instruments and services.
The supply of technology may either directly or indirectly affect the tourist enterprises. The dramatic spread of Internet in the last years is a clear example of a direct effect. The chance to reach the world market of potential consumers has surely stimulated firms to invest in new instruments for the electronic marketing and promotion of their product on the Web, sometimes just to follow fashion.
On the other hand, the development of new services by service providers such as banks (e.g. home banking systems) may be an example of an indirect effect. The opportunity to manage the enterprise's accounts simply through a PC connected to a modem may indirectly stimulate tourist enterprises to spread the use of ITT in other internal activities.

Comparing demand and supply, it is possible to evaluate the gap which characterises tourist enterprises. Apart from the influence of steady progress in technology, when and how this gap will be filled in depend on two main forces:

1. the *awareness* of the management and their willingness to recover the existing gap;
2. the *structural and organisational changes* which are characterising the single tourist enterprise and the market in which it acts. The analysis of the megatrends is crucial to design the scenarios for the future spread of ITT in the tourist industry. These changes stimulate increasing technology needs and are partly induced by the evolution of Information Technology (feedback).

With the help of log-linear analysis[1,2], a technique that translates multi-dimensional cross-tables in variable loads, the qualitative model of the investment process was reconstructed, enabling us to verify what variables effectively determine its outcome.
The model was applied to the results of a survey of 15,000 Italian tourist enterprises. Along with the evaluation of the actual equipment, the collection of the operators' opinions both on utility perceived and on future use of such instruments and services allowed us to evaluate the existing gap between desired and actual equipment and consequently the probability of growth of demand for ITT.

3. The demand for ITT of the Italian tourist industry

The implementation of the qualitative model has first of all confirmed that structural characteristics of the firms (location, size, sub-sector, legal status, etc.) are the most important factors in determining their willingness to employ ITT, considering both the present use of technology and its future applications.
The Italian tourist sector is strongly polarised. On the one hand, there is a small group of large tourist enterprises (air companies, international hotel chains and tour operators) which are the leaders in the use of these new tools and services. Technology is so much incorporated in their organisation that they tend to create independent and tailor-made systems for the management and marketing of their products/services. On the other hand, there is a great number of small and medium-sized enterprises, representing the majority of the tourist firms, which are lagging far behind. They represent the "tail-light" at the eve of the Technology Era.
The internal organisation (no. of departments/units, type of information flows, etc.) and the level of management culture are strictly related to the type of structure which characterise the same firms. In other words, these factors reinforce the negative attitude to invest in ITT in case of small-sized enterprises and the positive one in case of large-sized firms.
A medium-sized business company has thus a higher probability to increase their technology equipment than a small-sized co-operative association. In the first case, it is the organisation complexity which requires the adoption of new technologies for the management of information flows, also independently from the management's aptitude.

On the other hand, a simple organisation does not imply the introduction of ITT. In this case it may only be the sensibility of the owner which can stimulate the investment.

The strongest hindrances to the spread of ITT in the Italian tourist industry are represented by:

- the high fragmentation of the industry. More than 80% of the enterprises interviewed have less than 20 employees and more than 40% has a turnover less than 250.000 ECU;
- the prevalence of a traditional, family management. More than 80% of these firms are directly managed by the owner, while a clear distinction between ownership and management is rare;
- the lacking of a technology-oriented management culture. The scarcity of professional skills along with a poor knowledge of instruments and services offered on the market still represent a psychological barrier to the introduction of new technologies. On the other hand, an "accumulation effect" may arise in those enterprises where ITT are already intensively used;
- the prevalence of direct, informal and uncoded information flows both between ownership/management and staff and between the enterprise and the market (suppliers, clients, banks, etc.). However, the level of formality and coding tends to increase in external contacts;
- the high cost of investing in new instruments or accessing the digital network met by small and medium-sized tourist enterprises. Given the rapid evolution of technology, small operators are often afraid that the tool/service purchased becomes obsolete well before they have been able to amortise the investment;
- the location near an important tourist attraction/site. It has been noted that firms located, for example, in the historical centre of famous art cities and which enjoy a monopoly of tourist flows, have often no incentive to improve their internal organisation and to invest in Information technology.

Given these elements, the potential demand for ITT does not automatically translate itself in a real demand. A large number of operators do not have or do not perceive the need to apply new technologies. And it is just from this need that, as discussed before, the stimulus to investment arise.

4. Future scenarios

In spite of this, there are however signals of change. Not only at the global level, but also in Italy, four megatrends are forcing the tourist industry to modify its attitude towards ITT:

- the transformation of tourist enterprises;
- the evolution of the tourist market as a whole, both on the demand and the offer side;

- the development of technology (e.g. the completion of the shift from analogue to digital information systems, the creation of an integrated System of Information Technologies (SIT));
- the evolution of the institutional environment (deregulation).

Apart from recognising the importance of the development of Information technology and the stimulus coming from the increasing role of national and international institutions (e.g. EC), the first two trends deserve to be examined more thoroughly.

As far as the single tourist enterprise is concerned, the slow but steady growth in size and the multiplication, in some case, of local units; the gradual development of a process-based organisation and the progressive use of yield management techniques are elements which may contribute to the spread of ITT. In detail, the replacement of single departments with a sequential flow of co-ordinated activities aimed at the satisfaction of the final consumer[3,4], and the implementation of an approach "to maximise profit by carefully monitoring and managing pricing, inventory availability and sales"[5] will need more rapid and frequent exchange of information within and among tourist enterprises and therefore an intensive use of new technologies.

These changes are also enhanced by the evolution of the tourism market, both on the demand and the supply side. The birth of a new tourist, more expert and quality-conscious, and consequently the development of demand for more flexible and "tailor-made" holidays, is causing a revolution in marketing and promotional strategies. There is an increasing integration among all the actors which produce goods and services for the fulfilment of the travel experience (means of transport, tour operators and travel agencies, hotels, banks, insurance companies, etc.).
In this context, "technology plays an important complementary role in engineering the new tourism"[6]. It is the platform for creating and implementing partnerships and networks among firms which belong to the same value chain and pursue the same objectives. For example, the growth of local networks and above all the spread of the "destination management" approach allow tourist enterprises located in the same resort/area to enjoy higher visibility in the market and economies of scale, overcoming the hindrance of the insufficient size of the single firm. Generally speaking, partnerships and agreements with international operators favour the importation of up-to-dated management models and consequently of advanced telematic and informatic systems.
Finally, a further stimulus to innovation may come from banks and other new service providers (insurance companies, financial companies, etc.). The coming into the tourist market of such firms, which belong to other sectors but offer the same tourist services by replacing or overlapping traditional suppliers, influence the behaviour of tourist enterprises. The action of these actors often force the operators to introduce new technology instruments, which probably influence their internal organisation and management.

5. Closing comments

For the time being, the development of Information Technology in the world tourist industry interests almost exclusively large operators, while small and medium-sized tourist enterprises -- which represent the majority of the tourist firms -- are lagging behind.
Using a demand-side approach, the paper has aimed at evaluating the aptitude of tourism operators for new informatic and telematic tools and services. In detail, this paper has analysed the economic and behavioural factors which may stimulate or restrain the use of ITT for the management of information flows within and among tourist enterprises and the market in which they act, by looking specifically at the Italian tourist industry.
The results obtained give an interesting insight on the future prospects of the application of new technologies in tourism, from which both the tourism industry and academics may draw useful suggestions.
Though forming on the Italian market, its results can be easily extended to other European countries where small and medium enterprises represent the major segment of the tourist market.
Furthermore, the identification of the main hindrances and opportunities encountered by small and medium-sized tourist enterprises to the application of ITT provide the basis for a discussion on how a Telecom company can anticipate these developments.

As far as the Telecom Italia company is concerned, the guidelines of its new strategy may be summarised in the following points:

- *promoting destination management systems.* As stated in Section 4, ITT may be used as a strategic tool for the development both of the tourist resort/area as a whole and of the single small and medium-sized tourist enterprises located there[7]. In this context, the Company aims at having a role of catalyst in the setting up of local networks among tourist enterprises;
- *promoting electronic ticketing and new forms of telematic reservation of tourist products.* Up to now, CRS are mainly used for flight and train reservations. It should be useful to extend these practises not only to all the other means of transport (ships, car-rental) but also to accommodation establishments and tourist attractions (museums, theatres, theme parks, etc.). Telematic reservations would be made by TO and travel agencies or directly by the tourist on his/her personal computer.

As a consequence, the key issues of the Telecom Italia's strategy are:

1. pricing;
2. product innovation in the core business;
3. entry in new multimedia markets.

Price reductions are gained with high volumes of traffic. In the tourism industry, these volumes are easily achieved by large enterprises (such as air companies and hotel chains), which usually have many locations spread on the territory. But also a group of small-sized enterprises organised in a network may produce a good volume of information flows.

Product innovation in the core business mainly concerns high speed communications, supplementary telephone services, intelligent network services and international calling card.

Finally, the marketing of on line services, videoconferencing and Internet access providing represent the first step of the Telecom Italia company in the new multimedia markets.

References

1. Haberman, S.J.: The analysis of qualitative data, Vol. 1, Academic Press, New York (1978).
2. Knoche, D., Burke, P.J.: Log-linear Models, Sage, California (1983).
3. Davenport, T.H.: Process innovation: Reengineering work through information technology, Harvard Business School Press, Boston (1993).
4. Simon, K.A.: From structure to process: a vision of a process-based organization. In: Schertler, W., Schmid, B., Tjoa, A.M., Werthner, H. (eds.): Information and Communication Technologies in Tourism, Proceeedings of the International Conference in Innsbruck, Austria, Springer-Verlag, Vienna-New York (1995).
5. Andersen Consulting : Yield Management in small and medium-sized enterprises in the tourist industry, European Commission, DG XXIII-Tourism Unit (1996).
6. Poon, A: Tourism, Technology and Competitive Strategies, CAB International, UK (1993).
7. Buhalis, D: Information and Telecommunication Technologies as a Strategic Tool for Tourism Enhancement at Destination Regions. In: Klein, S., Schmid, B., Tjoa, A.M., Werthner, H. (eds.): Information and Communication Technologies in Tourism, Proceeedings of the International Conference in Innsbruck, Austria, Springer-Verlag, Vienna-New York (1996).

MINOTAURUS: A Distributed Multimedia Tourism Information System

S. Christodoulakis, P. Kontogiannis, P. Petridis, N. Moumoutzis, M. Anastasiadis, T. Margazas.

MUltimedia Systems Institute of Crete (MU.S.I.C.) Technical University of Crete, Chania, GREECE

e-mail: {stavros, pkonto, pantelis, nektar, manos, tm}@ced.tuc.gr

phone: +30-821-64803, fax: +30-821-64646

Abstract

The rapidly expanding technology of mobile computers, wireless data networks, vehicle navigation, multimedia and database systems has caused the development of powerful mobile information systems. These systems, consisting of a portable computer (Laptop, PDA), have large storage capacities, capabilities of wireless connection to a worldwide information network and provide to their users many functionalities like access to WWW, shopping, banking, reservations and other transactions.

This paper presents such a mobile system named **Minotaurus** which combines the above technologies to achieve the development of a mobile, multimedia tourist information system.

1. Introduction

The system's objective is to overcome the existing problems of traditional tourist information systems using modern technology, aiming to eliminate and resolve most of them. The baseline functionality that the system provides to its users has been categorized in:

- Presentation of historical, tourist and geographic information about the areas a user wishes to visit.
- Providing road-level navigation assistance through the display of the vehicle's position onto a geographical map when the user travels.
- Presentation of data stored in Remote Information Servers accessed through the Internet from wherever the user might be, using location-transparent methods.

In contrast to existing Tourist Information Systems, **Minotaurus** has substantial advantages. The most important issue is the *ease and speed of access* to the tourist-related information since the platform being used allows the user to carry the system inside his car in full effect.

The wireless modem being used over Mobile Telephony network enables the user to obtain access to a vast amount of data through a connection to a remote Information Server, and the WorldWideWeb (WWW). Thus, the system is *unrestricted with respect to the amount of information it can provide* to the user, while still *being portable*.

Yet another advantage of **Minotaurus** is the usage of a GPS device, that introduces the notion of vehicle navigation in Tourist Information Systems. The

GPS device is used for the display of the user's geographic position, assisting him during his excursions.
Last but not least, the system uses a well-defined structure based on Relational Data Base Technology and also provides a visual presentation using Multimedia Technology.

In the rest of this paper we will describe the architecture of **Minotaurus**, its basic characteristics, the functionality it provides and finally, the Client and Server components. We will mention the contribution of this work and we will conclude with plans for future work and possible improvement.

2. The *MINOTAURUS* architecture

In this section we'll describe the architecture of the **Minotaurus** system, its components and their attributes. **Minotaurus** consists of two parts: the **Information Service System** and the **Mobile Information Stations [6]**.
The **Mobile Information Stations** are portable systems which travelers and tourists can use during their vacations or tours in order to obtain access to valuable information. This portable system contains 3 components (Figure 1):

- A portable multimedia computer (Laptop).
- A Global Position System (GPS) receiver.
- A wireless modem.

The portable multimedia computer is the basic component of a Mobile Information Station. It runs the application software (**Minotaurus Client**) which is responsible for the presentation of the tourism information, manages the interface with the other peripheral components and handles the interaction with the user. It also contains large storage mediums (Hard Disks, CD-ROM) for storing the tourist multimedia data.
GPS is a satellite positioning system based on the NAVSTAR/GPS [8][13] project of the U.S. NAVY. In a few words, the GPS receiver consists of a PCMCIA card and a special antenna that receives specific signals from the GPS satellites. The PCMCIA card processes the satellite signals and can compute quite accurately (100 meters precision) the geographic position (latitude, longitude, height) of the Mobile Information Station every 1 sec.
Finally, the wireless modem consists of a mobile telephone and a PCMCIA cellular data card. It is used to establish communication connections between Mobile Information Stations and Information Service System through modern cellular data communication networks. **Minotaurus** uses these connections to successfully access dynamic information stored in remote service systems. Dynamic information is time-dependent (changes continuously) and cannot be stored in the storage media of the Mobile Information Stations.

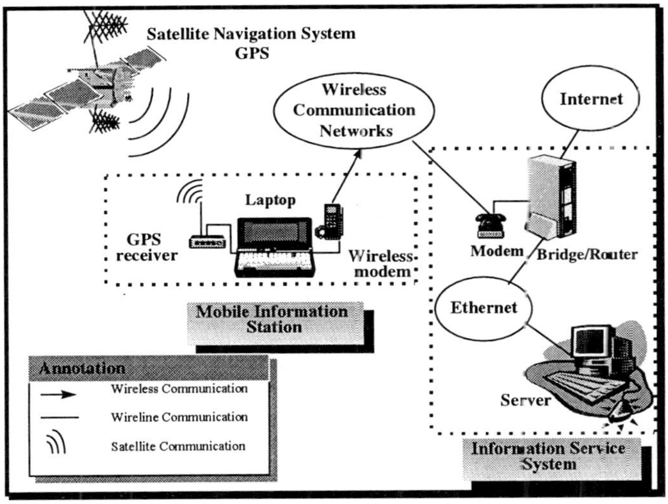

Figure 1. The **Minotaurus** architecture

The **Information Service System** is a centralized information system responsible of processing and servicing of the requests sent from Mobile Information Stations for presentation of dynamic tourist information.

The Information Service System consists of three components :

- A bridge/router.
- A master desktop computer.
- A set of secondary desktop computers.

The bridge/router is a **base station [6]** responsible of establishing the wireless communication between the Information Service System and the Mobile Information Stations.

The master desktop computer is the basic component of a Information Service System. It runs the software (**Minotaurus Server**) which is responsible of receiving requests from the Mobile Information Stations, and servicing them appropriately. Each request is transformed to one or more SQL queries which are then forwarded to a relational DBMS for processing. The results are retrieved and sent back.

The set of secondary computers is used to constantly update the database with the dynamic information. It constitutes a local network along with the master computer.

Finally, the communication between the three components mentioned above, is established using a traditional Ethernet-based LAN. Some alternative solutions could be ISDN, ATM or other networks. On the other hand the communication between the Information Service System and the Mobile Information Stations is established using cellular communication networks like GSM-900 and DCS-1800.

3. The *MINOTAURUS* Client

The **Minotaurus Client** is the software application running on the Mobile Information Stations. It exports the functionality of **Minotaurus** to the users.

The Minotaurus Client consists of three subsystems :

- The Geographical Position Display (GPD) subsystem, which is responsible for the accurate display of the geographical position of a Mobile Information Station onto a digital map of the area where the user is currently moving.
- The Static Tourism Information Presentation (STIP) subsystem, which is responsible for the presentation of multimedia tourist information to users.
- The Dynamic Information Management (DIM) subsystem which is responsible for the management of the communication establishment between the Mobile Information Stations and the Information Service System for access to dynamic tourist information and of presentation of information to users.

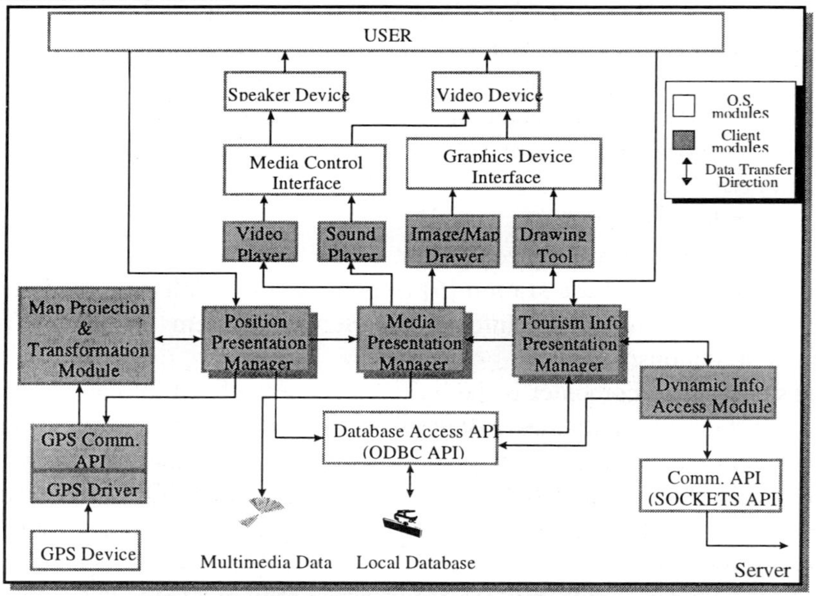

Figure 2. The Minotaurus Client architecture

In the following sections will describe the three subsystems and the software architecture (Figure 2) of Minotaurus Client.

3.1 The Geographical Position Display subsystem

The Geographical Position Display (GPD) subsystem is the component of Minotaurus Client which undertakes the display of the position of a Mobile Information Station onto a digital geographical map of the area the user is currently moving.

In order to display the user's position, GPD uses the GPS receiver and executes the following procedure (Figure 3):

- GPD communicates with the GPS receiver every 1 sec and receives the computed position of the Mobile Information Station. This position is described in an elliptic geographical coordinate system named World Geodetic System - 84 **(WGS-84)** [14].
- GPD transforms the position's value as measured by the GPS receiver from elliptic to Cartesian coordinates system using the *Universal Trancator Mercator* **(UTM)** projection method [2].
- GPD transforms the value of the user's position from the Cartesian **UTM** coordinates system to the Cartesian image coordinate system (Pixel coordinates) using the *Mapping Polynomial with Control Points* algorithm and *Resampling* methods. These methods improve the quality and the precision of the digital map removing the distortions caused by scaling, rotation, shearing, reflection, warping but not the ones caused by random errors.

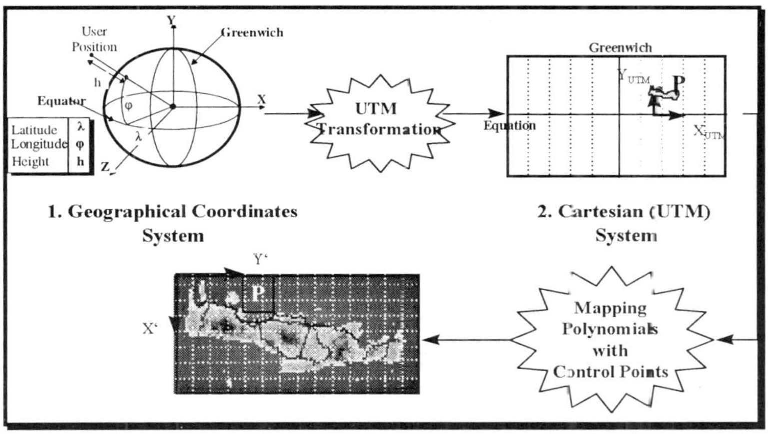

Figure 3. The Geographical Position Display on Digital Map Method

The whole procedure of Mobile Information Station position display is managed by the *Position Presentation Manager* (PPM) module. PPM not only controls the communication between Minotaurus Client and GPS device via the *GPS Comm. API* module and the projection and transformations of the initial values from GPS receiver via the *Map Projection & Transformation* module but it is also responsible for the access of digital maps from the Mobile Information Station local multimedia database and the display on the screen of the mobile computer.

3.2 The Static Tourism Information Presentation subsystem

The Static Tourism Information Presentation subsystem (STIP) undertakes the management and servicing of the user requests for *static* tousist information presentation. This information which describes historical and tourist sites or geographical areas of general interest via multimedia (texts, photos, maps, audio and video), doesn't require continuous updates and is stored in the secondary storage of the Mobile Information Stations (CD-ROM).

The STIP's basic operations are :

- The processing of the user requests for tourist information presentation.
- The formulation and execution of the appropriate SQL queries to the relational DBMS of the system.
- The final visual presentation of the information to the user.

The *Tourism Info Presentation Manager* (TIPM) is responsible of the management of the procedure for tourist information retrieval. Specifically, after receiving the user requests for information presentation, TIPM processes it and creates the appropriate SQL queries to retrieve the information from the local database in Mobile Information Stations. The SQL queries are executed via the *Database Access API* module and the SQL results are returned to the TIPM.

Then, TIPM informs the *Media Presentation Manager* (MPM) to undertake the final visual information presentation to the user. The MPM uses one of the modules *Video Player, Sound Player, Image/Map Drawer, Drawing Tool* to handle different media types. In the first three cases, MPM accesses the multimedia database of the Mobile Information Station and provides the corresponding data files to the Multimedia Players. In the last case, text or the attributes of the graphical objects are sent to the MPM by TIPM in order to be displayed.

3.3 The Dynamic Information Management subsystem

The Dynamic Info Access Module subsystem (DIAM) is responsible of managing the communication between the Mobile Information Stations and the Information Service System. Furthermore, it is responsible of the presentation of *dynamic* tourist information to the user.

The term *dynamic* is used to describe the time-dependent and time-varying tourist information such as news, weather forecasts, cultural and special events, time schedules of excursions with various transportation media, reservations in hotels and flights etc.

This information is stored into the database of the Information Service system in order to be easily updated. The Mobile Information Stations retrieve it using wireless, cellular communication networks (e.g. GSM 900). Cellular communication networks do have several disadvantages [1][9]:

- Low transmission rates (up to 9600 bps).
- Relatively high cost, compared to traditional networks (wired).

- Low connection quality (It is quite common to face difficulties in connecting to the server, while the connection itself suffers from noise which can frequently cause disconnections).

As a partial solution to these problems, we propose a caching technique: small-size databases are kept locally (on the client's side) and their content is updated by downloading the dynamic data from the centralized database located at the Information Service System. Each time the Client Application is executed on the Mobile Information Station, a connection is established between the Mobile Information Stations and the Information Service System, through which the local database's content is updated to reflect the latest state. This method allows us to:

- Minimize the duration of the connection to the Information Service System in order to minimize the related economic cost.
- Minimize the time delays related to the service of the user's requests, since the network is bypassed and data is now accessed from the local database.

However, this method may face data inconsistency problems. The majority of the dynamic data mentioned –such as weather forecasts, sport and cultural events, trip schedules etc.– changes slowly, allowing us to update the local caches on a regular yet in-frequent basis. By convention we've decided to update the local cached on a daily basis, during low-rush or no-rush hours as far as the system's workload is concerned (e.g. early in the morning).

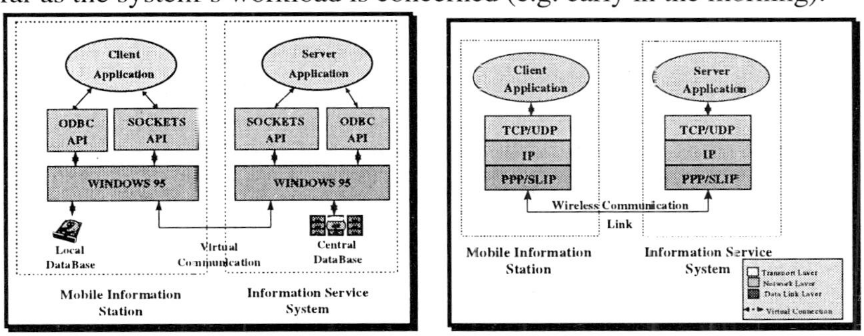

Figure 4. The Client - Server Communication

On the contrary, this method is not sufficient for frequently changing information, such as reservations-related information. To solve this issue, the system has no alternative to accessing the Information Service System directly, in order to serve specific user requests.

Access to the dynamic information is managed by the *Dynamic Info Access* module. When the DIAM receives a request for dynamic data by the Tourism Info Presentation Manager, either retrieves the related information from the local (to Mobile Information Station) database if it is up-to-date, or manages

the remote communication establishment and the transmission/reception of data between the Mobile Information Station and the Information Service System.
The fundamental characteristics of wireless communication between the Mobile Information Stations and the Information Service System (Figure 4) are:

- Client- Server architecture [4][7].
- TCP/UDP transport protocols for the end to end communication management [13].
- IP network protocol for the transmission over heterogeneous communication networks.
- SLIP/PPP data-link protocols for point to point communication between the Mobile Information Stations and the base station in Information Service System [12].
- Cellular PCMCIA Data Card and modems for the physical transmission of data.

4. The MINOTAURUS Server

The Minotaurus Server is the software application running on the master desktop computer of the Information Service System. Its main purpose is to serve the information requests sent by the Mobile Information Stations to the Information Service System. In this section we briefly discuss the fundamental components of the Server's architecture and their functionality.
The procedure for serving the information requests is comprised of the following steps:

- Reception of the information requests sent by the Mobile Information Stations.
- Analysis and translation of these messages.
- Construction and execution of the appropriate SQL requests to the central Information Service System's database [5].
- Retrieval of the results, construction of the communication messages, and transmission of these messages to the recipients (the Mobile Information Stations).

The *Dynamic Info Access Module* (DIAM) of Server application manages the communication and the message interchange between the Information Service System and the Mobile Information Stations. When DIAM receives an information request from a Mobile Information Station, forwards it to the *Message Translator-Creator* (MTC) module. This module's basic task is to analyze and translate the received messages to a *Message Language*. This language specifies the structure of the messages interchanged between the Mobile Information Stations and the Information Service System.

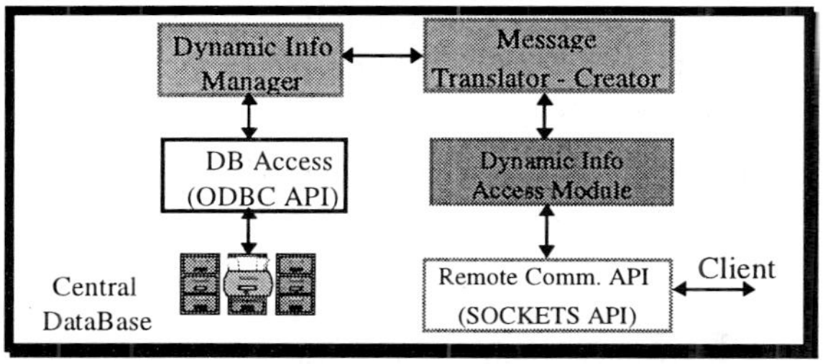

Figure 5. The Minotaurus Server architecture

In the next step, MTC requests the Dynamic Info Manager (DIM) to initiate the information retrieval procedure. The DIAM creates the appropriate SQL queries, executes them via the *Database Access API* module and the query results are returned to the DIM. The DIM forwards the results to the MTC module in order to construct the answer message (in order to Message Language). Finally, this message is sent to its recipient via the DIAM.

5. The functionality of MINOTAURUS

In this section we will describe the main features of the **Minotaurus**'s functionality. Specifically, we will present in depth the potential information capabilities provided to the users, as they travel with their vehicles. As previously stated, the **Minotaurus**'s main objective is the presentation of tourist information for a geographic province, an area which is characterized by similar attributes (history, weather, traditions, geographic neighborhood etc.). Such an area is the island of Crete which was used as a pilot-area by the **Minotaurus** project.

The geographic area **Crete**	The area of interest **Chania**

Figure 6. The presentation of the geographic province **Crete**

According to the model of information presentation **Minotaurus** uses [3][10], the geographic province is presented to the users via a set of *Areas Of Interest* and *Sites*. The *Areas Of Interest* can be smaller geographic parts with their own distinct attributes (e.g. towns, villages etc.), while the *Sites* are usually areas with special tourist, historical or geographical interest (e.g. monuments, museums, mountains, gorges etc.).

The *Areas Of Interest* are presented using the map of the geographic province where they belong. The geographic positions of the *Areas Of Interest* are displayed on this map as graphical objects (e.g. circles, polygons). This objects are hotspots. The user can select them with his/her mouse, to visit them and see the information that the system provides for them.
An *Area Of Interest* is usually composed of a set of smaller *Areas Of Interest* and/or *Sites*. For example, a city consists of a set of districts, roads, squares, described as *Areas Of Interest*, and a set of museums, churches, archeological grounds, described as *Sites*. They are presented in the map of the city in the same manner as described above (Figure 6).
In short, a tree hierarchy is created with root being the geographical province, nodes the *Areas Of Interest* and *Sites* as leaves.
The complete presentation of the information regarding the *Areas Of Interest* and *Sites* is achieved by using Multimedia (combination of text, photos, sound, video).
In addition, the **Minotaurus** system displays the location of the user as he/she travels along with the tourist information of the area. The user's location is presented as a blinking dot on the digital map of the area (Figure 8).
In summary we achieved:

- Simultaneous display of the user's position and the nearby *Areas Of Interest* on the common geographical map of the area.
- Automatic map interchange when the users arrive/leave at an *Areas Of Interest* which has its own geographical map.
- Automatic notification of the users using text and sound when they approach an *Area Of Interest.*

Also, the system has the ability to access dynamic information stored in remote database systems. The stored information can be:

- Structured data, dynamic information related to the *Areas Of Interest* and *Sites* according to the system's information presentation model. Such information could be cultural, sport, special events etc.
- Unstructured data, general information. These are not related to the geographical province presented and are displayed by a WWW browser.

Finally, the user interaction model [3][10] is used for accessing the tourist information. This model gives significant help to the user as it achieves to:

- Present the information on different levels of detail (e.g. using maps with different scale).
- Present independently the *Areas Of Interest* in the same format with the traditional tourist guides.
- Access information with different ways.
- Show the information perspectives at any point of the user navigation.

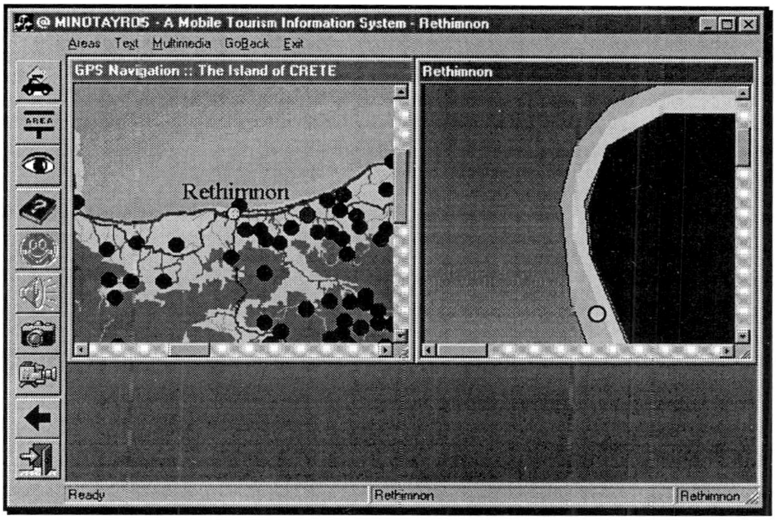

Figure 7. User's Position and Tourist Information Simultaneous Display

Conclusions

In this section we will mention the contribution of **Minotaurus** and in the tourist industry and the computer science, and we will suggest some future extensions.

The most important advantage of **Minotaurus** is the effective integration of three different heterogeneous systems: a GPS-based vehicle navigation system, a multimedia tourist information presentation system and a wireless communication system able to access data from remote database systems. The result of this integration achieves the development of a mobile information system able to replace the traditional tourist information systems. Its main characteristics are:

- The portable platform. In this way users can continuously carry the system inside their car in full effect during their trips.
- The visual information presentation via multimedia (text, photos, sound, video).
- The ease and speed of access to the tourist information.
- The well-defined structure and the independent presentation of the information.
- The unrestricted amount of information which is available to the users via wireless connections with remote databases.

Finally, **Minotaurus** has been developed in a platform which admits many future extensions. The most important of them are the increase of the precision in the GPS-based vehicle navigation system, the use of voice in the user interface, the multilingual support etc. Finally, there are plans for the exportation of the system's functionality into more light-weight platforms (e.g. PDAs).

References

[1] E.Brewer, T.Burd, F.Burghardt, A.Burstein, R. Doering, ***"Design of Wireless Portable Systems"***, IEEE Transactions on Communications, 1995.

[2] Clarke, ***"Analytical and Computer Cartography"***, 2nd edition, Prentice Hall.

[3] S.Christodoulakis, F. Kazasis, N.Moumoutzis, A.Servetas, P.Petridis, ***"A Software Bench for the Production of Multimedia Tourism Applications"***, Information and Communication Technologies in Tourism, Springer 1996.

[4] G.F.Coulouris, J.Dollimore, ***"Distributed Systems : Concepts and Design"***, International Computer Science Series, 1988.

[5] C.J.Date , ***"An Introduction to Database Systems"***, Volume 1, 5th edition, Addison-Wesley Publishing Company, 1990.

[6] M.H.Dunham, A.Helal, ***"Mobile Computing and Databases: Anything New?"***, SIGMOD Record, Vol. 24, No. 4, December 1995.

[7] A. Elmagarmid, J.Jing, T. Furukawa, ***"Wireless Client/Server Computing for Personal Information Services and Applications"***, SIGMOD Record, Vol. 24, No. 4, December 1995.

[8] B.Hoffman-Wellenhof, H.Lichtenegger, J.Collins, ***"GPS Theory and Practice"***, Springel-Verlag Wien New York, 2nd edition, 1993.

[9] T. Imielinski, B.R.Badrinath, ***"Data Management for Mobile Computing"***, SIGMOD Record, Vol. 24, No. 4, December 1995.

[10] F.G.Kazasis: "A Development Model for Multimedia Tourist Information Systems", M. Eng. Thesis Technical University of Crete, 1995.

[11] A.Reinhardt, ***"Special Report: Mobile Computing, From Here To Mobility"***, BYTE Magazine, June 1995.

[12] N.J. Muller, ***"Wireless Data Networking"***, Artech House Publishers, 1995.

[13] Andrew S. Tanenbaum,***"Computer Networks"***, 2nd edition, Prentice Hall International, 1991.

[14] David Wells, ***"Guide to GPS Positioning"***, Canadian GPS Associates, Fredericton, N.B, Canada, 1986.

WORKFLOW TECHNOLOGY: AN APPLICATION FOR TOURISM MANAGEMENT

A. Aguayo, J.L. Caro, A. Guevara
e-mail : <aguayo, caro, guevara>@lcc.uma.es
e-mail: jlcaro@ccuma.uma.es
Departamento de Lenguajes y Ciencias de la Computación
Universidad de Málaga. Spain

L. González
e-mail: cst@ccuma.uma.es
Departamento de Economía Aplicada (Estructura Económica)
Universidad de Málaga. Spain

Abstract. Workflow management is of great value for companies since today they have to confront a highly competitive environment, and are in need of reducing costs and swiftly developing new services and products. These demands cannot be met without reconsidering and optimising the way of doing business, developing policy and modifying information systems and applications. Workflow technology facilitates this constant evolution, providing methodologies and software to support: (i) the modelling of processes as workflow specifications, (ii) reengineering business processes in order to optimise them and (iii) workflow automation to generate an implementation from applications.

Tourism related organisations and business, like any other type of enterprise, can benefit from this technology and obtain greater efficiency. This will translate into increased customer satisfaction and better efficiency in the developed processes. In this work we intend to apply workflow technology to the management of a hotel by identifying processes, tasks and methods employed in the hotel industry. First we will gain an understanding of the processes involved in hotel management and then model them, from the moment of booking in, to the customer checking out. Thus, our objectives are: (i) to introduce workflow technology and (ii) to apply it to the tourism sector, specifically to a hotel.

Keywords: workflow, computer supported cooperative work, CSCW, reengineering, tourism information system, reengineering business processes, BPR.

1 Introduction.

The concept of workflow evolved from the world of industry and business to refer to the processes taking place in manufacturing and the office. Such processes existed from the time industrialisation began and administration began to look for ways to increase efficiency in the routine aspects of work activities. Initially, these processes totally were carried out by people, who manipulated physical objects. However, with the introduction and the advance of new information technologies, many processes, previously accomplished manually, are now partially or totally automated by information systems, that is to say, they are carried out with computer programs able to develop tasks and put into action a series of rules previously implemented by a human being.

Within a given organisation these processes can be categorised as follows: (i) Material processes. These involve tasks carried out by humans and include such things as the assembling of components and their distribution; (ii) Information processes. These relate both to fully automated tasks, that is, tasks performed by programs, and partially automated tasks (e.g., tasks performed by people interacting with computers); and (iii) Business processes. These refer to market-centred descriptions of the activities of an organisation which are implemented as information processes and/ or material processes.

All these processes, which can be found in any organisation, can be reengineered in order to improve or adapt them. Reengineering existing processes provides many advantages, among which we can outline the following: increase in customer satisfaction, improvement in efficiency, betterment of product quality, cost reductions and encouraging the finding of new business opportunities and challenges by changing existing services or introducing new ones.

Workflow is a concept closely related to reengineering and automating businesses and information processes within a given organisation. Workflow management technologies involve: (i) Defining workflows, that is to say, describing those aspects of a process that are relevant to controlling and coordinating the execution of tasks (and possibly the skills required by the people or information systems in charge of performing them) and (ii) providing rapid redesign and reimplementation of processes as business needs and information systems change. Thus, a workflow could describe business process tasks at the conceptual level needed for the evaluation, understanding and design of such business processes., as well as capturing information process tasks at a level that describes such process requirements for information systems and human skills.

Therefore, workflow management (WFM) supports the reengineering of information and businesses processes. In more concrete terms workflow management includes the following aspects: (i) process modelling and workflow specification; (ii) process reengineering; and (iii) implementation and automation of workflows. These three aspects range from the mentioned process modelling, to the implementation and automation of these processes by information systems. In order to tackle these tasks we require: a) models and methodologies for capturing and optimising processes; b) technologies capable of using information systems; and c) human developers to implement, plan, execute and control the workflow tasks.

In addition to this, the direct relationship between workflow and computer supported cooperative work (CSCW) is clear. We can envisage workflow as a continuum that moves from human-oriented to system-oriented workflow. At one end, human-oriented workflow implies human collaboration as regards the development and coordination of tasks; at the other end, system-oriented workflow involves computer systems which perform computation intensive operations and specialised software tasks. The implementation of human oriented workflow aims at controlling and coordinating human tasks while the latter aims at the control and coordination of other tasks with little or no human intervention. As can be intuited, CSCW belongs between these two poles, whether workflow carries the automation processes further.

2 Modelling a workflow process.

There is a great variety of techniques for process modelling within the workflow approach. We can classify them into three main categories [3]:

1) *Communication-based Methodologies* [4]. These assume that the main aim of reengineering a business process is to increase and improve customer satisfaction.

2) *Activity-based Methodologies.* These focus on modelling the task or job rather than the interaction between people.

3) *Object-oriented Methodologies*. This works with "actors" taking different roles in the workflow and their relationships in order to apply object oriented techniques to matters such as inheritance, etc.

In this article we will use a communication-based methodology to model the processes taking place in a hotel. Communication-based technologies translate any action into a workflow with four steps which are based on the communication between a client, requiring a service, and a server, offering the service.

The four steps are as follows:

1. *Initial request or preparation*. The client asks for a given service.
2. *Negotiation*. The conditions for satisfaction are defined and an agreement is reached .
3. *Development*. The action is carried out according to the terms agreed upon.
4. *Acceptance or satisfaction*. The client reports his level of satisfaction as regards the action carried out.

Each of the loops within a workflow can be linked to other loops creating a network in which each group of people take on certain roles within the business process in order to carry out the action successfully.

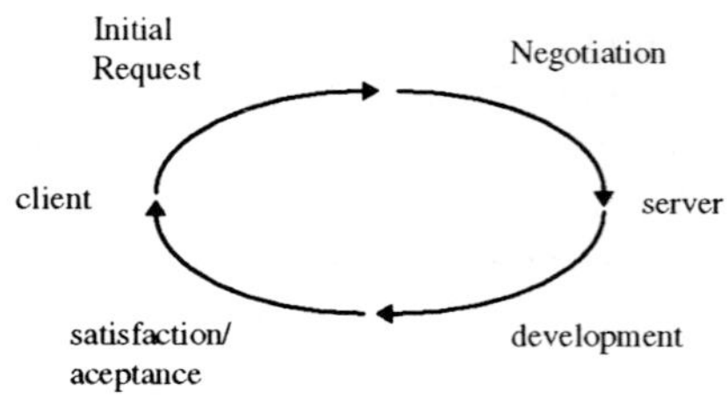

Fig. 1. Workflow loop.

3 A workflow model of a hotel information system.

Information and knowledge are indispensable resources in the operation of any information system. In our case, the system we are going to study is a hotel. The aim of this is to create a suitable information system, which will result in increased efficiency in the hotel itself. In order to develop an information system based on workflow technology we need a solid understanding of the services offered by the hotel and determine the relationships existing between the different facilities in the form of workflows.

A hotel can be understood as a large system consisting of various groups of people and information systems (e.g. reception) that work in cooperation to obtain a common aim: *the satisfaction of the client.*

In order to properly model the full system using WFMS, we will have to understand the workflows and information established between these groups of people or systems. The starting point is several groups cooperating and relating with each other. Also, as regards processes, we can see that a hotel is no more than a given number of business processes related or linked to each other by a client-server relationship, in which each process provides a set of services.

We will take the case of a tourist couple requesting the services of a hotel as an example to illustrate process modelling. We will begin modelling from the moment the clients arrive at the hotel. Therefore, the architecture of our workflow, at the highest level, will be a workflow loop called *client accommodation.* In this workflow loop the role of the clients will be played by the tourist couple and the hotel will be the server. The booking stage belongs to another workflow that we will not address here.

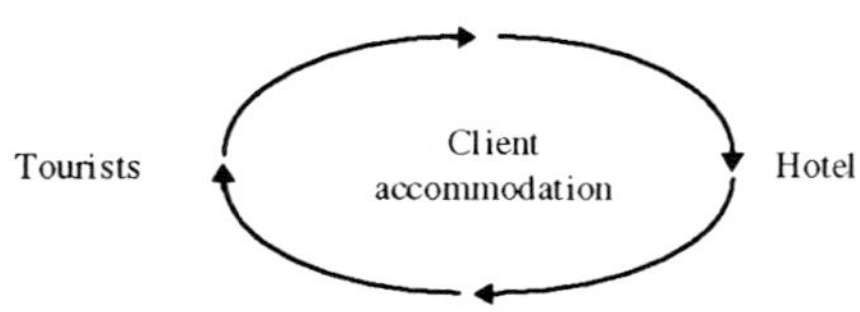

Fig. 2. Workflow loop: Client Accommodation.

This first loop can be broken up into three more workflows: *checking-in, using the hotel services and checking-out.* These encompass the server-client relationships that will be established while the clients stay in the hotel. The workflow loops show what acts as a server at the different stages. In our particular case the servers are the different facilities of the hotel.

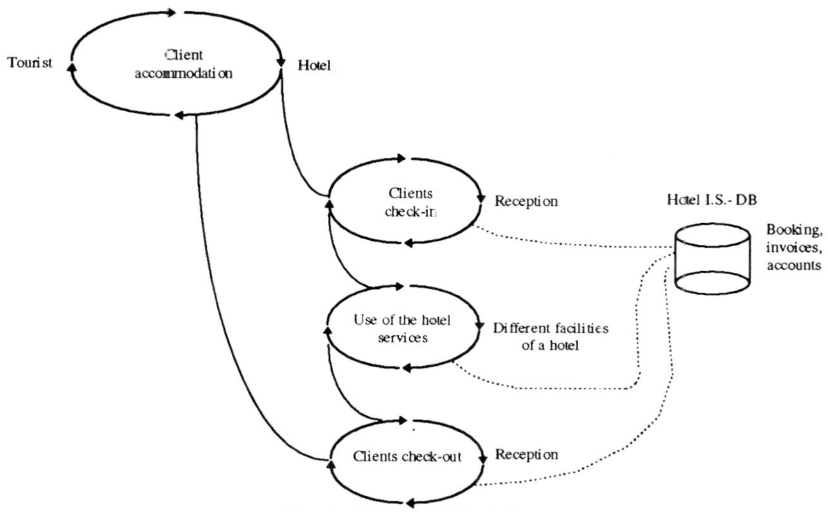

Fig. 3. Unfolded Workflow.

So, the first stage is the arrival of the clients at the reception desk in order to register and get a room. Then, during their stay they will use the different services available at the hotel, such as the swimming pool, restaurant, bar, etc. Finally, at the end of their stay, an invoice is made up to charge for all the services received. This is represented in figure 3.

In this figure, we can see how the three secondary workflows generated are linked one after the other, as they represent actions taking place in a sequential manner: arrival at the hotel, registering, the actual stay and finally the check-out.

In the first loop, *client check-in*, the tourists will be the client and the hotel reception the server. The reception can be considered a subsystem cooperating with the rest of hotel subsystems.

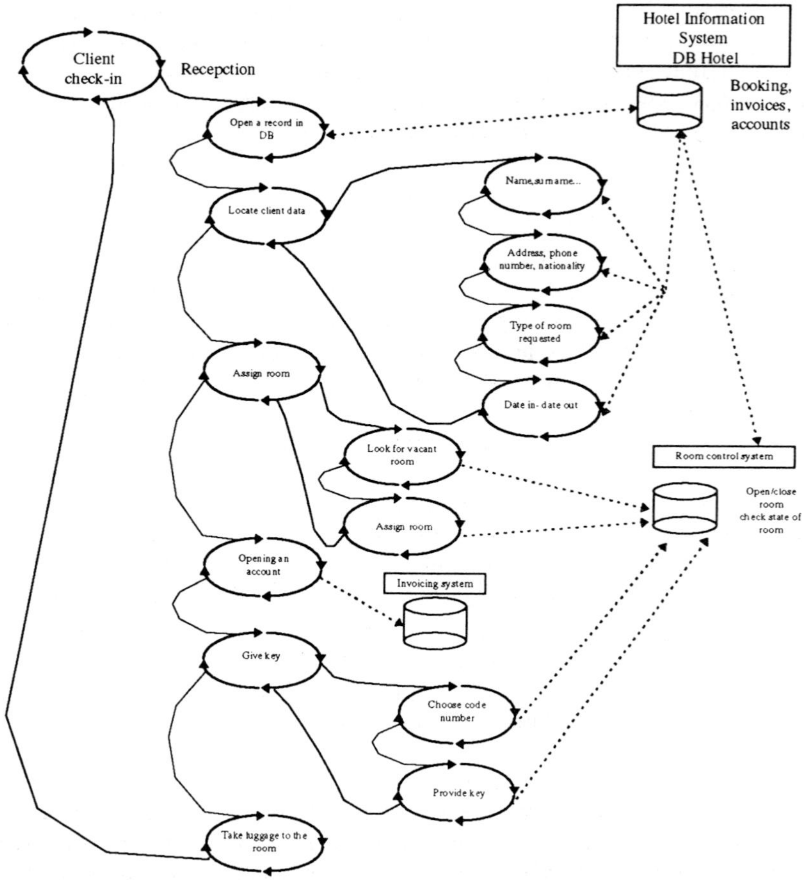

Fig. 4. Check-in Workflow.

The arrival of clients requesting the services of the hotel results in a series of linked workflows: from the registering of a new client to the assignment of the room so that they can begin their stay. Let us look at each workflow.

First of all we open *a record in the database* (if the clients do not already have one). Then, we *locate client data*. In this task the reception becomes the client while the server is the hotel's information system (i.e. the database management system storing clients data). This workflow opens up a series of workflows whose task is capturing new client data. This information might not always come via the receptionist directly asking the clients at the desk. For example, if the clients had visited the hotel before, the data will already be stored in the database, as it would be if a client had booked the room over the phone.

The next step, once the previous workflow has ended, is to *assign a room* to the couple. This generates two more workflows. The server will be the information system holding data about the state of the rooms (that is, vacant/occupied, double room/single room, etc.) We assume that the room system is fully automated and we can check whether they are occupied or vacant from the IS. Also, we assume that we can open and shut them electronically with, for example, a magnetic card system (electronic keys). These two workflows will first look for *vacant rooms* (by checking both the room system for vacant rooms and the clients' record to find out what sort of room they have requested). Subsequently, a room is assigned to the client and so it becomes occupied.

After assigning the room, we will carry out the workflow *opening an account*. We will assume that our hotel has facilities such as a restaurant, a bar or a shop, and that clients will be able to charge to their account any expenditures made at the hotel. In this way, when the client is ready to leave we will be able to make up the total bill for his stay. In this workflow we will simply inform the other information systems about the existence of a new client.

The last two workflows consist of giving the key to the couple and taking the luggage into their room. The *giving-the-key* workflow -assuming a fully automated system for opening and shutting the rooms- will result in two new workflows carried out in a sequential way (Fig. 4). The first one will be the selection of a code number for the room, which will involve access to both the database where we have stored the clients' record and the automated room system. Secondly, we will "create" a key for the client. This could be in the form of recording a code on a magnetic card so that the clients can open their room.

At this stage, the check-in process ends and the client is ready to enjoy the services on offer. The workflow *using the hotel facilities* will represent the model of the tourists' stay at the hotel (Fig. 5) and will depict the use the clients make of the different services available (Fig. 3). This workflow can be split into a number of workflows corresponding to the different hotel facilities. Note that these are not linked, as they are not sequential in time and can be required by the client at any moment. Each one of these workflows will have the tourist couple as the client and the different hotel facilities as the server.

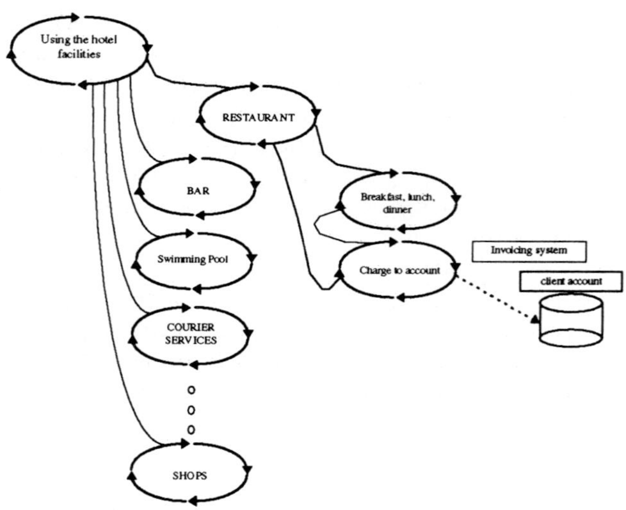

Fig. 5. Workflow using the hotel facilities.

Each of these services will generate several workflows that are not presented in this example. We will only show the workflow for the restaurant and in very general terms. For our purposes, this workflow can be broken up into an initial workflow consisting of, for example, *breakfast, lunch and dinner*, and a second one called *charge to account* in which we will access the client invoicing system to charge these services to their account.

The last step in our main workflow, *client accommodation*, refers to *the check-out* that will once again have the hotel reception as the server. The *check-out* workflow is depicted in figure 6.

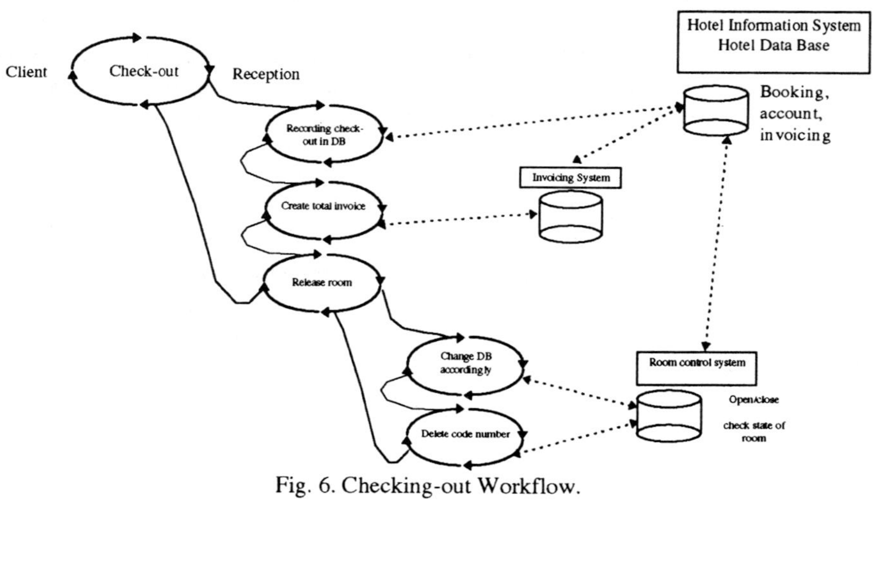

Fig. 6. Checking-out Workflow.

So, this last workflow represents the point at which the client is ready to leave the hotel. Three linked workflows are generated. And so, as one ends the other begins. We call the first workflow *recording check-out in database* which can include information like the check-out date, etc. Then, in the second workflow we will make out an invoice for all the expenditures of the tourist couple at the hotel. In order to do so we will check the date of arrival and the different charges to their account from the bar, restaurant, etc. Finally, we will release the room the couple had, which involves both changing the room status to vacant by using the automated room system and erasing the lock combination.

When the clients leave the hotel, the main loop *client accommodation* is finished and our workflow has ended. The last step in our workflow model will be to find out the level of client satisfaction with the hotel service.

4 Conclusions.

Workflow is a potent tool for the modelling of business processes, especially when customer satisfaction is an important factor. In the tourist industry the satisfaction of the client is directly involved in the accomplishment of the business processes. Therefore, workflow technology applied to tourism has great potential for the optimization of processes typically found in this industry.

In this paper we have carried out a process modelling using communication methodology which assumes that the main objective of the business is to fulfil as much as possible all the client's expectations. Workflow technology is essential in re-engineering processes since once we have a preliminary model of the business processes developed by the company, we can optimize the workflow model and thus obtain a greater level of satisfaction for all parts involved.

The processes of optimization and client satisfaction are two fundamental aspects for any company today, since the business world is a highly competitive environment which forces them to think of new ways of organising processes as to increase efficiency and reduce costs. But although greater efficiency and lower costs are very important, customer satisfaction is equally essential, especially for businesses directly related with tourism.

We are in favour of workflow technology because it offers a solid basis for process modelling oriented to achieve the full satisfaction of the customer and provides a good framework for future redesigns and optimizing processes before implementing the workflow management system (WFMS).

5 References.

1. D. Georgakopoulos, M. Hornick y A. Sheth, "An Overview of Workflow Management: From Process Modeling to Workflow Automation Infrastructure", Distributed and Parallel Databases, 3, 119-153 (1995).

2. R. Medina-Mora, H. Wong and P. Flores, "Action Workflow as the Enterprise Integration Technology", Bulletin of the Technical Commitee on Data Engineering, IEEE Computer Society, Vol. 16, Nº 2, June 1993.

3. R. Marshak, "Software to Support BPR - The Value of Capturing Process Definitions", Workgroup Computing Report, Patricia Seybold Group, Vol. 17, Nº 7, july 1994.

4. T. Winograd and R. Flores, "Understanding Computers and Cognition", Addison-Wesley, 1987.

5. P. Korzeniowski, "Workflow Software Automates Processes", Software Magazine, Feb. 1993.

Edite - A Natural Language Interface to Databases
A new dimension for an old approach

Paulo Reis João Matias Nuno Mamede

INESC

Av. Duque d'Ávila 23, 1000 Lisboa, Portugal

{Paulo.Reis, Joao.Matias, Nuno.Mamede @inesc.pt}

Abstract

This article presents the Edite system, a Natural Language Interface for Databases (NLIDB), that tries to explore the advantages of joining natural language processing with the expressiveness of graphical interfaces. In order to guarantee a permanent adaptation of this type of solution to a dynamic domain one should consider two critical fundamental factors: extensibility and portability.

An overview of the system architecture is presented, emphasising those choices that were imposed by the demands of portability and extensibility. Several general problems of natural language processing that were faced in constructing the system are discussed. Future work is highlighted.

Keywords: tourism, natural language processing, intermediate representation languages, database querying.

1. Introduction

The importance of the tourist industry for the Portuguese economy has significantly increased in the last 10 years. It accounts nowadays for approximately 8% of the GDP and equals that of the financial sector. The main reasons for the success of tourism are: the climate, the historical and cultural heritage, the tourist infrastructure, the hospitality of the Portuguese people and the relative close emissary markets.

It is important to create conditions to consolidate the registered growth, which can be achieved through a firm development strategy focusing on the quality of supply and human resources and on diversification of markets and products.

National tourism has adopted new instruments in order to reinforce their performance focusing on the new information and communication technologies.

The Inventory of Tourist Resources (IRT) emerges as the largest R&D Portuguese project in this area, actually exercising fundamental support on tourist ordering and planning and on forming a global reference point our tourism. The IRT initially emerged to eliminate the shortage of institutional information, positioning itself as the largest data repository of tourist resources on Portuguese tourism.

IRT's vast variety of purposes determined the adoption of an architecture able to support the integration of a large informational universe on a stage of multiple user segments and on the various applicational objectives (e.g., planning, promotion, distribution, management ...). Nowadays, IRT's informational domain uses Internet, multimedia "kiosks", GIS, graphic user interfaces and "natural language interfaces".

The supporting technologies on the diverse functions of IRT are in well marked, distinct stages of development, from a scientific and technological point of view as well as from a commercial point of view.

This article intends to approach and deal with the problem of integrating an advanced technology, the so-called plain language interfaces for databases, into systems with a complex architecture.

2. Motivation

One of the main characteristics of multimedia kiosks is their familiar visual appearance, reducing the complexity of communication between man and machine to a minimum. The anthroponomical synchronisation of Image, Video, Audio and Text is one of the crucial factors to "seduce" the user into wanting to experiment the system. Another fundamental characteristic is the emergence of the utility. The user must feel, when using the system for the first time, that it can be useful. This is only possible with a well designed interface where the information can be easily accessed without having to learn another vast and complex communication language (the one used by the system).

In spite of the large variety of existing systems there is not yet a standard for these interfaces. This causes the user to fully understand what the system does, only after a certain amount of time. Another critic one could have on this method of interaction has to do with the fact that in a traditional system, with navigation through several successive windows, one can not always get the needed information. This occurs either because it does not exist (and the system is not able to inform the user on this) or because the user does not know the system language well enough to extract the information in question (it may take too many steps to get there).

What could be the best browsing alternative that passes beyond the above mentioned limitation? The answer could be a NLIDB.

The evolution of technology has caused a continous development of NLIDBs, especially in the area of natural language processing, exploring architectures that transform the NLIDBs into relational agents, and integrating languages and graphics that explore the advantages of both modalities [1,6].

In order to guarantee a permanent adaptation of this type of solution to a dynamic domain one should consider two critical fundamental factors: extensibility and portability [9]. On the one hand, a product like a NLIDB has to guarantee the ease to increase the linguistic coverage related with the domain, without requiring a vast technical knowledge of the natural language. On the other hand, the system has to guarantee the portability to other databases, minimising the number of changes.

If the above mentioned factors are indispensable for the development of the system whilst a product, we will have to guarantee the emergence of the utility. We believe that in order to achieve this, three factors are indispensable: a good linguistic coverage, the comprehension of the domain (ideally an interface of this type can use all of the existing information in the database) and great interaction with the user.

Bearing this factors in mind, INESC has been developing Edite, a multimodal interface, since 1995. This interface tries to combine the advantages of the natural language with the expressiveness of the graphical interfaces.

Chapter 3 presents the NLIDB architecture of the Edite system, describing the principle modules. Chapter 4 will present some of the options for a future development. Chapter 5 will present the main conclusions of this work.

3. An overview of Edite

3.1. Introduction

Edite is a *multi-lingual* (Portuguese, French, English, Spanish) natural language front-end for relational databases. It answers written questions about tourism resources by transforming them into SQL queries. The answer depends on the type of question. It can be a nominal list of resources, text, images or graphics. At present, the database contains 53000 tourism resources, arranged on 253 distinct types, corresponding to 209 tables.

The main goal of a NLIDB is to provide users with the capability of, in an efficient alternative way, obtaining information stored in the database [5]. The user is not required to learn an artificial communication language being possible to formulate the question in his own native language. The system building up was driven by two main objectives: (a) the exploration of a new technology in an existing context, *i.e.* by exploring how this technology can be used to increase the efficiency of current processes. The technology should also work as a new motive of attraction. We could integrate this into multimedia kiosks, but we can foresee the adequacy to other information supports, like *e.g.* Internet; (b) the advantages NLIDBs have, compared to others interfaces like formal query languages, form-based interfaces and graphical interfaces: (1) the user is not required to learn an artificial communication language to use the system's potentiality. This does not mean that there is no need for some information (training) about the system's functionality (linguistic coverage, language's domain); (2) there are kinds of questions (*e.g.* questions involving negation or quantification) that can be easily expressed in natural language, but that seem difficult (or at least tedious) to express using graphical or form-based interfaces. For example, "*What Lisbon's hotels are rated over 3 stars?*" (numerical quantification), or "*Which are the Algarve's golf courses without a driving range?*" (negation), can be easy expressed in natural language, but they would be difficult to express in most graphical or form-based interfaces; (3) the system will support anaphoric and elliptical expressions. NLIDBs of this kind allow the use of very brief, under-specified questions, where the meaning of each question is complemented by the discourse context. In other interfaces this notion of discourse context is usually not supported.

3.2. System Architecture

The system architecture (see Figure 1) is based on *intermediate representation languages*, where the natural language question is transformed into an intermediate logical query (LIL), before the final translation to a SQL query. This language expresses the meaning of the sentence in terms of high-level concepts, independent of the database structure [3,4].

The system architecture can be seen as being made of two big modules. The first module controls the natural language processing (linguistic component), where a question is submitted and successively transformed (morphological, syntactic, and semantic analysis), obtaining at the end of this process one or more LIL expressions. These expressions correspond to the possible interpretations of the initial question. Given the dimension of the domain and the flexibility of the natural language, there exist usually, several interpretations for the same question (the same happens with the spoken language).

The second component is in charge of the connection with the database, translating the LIL expressions to Structured Query Language (SQL) expressions (using mapping tables, see 3.3.4) and sending them to the Data Base Management System (DBMS) to produce the answers.

The main advantage of this architecture is the complete separation between the linguistic component and the database knowledge. The portability of the system to another relational database is guaranteed by the configuration of the mapping module. Of course, there will always be other modules to configure, because of the linguistic coverage but the main work remains on the mapping tables.

3.3. System Components

Morphological Analysis

The morphological analysis module is an extension of Ispell (dictionary support usually employed by UNIX users) and allows one to obtain information about the words that form the sentence. For this purpose, it uses a dictionary with the root form of words and one affix file which contains the construction rules. Thus, it is possible to get the morphologic and semantic information of all the words of a family; however, this information does not have an explicit representation on the dictionary [2,10].

Each entry of the dictionary holds, besides the root, the construction rules that can be applied, syntactic and semantic information. *E.g.* the words `hotel` and `lodge` have the following entries:

```
hotel/CAT=nc,SEM=hotel(X),TYPE=HOTEL/p/
lodge/CAT=v,SEM=lodge(X,Y),TYPE=IND,T=inf,TR=t/XYPLSM/
```

About the word `hotel`, we have the information that it is an ordinary name that could be conjugated in the plural form. Its semantics is `hotel(X)` where `X` is a variable of the type HOTEL.

About the `lodge` word we have the information that it is a transitive verb, and one can apply the construction rules XYPLSM; the semantics of the word is `lodge(X,Y)` where `X` and `Y` are variables of an indeterminate type.

The syntactic and semantic information are used in the correspondent analysis. It is important to remark that there may exist equal words in the dictionary that have different semantics. This fact is extremely important for the semantic analysis.

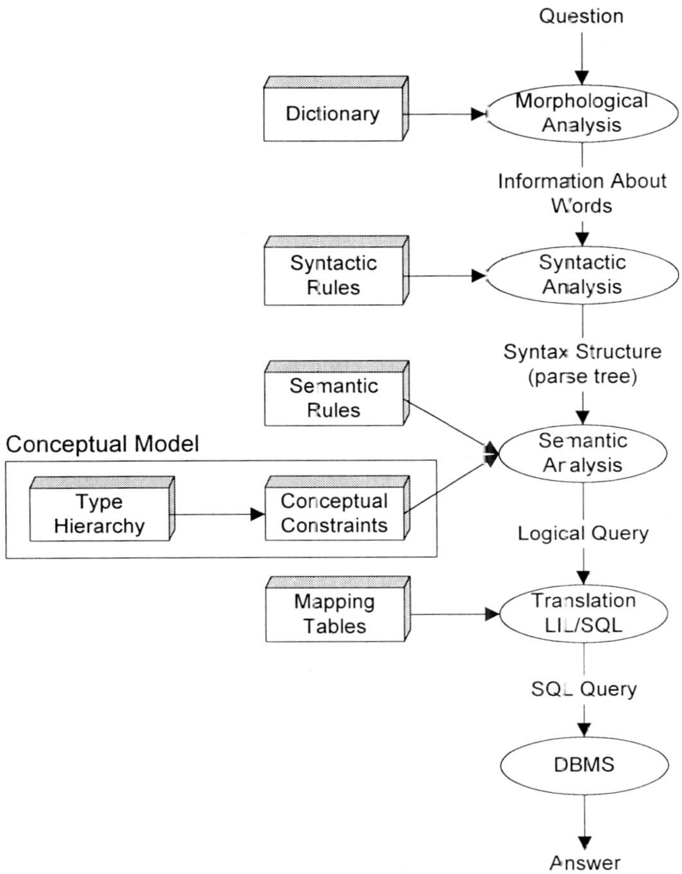

Figure 1 - The Edite Architecture

Syntactic Analysis

The syntactic analysis is based on the chart-parser technique, using a context free grammar and the Earley algorithm [1,7]. The result of this module is a set of parse trees, corresponding to the possible syntactical interpretations of the sentence.

One remark about this phase is the relaxation of some grammatical rules. This means that certain "less correct" sentences are also accepted and interpreted by the system. This was motivated by the evidence that those are currently used sentences.

Semantic Analysis

The final goal of NLIDB is the determination of the meaning of a sentence. In our case, and in the context of a database, that meaning is given by the SQL expression obtained after processing the initial statement (a question).

To obtain that expression, it is necessary to determine its meaning, usually represented in a logical form. In our case, we use the LIL language [14], similar to the First Order Predicate Calculus.

The sentence "*Which is the telephone number of the Ritz hotel*?"[1] is represented in LIL as:

```
hotel(Ritz-hotel), telephone (_11), of(_11, Ritz-hotel).
```

We will describe, very briefly, the LIL generation process (the space available being insufficient for a full description), concentrating on the interaction with the conceptual model. [10] The translation to LIL is similar to the syntactic analysis, using a dictionary and rules.

The semantic dictionary contains the words sense, called unit of meaning. When the meaning of a word is domain dependent it has to be defined using a logical predicate [14]. As an example, the meaning of the word "*city*" can be described by the predicate *is_city*(X), and consequently the question "*Is Lisbon a city?*" should be translated into: *is_city*(*Lisbon*).

Besides the predicates related with the words that are domain dependent , there exist another class of predicates, domain independent, such as the logical operators (*AND*, *OR,* and *NOT*) and the predicates *min*, *max*, *inf*, *sup*, and *exact*.

The semantic rules define the relations between the meanings of the words in a sentence. Each syntactic rule has a correspondent semantic rule, and for that reason this process is called rule-by-rule style.

Like syntactic analysis, semantic analysis is compositional, which means that the meaning of a constituent is derived solely from the meanings of its subconstituents. Compositional theories have some attractive properties. In particular, interpretations can be built incrementally from the interpretations of subphrases [1].

The semantic analysis process generates several LIL expressions, however, not all relations are conceptually acceptable in a given domain. The **conceptual module** is used to help ruling out the invalid LIL expressions. We say that a sentence is conceptually well defined when there are no violations of the constraints of the *domain of discourse*. If a sentence does not respect the conceptual constraints, then it is conceptually wrong (nonsense). In [13] we present a formalism for modelling conceptual constraints. The conceptual module is extremely important because it invalidates the wrong translations performed by the semantic analysis, reducing the number of LIL expressions that are generated.

The following example shows the importance of the conceptual module. Suppose that the dictionary contains three resources named Lisbon, a city, a hotel, and a dancing:

```
Lisbon/CAT=np,SEM=city(X),TIPO=city/p/
Lisbon/CAT=np,SEM=hotel(X),TIPO=hotel/p/
Lisbon/CAT=np,SEM=dancing(X),TIPO=dancing/p/
```

If asked "*Is Lisbon a city?*", during the semantic analysis the system would try to generate the following expressions: *is_city*(*Lisbon-city*), *is_city*(*Lisbon-city*),

---

[1] In this paper, all the english questions and vocabulary were adapted from the portuguese system module.

is_city(*Lisbon-city*). However, when the conceptual module is requested to validate the previous expressions, two of them would be eliminated due to the constraint *is_city*(*city*) (city represents the type of the argument).[2]

In the Edite system types are organised in a hierarchy: a forest of trees, where each ascendant subsumes its descendants types.

The conceptual model is also used to solve other problems related with the scope of the modifiers (adverbial phrases and prepositional phrases). The question "*Which are the hotels of Lisbon with air-conditioning?*" only has one interpretation, because "*Lisbon does not have air-conditioning*". During the syntactic analysis, and due to the second modifier, two syntactic trees are generated. In one, "*with air-conditioningr*" is a modifier of the noun "*hotel*" (syntactic rules NP→ Det N NP PP, PP→Det N), in the other, as a modifier of the noun "Lisbon" (syntactic rules NP→ Det N PP, PP→Det N PP, PP→ Det N). However, during the semantic analysis, when applying the correspondent semantic rules, the second interpretation fails, since the predicate *with(Lisbon, air-conditioning)*, is not validated by the conceptual model. Consequently, a unique LIL expression is generated at the end of the interpretation process. Note that the question "*Which are the hotels with a bar in Lisbon?*" has two possible interpretations.

LIL-SQL Translation

The LIL-SQL translator [12] is based upon several mapping tables, which are highly dependent on the database organisation. This process is very efficient and, more importantly, it allows Edite to be used with any other relational database, being only necessary to define new mapping tables.

The question "*Has hotel Berna a swimming pool?*" is translated to the following LIL expression: `hotel(Berna-hotel),` `have(Berna-hotel,swimming-pool).` This last expression is then transformed into SQL:

```
SELECT FROM hotel B
        WHERE B.cod_type_abord=16      AND   B.cod_hotel_gr=1 AND
              B.cod_rec_turist=12234   AND   B.swimmpool > 0"
```

The answer is produced by another system, in which the natural language package is included. It is this system that presents the answers using nominal lists of tourism resources, texts, and graphics.

The relations used in LIL cannot be directly translated into database relations. In the systems that use a similar architecture, the database is defined taking into account the natural language processing. Since the possibility of expressing questions in natural language was only considered when the database was defined and full loaded, we were forced to define meta-relations to map the LIL relations into the database organisation. This approach does not degrade the response time, because the time

---

2 The arguments of a logical predicate correspond to a given semantic. In the previous example, *is_city* would have to be declared, in the conceptual module, as the constraint *is_city*(*type_city*), denoting that the argument of the predicate has to be of *type_city* [11,13].

spent during this last translation step is very small when compared with the previous modules.

4. Future Work

At the present moment the project is in the final stage of conception and prototyping.

In fact the Portuguese prototype is ready to be integrated in the experimental version of the entire system. On the other hand, the dictionaries necessary for the multi-language processing are being embedded in the application. Therefore, in the beginning of 1997 the product will be available for testing in the WEB and in the Portuguese Tourism Information Network.

In the future, we intend to extend Edite in order to achieve the following goals:

- Resolution of several language phenomenons such as anaphora and ellipse with the purpose of increasing the communication speed. *E.g.* "*Which are the country clubs in Albufeira area? And which are the hotels?*".
- Treatment of negation. *E.g.* "*Which campings do not require a camping licence?*"
- Handling temporal questions. In this type of questions, the numbers, dates and times have to be carefully treated. *E.g.* "*Is the Jeronimos monastery open on Saturdays?*" , "*In Lisbon, what are the churches open between 7 am. and 10 am.?*"
- Semi-automatization of the fulfilment data. The goal is construct a software tool, Domains Editor, that partially automates the information augmenting process; by information we mean dictionaries, conceptual model and mapping tables. This acquisition component is crucial to the success of a portable system. This tool allows a better management with fewer demands on (i) the knowledge of the system's internal workings; (ii) the intricacies of the grammar; (iii) computational linguistics in general.
- The development of techniques for generating co-operative responses. Nowadays, if the user asks "*What are the hotels in Marinhais?*", the system will return an empty list. In the future, we expect that the system answers with: "*There are no hotels in Marinhais.*"; next it can present information related to other types of lodging. The idea is to progress from the actual answers to co-operative ones.
- Natural language generation.
- Due to the evolution of signal processing technology, namely digital speech recognition, it will be possible to integrate, in the future, a voice analyser in this system. This will allow the establishment of a direct oral dialogue between the final user and the system.

5. Conclusions

In spite of the linguistic limitations, the system is powerful, portable and user-friendly enough to be used in multimedia kiosks and the Internet. The main conclusions are the followings:

- After being duly tested, one can conclude that Edite can be used for various purposes. Its advantages were laid out along this paper: the user communicates in his own language and the system embraces the whole domain.
- One has to fit this type of interface in a class that still has limitations because of the computational demands and the complexity of the natural language. However, we think that with perfectly user defined functionalities and linguistic coverage, the NLIDBs can play a complementary role in current systems.
- This type of systems are important for the evolution of its informational domains as the characteristics of the natural language do not impose any restrictions on the information that the user might want to request.
- The system architecture allows this software to be ported to other databases. This has been one of the main guidelines during the software design phase and has resulted in an independence between natural language processing and linkage to databases. This feature makes possible to apply this module to other databases or Universes of Discourse with low development overheads and costs and high efficiency.
- The system has some limitations that we address here. First of all, regarding the linguistic coverage, it only accepts questions. Imperative or declarative statements are not allowed. With this approach it was possible to perfectly identify the subset of the natural language universe of the system. This is also an indication to the user, that the system only responds concerning the information of the database domain. It is not useful to have a chat about football games (typical expectations generated by the natural language systems). Another limitation is the set of restrictions imposed on the design and conception of the database.
- Despite the fact that the system has a good linguistic coverage, it is important to foresee the future extensions of the database domain. Thus, there should exist mechanisms to log the failures, so that the system could be properly extended. This mechanism should support vocabulary enrichment and the extension of the conceptual model and mapping tables.
- This project does not expect to demonstrate that NLIDBs are the future, but that they can perfectly complement the other types of interfaces.

6. Acknowledgements

Ministério da Economia de Portugal, Secretaria de Estado do Comércio e do Turismo - for project financial support and for the wise decision of linking National Information Technologies & Telecommunications to Portuguese Tourism Strategic Development Plan.

Dr. José Sancho Silva - Director of Gabinete de Estudos e Planeamento da Direcção Geral do Turismo and IRT Chief Executive Project - for highly professional competence, availability, persistence and courage for sustaining the project unreservedly, beyond all the adversities met.

Dra. Vanda Boavida (ICEP), Dra. Elisa Almeida (DGT), Dr. Luís Costa (ICEP) for the precious ideas, encouragement and assistance.

INESC, Instituto de Engenharia de Sistemas e Computadores - Director Prof. José Alves Marques - for skilful management and project belief. INESC, Centro de

Sistemas Computacionais, Grupo de Sistemas e Serviços Telemáticos - for teamwork and for some sleepless nights.

Finally we would like to show our special gratitude to Ana Matos, Luisa Marques, Eduardo Augusto and Nuno Santos.

7. Bibliography

[1] Allen, J. 1995. "*Natural Language Understanding*". The Benjamin/Cummings Publishing Company, Inc.

[2] Almeida, J. J. D., Pinto, U. 1994. "*Jspell - um módulo para análise léxica genérica de linguagem natural*". Technical Paper, Dept. de Informática, Universidade do Minho.

[3] Androutsopoulos I., Ritchie G., Thanisch, P. 1993. "*An Efficient and Portable Natural Language Query Interface for Relational Databases*". Proceedings of the 6th International Conference on Industrial & Engineering Applications of Artificial Intelligence and Expert Systems, Edinburgh, U.K., pages 327-330. Gordon and Breach Publishers Inc., Langhorne, PA, U.S.A., June 1993.

[4] Androutsopoulos, I. 1993. "*Interfacing a Natural Language Front-End to a Relational Database (MSc thesis)*". Technical paper 11, Dept. of AI, Univ. of Edinburgh.

[5] Androutsopoulos, I. 1994. "*Natural Language Interfaces - An Introduction*". Journal of Natural Language Engineering, Cambridge University Press.

[6] Cohen, P.R. 1991. "*The Role of Natural Language in a Multimodal Interface*". Technical Note 514, Computer Dialogue Laboratory, SRI International, 1991.

[7] Earley, J. 1970. "*An efficient context-free parsing algorithm*". Commun. of the ACM 13, 2:94-102. Reprinted in "*Readings in Natural Language Processing*".

[8] Godbert, E. 1994. "*Modelling domain and connectivity constraints in natural language processing*". Current Issues in Mathematical Linguistics, C. Martin-Vide (Ed.), Elsevier Science B. V.

[9] Grosz, B. J., Appelt, D. E., Martin, P. A., Pereira, C. N. 1987. "*TEAM: An Experiment in the Design of Transportable Natural-Language Interfaces*". Artificial Intelligence 32, pages 173-243. Elsevier Science Publishers B.V. (North-Holland).

[10] Marques, Luisa 1996. *M.Sc. Dissertation*. Instituto Superior Técnico, Universidade Técnica de Lisboa.

[11] Milhaud, G. 1994. *Ph.D. Dissertation*. Université d'Aix-Marseille II.

[12] Reis, P., Mamede N. 1996."*LIL-SQL. Processamento de Interrogações LIL por Tradução para SQL*". Technical Report. Grupo de Sistemas e Serviços Telemáticos, INESC. (in preparation)

[13] Reis, P., Mamede N. 1996."*Modelo Conceptual. A Hierarquia de Tipos*". Technical Report. Grupo de Sistemas e Serviços Telemáticos, INESC.

[14] Reis, P., Mamede N. 1996. "*LIL - Linguagem de Interrogação Lógica*". Technical Report. Grupo de Sistemas e Serviços Telemáticos, INESC. (in preparation)

Some Applications of the Multimedia Geographical Information Systems in Tourism

Dejan Rančić, Slobodanka Djordjević-Kajan, Leonid Stoimenov and Vladimir Tošić
Faculty of Electronic Engineering, University of Niš
Beogradska 14, 18000 Niš, Yugoslavia
e-mail: drancic@efnis.elfak.ni.ac.yu

Abstract: Multimedia enters all computer applications bringing new and powerful capabilities. It has also found its natural place in Geographical Information Systems (GIS). In this paper, basic principles of including multimedia into GIS and several capabilities for applying multimedia and GIS in tourism are discussed. Also, MS Windows multimedia support mechanisms and their application in GINIS-Agency, a multimedia GIS software for tourism, is presented.

Key words: multimedia, GIS, database, cartography, MS Windows

1. Introduction

One definition of multimedia says that multimedia is a variety of analogue and digital forms of data that come together via common communication channels. However, in everyday terms, multimedia refers to the technology of data integration and presentation. There are five basic data types forming multimedia: text, graphics, animation, video and sound. Multimedia is not a novel technology, but only several years ago it became possible to apply it on personal computers (PCs) and, hence, bring it closer to a wide range of users. The multimedia technique has been primarily applied in the fields of entertainment. At present, its application has being come into the professional fields. The most important role of multimedia is in educational software and, among them, geographical educational software [Krygler, 1995] bringing new qualities to computer aided learning. However, one must constantly keep in mind that educational system must not be organized as a pure knowledge presentation to a user, but rather with an active user during learning process. According to research of USA Department of Defense, only 20% people memorize information presented as a video clip. However, if the user gets the information through the system interaction that percentage rises to 60%. Apart from educational systems, multimedia also enters other fields of information technologies, including Geographical Information Systems (GIS). Recent fast development of GIS as well as the fast development of hardware and software have made possible including new technologies in GIS and their accessibility to the wide range of users. Due to their demands, up to recent past GIS applications were available only to the users of the powerful workstations. However, nowadays not only that they are accessible to PC users, but also gain new capabilities when enriched with multimedia. Multimedia has found its natural place in GIS since the need for multimedia in GIS is well known since the beginnings of GIS development. Photos and videos are natural way of displaying spatial objects. On the other side, visual component is the most important in human understanding of space and his referencing in it. Apart from that, it has been said that a picture speaks a thousand words, and one need look no further than (digital) maps which are base of every GIS application. A possibility of including sound in GIS is of special importance because they naturally supplement video, and also have independent applications in forms of some reports to the user or music.

Multimedia enables fast and effective displaying of vast amount of data, but those data have to be stored which demands adequate hardware support. CD ROM enables storing vast amount of data with acceptable access speed. Apart from hardware, a software support is needed not only from operating system, but also from the independent programs for multimedia support. Also, adequate software techniques for displaying vast amount of data are needed. In the PC world, MS Windows has gone the furthest in that way providing complete support for multimedia devices.

2. Multimedia Component of GIS

It is evident that GIS systems were and still are tool for the specialists. So what then are the barriers preventing GIS from becoming the everyday tool for everybody? It would appear that at the core of the matter are two issues:

1) system interface and
2) data presentation and interpretation.

There is no standard GIS interface, and there probably never will be. There are a lot of different user interface forms used by different commercial GISs. However, it would appear that the Windows graphical user interface (GUI) from Microsoft has become the de-facto standard for the office desktop. There are about 50 millions Windows licences in the world today, and this is increasing at a rate of 2 millions per month. Also, that process was helped by the spanning of PC and workstation capabilities. According to the Dataquest research [Barnes, 1994] in 1997 the market share will be:

Windows/DOS	54%
WindowsNT	17%
UNIX	6%
MAC OS	4%
OS/2	1%
others	18%.

On the other hand, interpretation of geographical data is often in unintelligible form for the wider range of users. An array of numbers or charts does not mean a lot to a common user. On the contrary, these data become very understandable if they are presented in an appropriate form. Clearly, what is needed is a GIS possessing the sophistication of being able to bring together and integrate disparate data sources, yet is as easy to use as any office tool. The solution could be found in the new generation of Windows based multimedia systems. For example, if a three-dimensional (3D) ground model is used instead of an array of numbers which represent altitude, then the data become understandable to every user. Including photos, satellite snapshots, videos and sounds (which is the task of the multimedia component of GIS) significantly contributes to the GIS understanding. Multimedia forms could be connected to any location covered by a GIS application enabling user to get adequate photo, video or audio for the chosen location. One of the functions of multimedia component is presentation of two-dimensional data, as well as creating and presentation of three-dimensional models and objects and three-dimensional data. It is also possible to add texture, illumination, shadowing and reflection to the three-dimensional objects and, hence, bring the picture closer to reality. Three-dimensional animation creates an illusion of movement. A final step towards a full, realistic animation is rendering - a process in which computers, using all available graphical capabilities, make the animation smooth, without jumps between two consecutive sequences.

However, it is often needed more from the computer made pictures. Including three-dimensional model is important for the presentation of three-dimensional ground or important objects, while three-dimensional animation in GIS enables three-dimensional movements through maps, i.e. space.

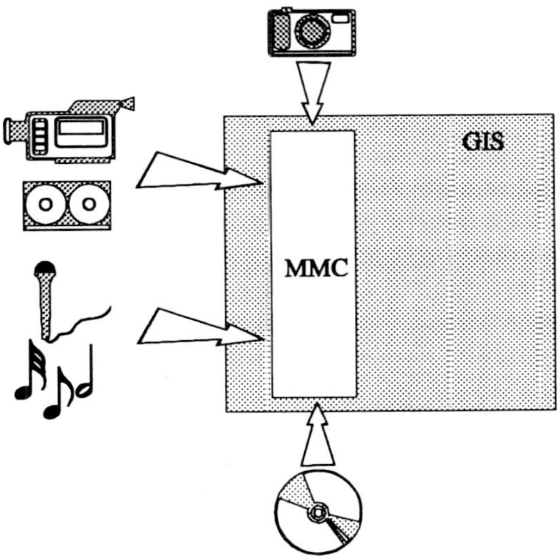

Fig. 1. Multimedia component (MMC) of the GIS

As a conclusion, it could be said that one of the solutions for the bringing GIS closer to a wide range of users is in new generation of Windows-based multimedia GIS.

3. Multimedia Support of MS Windows

In order to discuss this issue, the basic condition of the realizing multimedia technique should be analyzed. In the software support for multimedia, Microsoft has achieved the most by founding a PC multimedia marketing council (a manufacturers group). This group has defined the Multimedia PC (MPC) standard for Intel-based platforms. This definition covers two levels of PC technology starting from 386SX processor. The definition includes memory specifications (2MB min.), hard disk space, audio and video display, as well as system software (MS Windows 3.1 and above with multimedia extensions). Multimedia extension defines a complete architecture for sound, animation and video handling. The main components of which are [Barnes, 1994]:

- Object Linking and Embedding (OLE) is a mechanism for linking and embedding objects which enables insertion of images, audio and video into the documents.

- Windows Sound System which defines sound capture and storage options.

- The Windows Multimedia Application Programmers Interface (API) which interfaces to the Media Control Interface (MCI) and allows programmers to play sounds, run animation and control external devices, such as CD players or VCR's.

Additionally, Windows Dynamic Data Exchange (DDE) can be used to share data between different applications.

Multimedia extension of MS Windows is built in several layers which isolate multimedia applications from physical multimedia architecture. In such a way, a high level of hardware independence is achieved. The foundation of the extension is MMSYSTEM.DLL (dynamic link library for multimedia) which contains external device control functions. Some of the devices are controlled directly through MCI drivers which are a kind of logical extension of Windows API functions.

On the other hand, MMSYSTEM.DLL contains functions for handling multimedia device drivers. For the most simple device control, only three operations are needed: OPEN, PLAY and CLOSE. For more complex operations an existence of additional functions is needed. As different devices use different methods for data compression, a hardware independent interface is needed which should release programs from concerns on details of specific data compression methods. It is solved in such a way that programs turn directly to ICM (Installation Compression Manager), which in communicate with drivers and various hardware resources, enabling data compression and decompression.

In such a way, by isolating applications from physical devices, MS Windows has built a foundation for simple creation of multimedia applications and, hence, brought closer multimedia to a wide range of programmers.

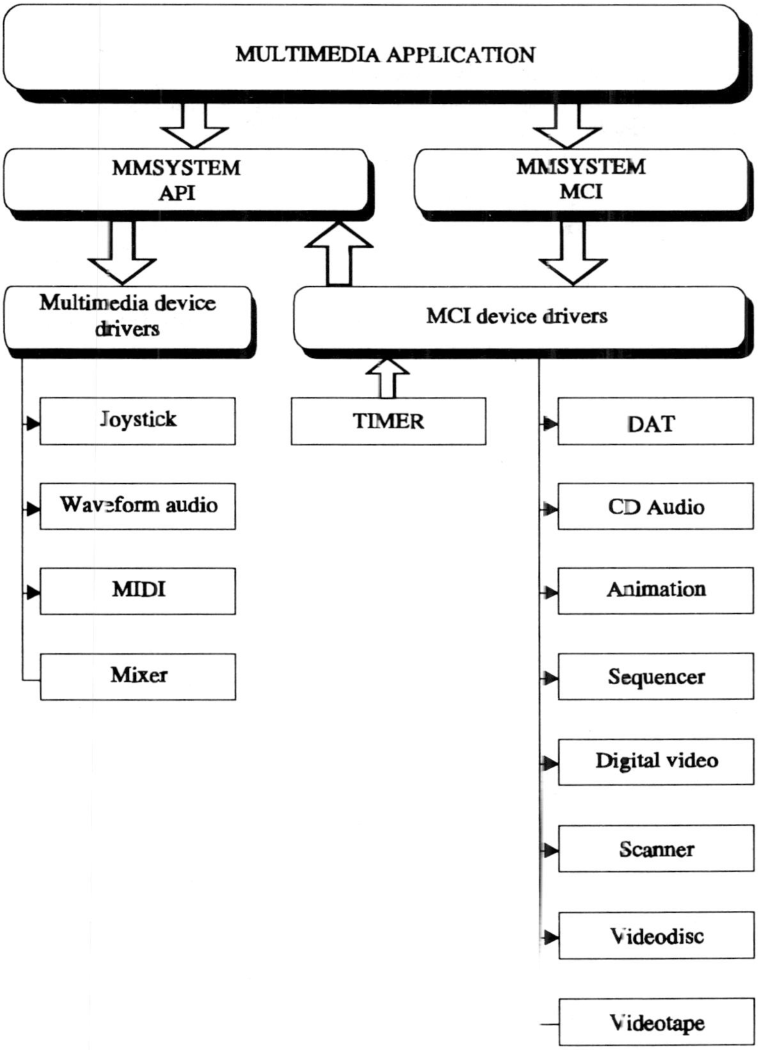

Fig. 2. Windows multimedia extension

4. Application of Multimedia and GIS in Tourism

Tourism is one of the fields with a strong need for integrating GIS and multimedia. There is a wide range of usage possibilities of multimedia GIS in tourism and a massive release of this kind of software for tourism is expected. First of all, they give tourist agencies an opportunity to present their tourist programs to potential clients in a simple, fast and effective way. Client can get all the information related with travels fast and simply. With the improvement of human living, the tourism has been developed and colorful touring maps, for example, scenic area, traveling lines, selling and historic cultural relics, etc. have coming into the human living. As a new kind of the modern map based on the GIS and multimedia support, electronic map is a new sort of the modern map that is displayed on the screen. On one hand, GIS enables to query for roads and distances between certain destinations and view interesting locations on the map. On the other hand, multimedia provides photos, audio and video records related with those places. In addition, we can establish the electronic map of scenery guide and consulting service of traveling can be set at the entrance to the scenery area. In this way, all the scenic spots can be displayed on the screen and the touring line (route) can be animatedly displayed as well. This kind of touring electronic map is more joyful than any traditional traveling paper maps. Therefore, it is sure to become an important aspect of the tourism developing.

Apart from tourist agencies, a need for integration of GIS and multimedia also encounters in museums. Such a system is the simplest way for a visitor to get an information about arrangement of museum exhibits and a picture, video and description in various languages of every exhibit.

A possibility of using information systems of this kind as city guides is also worth mentioning. Deposition of digital network of roads and important objects onto the scanned city map with corresponding data base enables visitors founding and getting information about important objects, e. g. taxi and bus stations, and streets.

It is obvious that there is a very wide range of possibilities for usage of GIS and multimedia in tourism and a massive release of this kind of software for tourism is expected.

5. GINIS-Agency

GINIS-Agency is a multimedia GIS application for tourism developed at the Computer Graphics and GIS Lab of the Faculty of Electronics at Niš, Yugoslavia. GINIS-Agency enables fast, simple and effective retrieving of information about tourist destinations and, hence, helping clients to make their mind about the travel. Program operates in MS Windows environment and uses all of its multimedia capabilities. The foundation of this application is georeferenced electronic map [Djordjević, 1995] with deposed layer of tourist destinations. As a part of this, spatial database covers various data such as: photos of destination and important objects in it, traveling departments, scenic spots, touring lines, selling, nearest airport, hotels, banks, restaurants, exchange rate data, natural conditions and social customs and others. These data can be managed and expressed in the form of graph, text, sound, video and animate figure on the base of the multimedia technique. For example, if hotel has been selected, client can get photos of hotels and rooms in them.

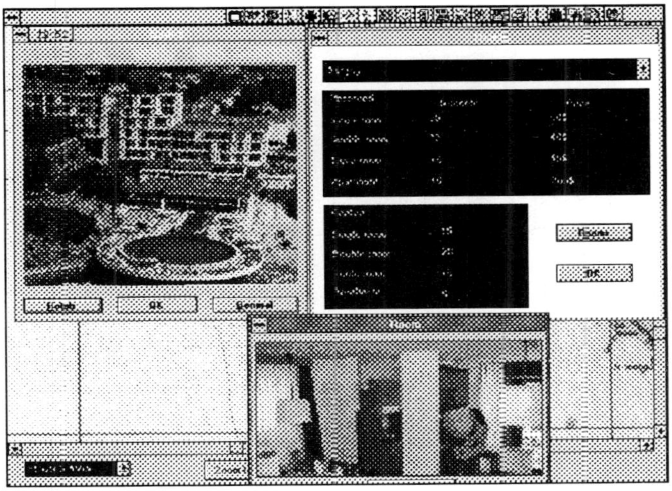

Fig. 3. Program "GINIS-Agency"

In addition to all that, it is possible to see a video and audio record of several characteristic events taking part at the selected destination. It is possible to choose one of available hotels and get information about it, e. g. available rooms and prices, and to do complete booking. Program enables map zooming up to the level of city maps. Also, a part of GINIS-Agency is an expert module (still under developing) which enables route planing functions. For example, if the client wants to go to the seaside at spring and also wants to see carnival, expert module will search, on the base of spatial query, all appropriate destinations and will plan and animate the route.

An application conceived in the described manner enables fast, simple and effective retrieving of information about tourist destinations and, hence, helps clients to make their mind about the travel. Also, an embedded data base enables complete booking and constant insight into the number of available beds, prices and other relevant information.

GINIS-Agency is developed using GINIS, a tool for developing GIS applications (also developed at the Computer Graphics and GIS Lab of the Faculty of Electronics at Niš, Yugoslavia). Electronic map is formed from scanned paper maps using a component of GINIS called MapEdit [GG&GIS Lab, 1995]. Electronic map formed in such a way consists of a number of electronic sections which solves problems of presenting huge maps. Hence the names of electronic sections correspond to geographical coordinates of the part they contain, only those electronic sections which are currently visible on the screen are loaded. Speed of work is improved by using cache technology, i. e. by loading electronic sections neighbor to currently visible on the screen. A relational database containing all relevant information about spatial objects which were originally on the paper map or were supplemented in the electronic map is created over a such electronic map.

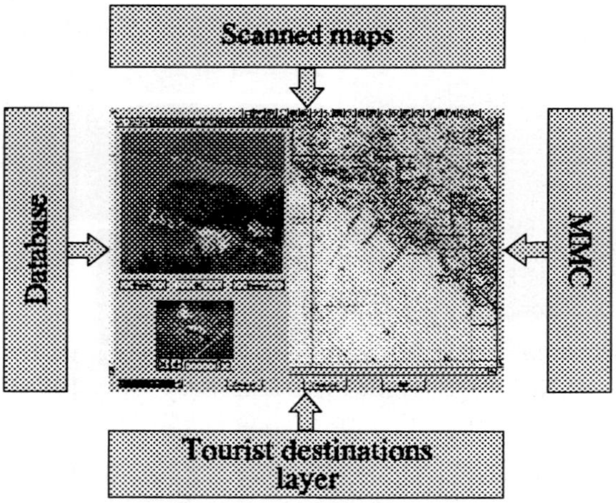

Fig. 4. Conception of "GINIS-Agency" program

6. Conclusion

Due to the great expansion of Geographical Information Systems, a need for converting existing spatial data into a form suitable for use and representing on the screen has arisen. Development of multimedia has, in turn, enriched a variety of possible representation methods. Photos, sound records, video and other forms of multimedia found their place in GIS. With adequate hardware and software support a relatively easy development of multimedia applications is enabled. This development had as a consequence entering of multimedia and GIS in the PC world and, hence, their availability to the wider range of users. Taking a fast hardware (especially components for storing huge amount of data with low access speeds) and software development into consideration it is realistic to expect that in near future these systems become even more powerful. Multimedia component will gain more and more importance bringing closer existing systems to virtual reality, whose time is coming.

Tourism, as one of the fields with a strong need for integrating GIS and multimedia, was gain new capabilities. They give tourist agencies an opportunity to present their tourist programs to potential clients in a simple, fast and effective way. Client can get all the information related with travels fast and simply. There is a wide range of usage possibilities of multimedia GIS in tourism and a massive release of this kind of software for tourism is expected. It is sure that GIS applications, such as GINIS-Agency, which is presented in this paper, become an important aspect of the tourism developing.

* This paper is partially financed by the Ministry of Science and Technology of Republic of Serbia.

References

CG&GIS 1995. The Methodology, Procedures and Software for Production of Digital Geographic Maps, Technical Report, Computer Graphics & GIS Lab, Faculty of Electronic Engineering, Niš, Yugoslavia, January 1995. (in Serbian)

Đorđević S., Petković M. i Rančić D. 1995. Georeferencing in Digital Cartography", Proceedings of the XXXIX Yugoslav Conference ETRAN '95, Zlatibor, Yugoslavia, June 1995., pp. 149-152. (in Serbian)

Mitrović D., Đorđević S., at al.,1994. MapSys - A Spatial Objects Control System, In: Proceedings of the III Int. Conf. on automation, robotics and computer vision, Singapore, 1994, WA6.1, pp. 185-189.

Visvalingam M. 1990. Trends and concerns in digital cartography. In: Computer Aided Design, vol. 22, pp. 115-130, no. 3, April, 1990.

Crisman N. R. 1990. Deficiencies of sheets and tiles : building sheetless databases. In: Int. Journal of GIS, vol. 4, pp. 157-167, no. 2., 1990.

Guptil, S. C. 1989. Speculations on seamless, scaleless cartographic data bases. In: Proceedings of the Auto-Carta 9, Baltimore, April, 1989, pp. 436-443.

Barnes P. 1994 Multimedia brings GIS to the masses. In: Proceedings of the AM/FM European Conference X, 18-20 October 1994, Heidelberg, Germany, pp. 190-194.

Hua W. J. and Rui Z. G. 1995. The application of the multimedia technique in the electronic map. 17th Intern. Cartographic Conference and 10th General Assembly of the ICA/ACI, September 3-9, 1995, Barcelona, pp. 2074-2080.

Krygler J. B. 1995. Multimedia in Geographic education. 17th Intern. Cartographic Conference and 10th General Assembly of the ICA/ACI, September 3-9, 1995, Barcelona, pp. 1611-1620.

INFORMATION AGE AND TOURISM

Jaro BERCE, SIP&A, Ljubljana, Slovenia
e-mail: jaro.berce@guest.arnes.si
URL: http://www2.arnes.si/guest/ljsipa1/index.html

The rapid advances in the fields of information technology and telecommunications are conducting the changes in economic, socio and business patterns of behaviour. Data and information are elementary to the new economic and trade foundations. Without them - and one can be driven out of business fast! Obviously the changes are fast and *knowledge economy* is taking over the leadership.
During the past few years, the generations of users of tourist services, as well as, the way they were using them, have undergone the drastic changes. With lifestyle changing on a daily basis, the vacations are becoming more pro-active and frequent. Tourists are trying to get the most of the several, but short trips.
Tourism has been always dependant on information data. Therefore, usage of Information Technology is a postulate for tourism. High quality services is expected, and so is an active participation bordering to adventure. These are the trends that follow the change of the generations. Thus the inclusion of the newer and more sophisticated services such as access to fax, E-mail, Internet, etc., is becoming a necessity within the mobile business community.

Background

Initially, computer technology in tourism was predominately used to speed up the process of calculations and paperwork. The *Information age* has begun and still most companies' internal business processes remain deeply rooted in the *Industrial Age*.

Intermediary in tourism and touring-travel companies were, due to bulk and need of fast response to customer requests, first to use the computer technology in order to remain competitive. Gradually, the information technology was introduced in hotels, tourist information centres etc.. Connectivity of tourist service providers and tourists are yet to be supported with new technology. Both parties still reside on the opposite ends of the "production line".

Not long ago, tourists were travelling mostly on business or pleasure. Individuality was lost in a mass of non-active vacations. On the other hand, providers were focused only to meet expectations and through mass production generated profit. There were no tailor-made tourism products for individual needs available. Therefore, huge tourist industry was built to cater for the enlarged appetite of the customer and the travel organisations alike.

With beginning of Global Information Networks and their interconnection - INTERNET a new paradigm evolved. *Open networks* allow businesses to share computerised data not only locally but mostly on global market scale with customers, intermediaries, and other entities. The tourist sector to be competitive is therefore forced to follow these new trends. A trend where information and representation of destination could be brought directly to customer is becoming crucial in the new information age. A real demonstration of desired destination should be within reach of perspective user. Not only picture and text data are the representation of tourism products, but with more advanced technology multimedia information spreads around the globe. This data is showing on-line weather reports (temperature of sea, air, humidity etc.), surrounding environment of destination, location of sport or other resorts, interiors of hotels and their rooms, (graphical) schedule of forthcoming events and much more.

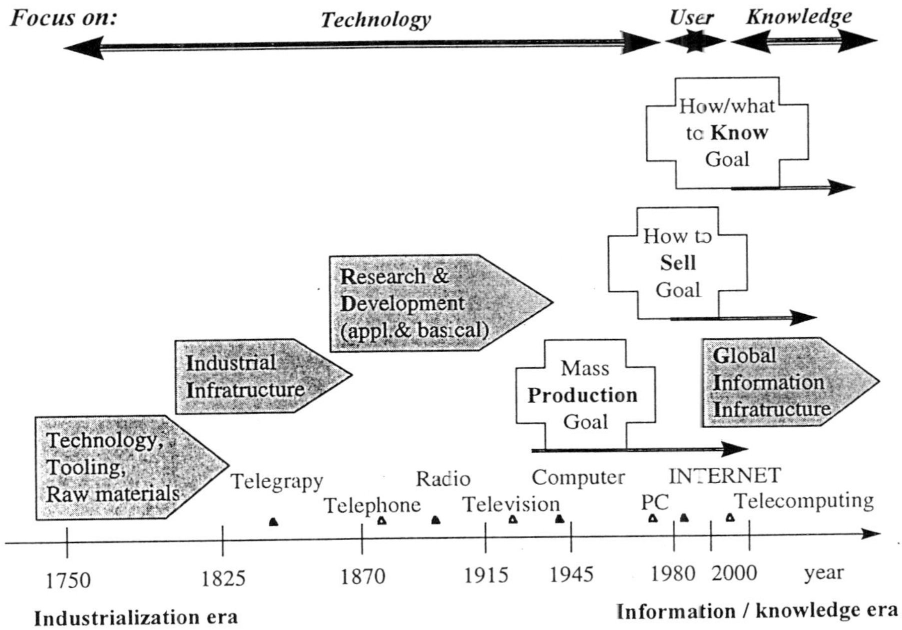

Figure 1: Evolution of human economy during last 250 years

Due to the shift into *knowledge era* where mass production replaces the problem "how to sell the product", also tourism is not protected by this influence. This effect is shown in tourism through advertising and type of services, products and tools that are offered to tourists (smart-card, laptop, modem, e-mail, ...) during their trips.

Virtual vs. real

Real world companies and tourist are facing "images" of so called "virtual world" through computer screens. Graphical user interfaces (world-wide-web /www/ browser) are helping new users on viewing basis, to get most out of the global contents of commercials. *Virtual era* is now in its evolving stage:

- Presentation of information and
- Advertising services and products.

On the other hand first initial period of:

- Research and development (DARPA, University level usage) and
- Spreading-interconnecting networks around the world

is slowly emerging the maturity, as dubbing (users, servers, networks) from one year to another cannot be forever. Therefore the virtual space is now however more about users seeking information than it is about providers proactively distributing it. No active real time interactions as known to everyday users of PC, exist on WWW pages. Home-pages are linked in (dis)organised passive technique. A new tool - Java that is promising to change intercommunication, is still in its introductory level. Although Java and INTERNET-embedded applications model will have the most significant effect on brand-new applications that resides on INTERNET. We will see significant transition from today's traditional client-server applications to this new model[1].

We should remember that at the moment the marketplace is accustomed to the notion of purchasing PC software for permanent use. The rent - as a new service, technically involve the vendor granting a licence to the user for which user has to pay fee only for each use of application instead for the whole software. These services are going to be maintained on promising Network-Centric-Computers (NCC) and should facilitate user's connectivity and problems with standardisation (release update). In parallel there are evolving intelligent computer programs - *Intelligent Agents*, that will be tailored and accustom to each user and help her/him to select destination, provider, or to make reservations and/or endorse of money transfer.

A random survey of WWW users reported by USA Today and Intelli-Quest's Web Satisfaction Study[2] shows highest desired changes:

- 74% - faster access,
- 55% - information that's easier to find,
- 53% - lower cost.

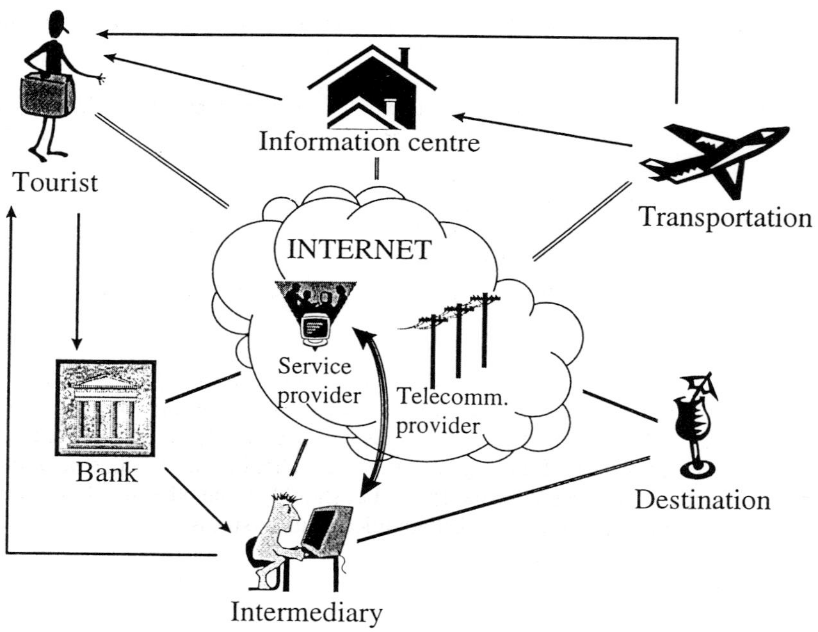

Figure 2: Shift to "virtual environment"

However, as technology becomes more widely used and standardised, its price declines. ATM, B-ISDN and faster infrastructure are build around the world. Global networks are nevertheless promoting business and as George H. Heilmeier current CEO of Bellcore once argued: "... information infrastructure is simply the entry fee for participating in the global economy..."[3].

New digital commerce is pushing toward transmission without delay and cashless business deals. Within these conditions INTERNET with its on-line distribution-access pipe model is having a revolutionary impact on the leisure and business travel market. Tourism services and products are one that cannot be pre-practised therefore are subject to financial and emotional jeopardy as "destroyed" vacations or business trip cannot be changed. Thus through Open-network access one can much easily, with help of real-time multimedia connection, take the correct decision. Therefore joining destination provider and user through this high-speed and real-time connection is making satisfaction on both sides. Also a new model that is taking place on INTERNET is becoming increasingly important. *Collaboration work* is not only tool for researchers and their environment but should be applied as well in business relations. So that proactive definition of one's trip can be built within real-time frame.

It is common known that investing in Open networks gain the competitive advantage. But on the other hand do these Open networks improve business productivity, efficiency and quality of work? As seen from market surveys and other sources'[4] companies using open networks are more profitable and effective and are having more fruitful relationship with their clientele. Through INTERNET companies can promote their activities, services and products and as well handle data transaction needed for business purposes. How companies are using networks generally depends on their individual competitive strategy and on the maturity of their market.

Survey guided by Georgia Institute of technology's Graphics, Visualization, and Usability Centre (October 1995 with 23.300 users responding) reported that INTERNET users are:

- 29.3% female (fast growing proportion),
- the average age is 32.7 (more than 75% of total users are 5 to 32.7 years old),
- having on average income of 63.000 US$,
- 78.4% are using browser daily - 63.6% for entertainment and only 11.1% for shopping[5].

One should remember that modern tourist wishes also to have on-time connectivity during the excursion, possibility to work (workforce mobility - virtual corporation) and gets link to hers/his data from different locations (service portability and transparency). Tourism as a part of very profitable global economy should take in consideration all these benefits that are showing in virtual space, but should on the other hand as a service provider thinks about human relations that are juice of its business.

Next, there exists also details that are regard to be important during the shaping of the virtual environment:

- Standardisation of representation of tourism data (how to show restaurant, ski lift, bus station, etc.), as all around understood icon,
- Privacy and/or tractability during the journey and
- Protection of past trips that tourist took as a source of advertising new ones.

Fulfilment of some other issue's that are waiting to help virtual commerce to take-off the ground is:

- Verification,
- Authentication,
- Security and confidence of information and transaction.

Conclusion

INTERNET is now mostly used in tourism as a new media for informing and advertising. Most of home-pages are having an e-mail option so that prospective user can ask some questions concerning information and data shown on the page. But this is far away what technology is capable to support. E-mail options should be final version of user interference as intelligently linked pages, active relational databases (unified procedures and data) and other tools should provide most of the answers to promised user - tourist. On the other side this type of intelligent pages or programming options should assist tourism organisations to provide up-to-date and verified information from different sources and places on real-time basis. In this way we can make integration of generalisation of the typical request-response computing model into *workflow* - a script of tasks that must be executed.

In Europe there are several actions taking place in parallel, to support SME and besides tourism sector to compete within growing opportunities in information society. INFO2000 aims at stimulating the emerging multimedia content industry to recognise and exploit new business opportunities. As the importance of multimedia content has been recognised in number of meetings. The central theme of INFO2000 is the development of a European information content industry. Which should be capable of competing on global market scale and able to satisfy the needs of Europe's enterprises and citizens for information content leading, on the one hand, to economic growth, competitiveness and employment and on the other hand to individual professional, social and cultural development. A consortium of several European countries with their companies and tourism entities was established to tender for project that will enable setting needed information structure to support virtual economy on INTERNET.

FECTO members are on the other side also setting up a working group for FECTO member presence on the INTERNET. The IT working group has recommended that a corporate site be developed that would maximise exposure of FECTO members' cities to growing market of INTERNET users. The WEB site would act as a gateway to city sites and provides general information about Europe and a common approach to standardised information organisation. It would enable browsers' to access individual member sites quickly and easily.

Hence business processes in tourism industry must be furthermore reengineered to fully exploit the new technologies that radically reduce transaction cycle time, remove distance as barrier to interaction, and improve the quality of information generated and received.

References

[1] Yourdon Edward: **Java, the Web, and software developmnet**, Computer, pp. 25-30, Vol. 29, No. 8, August 1996.

[2] Fox Robert: **News Track**, Communications of the ACM, pp. 11, Vol. 39, No. 8, August 1996.

[3] Heilmeier H. George: **Global begins at home**, IEEE Communications, pp. 50-56, Vol. 30, No. 10, October 1992.

[4] Lynn A. Streeter, etc.: **How open data networks influence business performance and market structure**, Communications of the ACM, pp. 63-73, Vol. 39, No. 7, July 1996.

[5] Pitkow E. James: **Emerging trends in the WWW user population**, Communications of the ACM, pp. 106-108, Vol. 39, No. 7, July 1996.

SpringerComputerScience

M. Göbel, J. David, P. Slavik, J. J. van Wijk (eds.)

Virtual Environments and Scientific Visualization '96

Proceedings of the Eurographics Workshops
in Monte Carlo, Monaco, February 19–20, 1996,
and in Prague, Czech Republic, April 23–25, 1996

1996. 169 partly coloured figures. VIII, 324 pages.
Soft cover DM 118,–, öS 826,–
ISBN 3-211-82886-9
Eurographics

X. Pueyo, P. Schröder (eds.)

Rendering Techniques '96

Proceedings of the Eurographics Workshop
in Porto, Portugal, June 17–19, 1996

1996. 197 partly coloured figures. IX, 294 pages.
Soft cover DM 118,–, öS 826,–
ISBN 3-211-82883-4
Eurographics

B, Urban (ed.)

Multimedia '96

Proceedings of the Eurographics Workshop
in Rostock, Federal Republic of Germany, May 28–30, 1996

1996. 71 figures. VII, 178 pages.
Soft cover DM 85,–, öS 595,–
ISBN 3-211-82876-1
Eurographics

SpringerWienNewYork

P.O.Box 89, A-1201 Wien • New York, NY 10010, 175 Fifth Avenue
Heidelberger Platz 3, D-14197 Berlin • Tokyo 113, 3-13, Hongo 3-chome, Bunkyo-ku

Springer-Verlag and the Environment

WE AT SPRINGER-VERLAG FIRMLY BELIEVE THAT AN international science publisher has a special obligation to the environment, and our corporate policies consistently reflect this conviction.

WE ALSO EXPECT OUR BUSINESS PARTNERS – PRINTERS, paper mills, packaging manufacturers, etc. – to commit themselves to using environmentally friendly materials and production processes.

THE PAPER IN THIS BOOK IS MADE FROM NO-CHLORINE pulp and is acid free, in conformance with international standards for paper permanency.

SpringerComputerScience